Nitride Phosphors and Solid-State Lighting

SERIES IN OPTICS AND OPTOELECTRONICS

Series Editors: **E Roy Pike**, Kings College, London, UK
Robert G W Brown, University of California, Irvine

Recent titles in the series

Nitride Phosphors and Solid-State Lighting

Rong-Jun Xie
National Institute for Materials Science, Japan

Yuan Qiang Li
Lightscape Materials Inc., USA

Naoto Hirosaki
National Institute for Materials Science, Japan

Hajime Yamamoto
Tokyo University of Technology, Japan

CRC Press
Taylor & Francis Group
Boca Raton London New York

CRC Press is an imprint of the
Taylor & Francis Group, an **informa** business

A TAYLOR & FRANCIS BOOK

Taylor & Francis
6000 Broken Sound Parkway NW, Suite 300
Boca Raton, FL 33487-2742

First issued in paperback 2020

© 2011 by Taylor and Francis Group, LLC
Taylor & Francis is an Informa business

No claim to original U.S. Government works

ISBN-13: 978-0-367-57695-0 (pbk)
ISBN-13: 978-1-4398-3011-6 (hbk)

Visit the Taylor & Francis Web site at
http://www.taylorandfrancis.com

and the CRC Press Web site at
http://www.crcpress.com

Contents

Preface

The history of inorganic luminescent materials (phosphors) dates back to the time when the first phosphor, barium sulfide, was produced in the nineteenth century. Since then, phosphors have been significantly developed for use in lighting, display, sensing, and labeling, covering a large number of materials with different crystal structures and compositions. Over the past decade, great advances in solid-state lighting technologies, especially in white light-emitting diodes (LEDs), which use phosphors as wavelength converters, have catalyzed the development of a novel family of luminescent materials: *nitride phosphors*. Because nitride phosphors make white light LEDs brighter, more reliable, and natural, many beginners in this field are looking for an introductory book on their crystal chemistry, synthesis, luminescence, and applications. This is the first objective of this book.

Although some useful books on luminescent materials have been made available in recent years, for example, Glasse and Grabmaier (1994), Yen et al. (2006), Ronda (2008), and Kitai (2008), nitride phosphors, as well as phosphors for LEDs, were not treated as a specialized topic and were sketchily described in all these books. Moreover, a number of novel nitride phosphors, together with other phosphors, have been discovered for solid-state lighting just in the past 5 years. Therefore, the second objective of this book is to introduce up-to-date nitride phosphor materials and technologies.

Since the field of phosphor materials is a multidisciplinary study that is involved in materials science, crystal chemistry, inorganic chemistry, solid-state chemistry, solid-state physics, optical spectroscopy, crystal field theory, and computational materials science, in this book we cover these subjects based on our extensive experimental data and published data by other research groups worldwide. This is the third objective of this book.

The present book is organized as follows: Starting with a brief introduction to solid-state lighting, Chapter 1 describes the working principles, semiconductor/phosphor materials, and characterizations of solid-state lighting, as well as white light-emitting diodes. Then, we summarize the prospective applications of solid-state lighting.

Chapter 2 is devoted to the fundamentals of optical and luminescence processes of optical centers in a solid, which enables the reader to have an idea of how the emission and excitation are affected by host lattice, concentration, and temperature, and how the nonradiative transition occurs in luminescent materials. Finally, the optical centers—rare earth and transition metal ions—are introduced in terms of their electronic configurations and characteristic luminescence.

Chapter 3 presents the photoluminescence properties of typical traditional phosphors for white LEDs. These include garnets, aluminates, silicates, sulfides and oxysulfides, phosphates, and scheelites. Although these phosphors have long been known about, they have been investigated as down-conversion phosphors for white LEDs over the past few years. This chapter describes in detail the crystal structure, excitation and emission spectra, thermal quenching, and quantum efficiency of these traditional phosphors.

In Chapter 4, we briefly introduce the crystal chemistry of general nitride compounds, and then we focus on the crystal structure and photoluminescence properties of newly developed nitride phosphors, classified into four groups according to their emission color. In this part, we also describe the synthetic methods for preparing nitride phosphors.

Chapter 5 presents details of the structural analysis of nitride phosphors. These include the application of x-ray powder diffraction and Rietveld refinement, electron microscopy observations, XANES/EXAFS techniques, first-principles calculations, and molecular orbital cluster calculations. The experimental and computational results of some typical nitride phosphors are given for a better understanding of their photoluminescence properties and the further design of nitride phosphors.

Chapter 6 discusses some key issues in generally understanding nitride phosphors, including significantly red-shifted excitation and emission spectra, small thermal quenching, and high quantum efficiency, compared with conventional phosphors. The effect of lattice imperfections on the photoluminescence of nitride phosphors is discussed with an interesting red phosphor that shows multiple luminescence bands.

Chapter 7 is devoted to the applications of nitride phosphors in white LEDs for general lighting and LCD backlight purposes. The general selection rules of LED phosphors, and the fabrication and optical properties of white LEDs utilizing nitride phosphors are mainly addressed in this chapter.

Rong-Jun Xie
Yuan Qiang Li
Naoto Hirosaki
Hajime Yamamoto
Tokyo, October 2010

Acknowledgments

We acknowledge many colleagues who have supported us in a variety of ways in the preparation of this book. They are Dr. Takashi Takeda (National Institute for Materials Science, Japan), Dr. Kyota Uheda (Mitsubishi Chemical Science and Technology Research Laboratory, Japan), Dr. Takayuki Suehiro (Tohoku University, Japan), Dr. Kousuke Shioi (SHOWA DENKO K.K., Japan), Mr. Kohsei Takahashi (Sharp Corporation Advanced Technology Research Laboratory, Japan), Dr. Ken Sakuma (Fujikura Ltd., Japan), Mr. Naoki Kimura (Fujikura Ltd., Japan), and Professor Kee-Sun Sohn (Sunchon National University, Korea). Also, sincere thanks go to Professor H. T. Hintzen (Eindhoven University of Technology, the Netherlands) for his longtime support and critical comments. These colleagues helped us to clarify our understanding of certain specialist topics, spent valuable time with us in fruitful discussions, and provided us published and unpublished data for use in this book. We also give our sincere thanks to Ms. Kazuko Nakajima, Ms. Yoriko Suda, and Mr. Tomoyuki Yuzawa for their technical assistance in the experiments.

R.-J. Xie, one of the contributors, gives his special thanks to Dr. Mamoru Mitomo (National Institute for Materials Science, Japan), who shepherded Xie into the colorful world of phosphor materials.

We greatly appreciate our other colleagues for their technical support in x-ray diffraction analysis, chemical analysis, and microscopy observation, as well as the supercomputer simulation facility.

We thank the following societies and publishers for granting permission to reproduce a number of figures and tables in our book: John Wiley & Sons Ltd., CRC Press, Oxford University Press, Springer-Verlag, the American Institute of Physics, and the American Ceramics Society.

We thank the publishers for their patience and support, and the reviewers and advisers who commented in a helpful and constructive way on this book.

Last but not least, all the family members of the authors are deeply acknowledged. Without their understanding and support, there would be no book.

1

Introduction to Solid-State Lighting

Lighting has a great impact on the lifestyle of humans, architectural design, and the quality of building, as well as entertainment environments. Starting with fire 500,000 years ago, lighting sources have consisted of gas/oil lamps, electric filament bulbs, and fluorescent tubes. The first electric filament (incandescent) lamp was invented by Edison in 1879, producing only 1.4 lm/W. Although today's incandescent lamp has the luminous efficacy of 15–25 lm/W, it generates more heat than light, with the efficiency of converting electricity into visible light being about 5%. The lighting industry was revolutionized by the introduction of the fluorescent lamp by General Electric (GE) in 1937; it now has a much higher luminous efficacy (60–100 lm/W) and conversion efficiency (~20%), as well as a longer lifetime (7,500–30,000 h), than the incandescent lamp (1,000 h). With increasing concerns about energy savings, environmental protection, and life quality, novel lighting sources that exhibit higher luminous efficacy, energy efficiency, and longevity are continuously pursued. *Solid-state lighting* (SSL) is considered the third revolution of the lighting industry, following the introduction of incandescent bulbs and fluorescent lamps. Solid-state lighting is a new lighting source that is based on semiconductor devices, with the light emitted by solid-state electroluminescence, which has great potential to significantly surpass the energy efficiencies of incandescent and fluorescent lamps, and thus promises huge energy savings and reduction of greenhouse gas emissions.

Solid-state light sources are much "smarter" than traditional light sources in controlling their spectral power distribution, spatial distribution, color temperature, temporal modulation (Kim and Schubert 2008), and polarization properties (Schubert and Kim 2005; Schubert et al. 2006). Therefore, they are used in a broader range of applications, such as general illumination, backlighting, traffic signals, medical treatment, agriculture, water purification, and imaging. In this chapter we briefly introduce how solid-state lighting works and the semiconductor materials usually used in LEDs (Section 1.1). Section 1.2 deals with white light-emitting diodes, including the approaches to fabricate white LEDs, characteristics of white light, and LED phosphors. In the final part of this chapter (Section 1.3), applications of solid-state lighting are overviewed.

1.1 Basics of Solid-State Lighting

1.1.1 Solid-State Lighting

Solid-state lighting (SSL), known as the next-generation lighting source, is a pivotal emerging technology that uses semiconductor materials and devices to convert electricity into light. The very first solid-state light source was demonstrated in 1907 by Henry J. Round (Schubert 2003), who reported light emission from a man-made silicon carbide crystal when a high voltage was applied to it. Over the last century, significant progress has been made in understanding the quantum physics behind semiconductors, and thus in developing solid-state light technologies. Even today, scientists and engineers are making great efforts in continuously improving the performance of solid-state light devices, making them more powerful, brighter, and more efficient and reliable.

The coming era of solid-state lighting is catalyzed by remarkable advances in semiconductor technology and materials science, together with much concern about our ecological environment and the earth's limited resources. As a novel light source, solid-state lighting offers the following advantages over traditional lighting sources, such as incandescent bulbs and fluorescent tubes:

1. *Energy savings.* Lighting generally consumes 22% of electricity, or 8% of total energy consumption, in the United States (Humphreys 2008). An incandescent bulb converts only 5% of its consumed energy into visible light, and fluorescent tubes, 25%. Solid-state lighting has the potential to significantly surpass the energy efficiencies of those conventional light sources. Therefore, solid-state lighting technology has great promise for significant gains in energy efficiency and huge overall energy savings; no other lighting technology offers us so much potential to conserve electricity at a time when we need bold solutions to achieve greater energy independence.

2. *Environmental friendly.* Solid-state lighting exhibits a longer lifetime (~50,000 h) than incandescent lamps (~1,000 h) and fluorescent tubes (~10,000 h), and is free of mercury. In addition, giant energy savings will result in the dramatic reduction of acid-rain-causing sulfur oxides and nitrogen oxides, carbon dioxide, fly ash, mercury, radioactive substances, and other pollutants emitted by fossil fuel power stations.

3. *Extensive applications.* Solid-state lighting allows one to control or tune the optical properties, such as color temperature, color rendering, and emission spectrum, with much greater accuracy, thereby enabling its use in a variety of applications, including general illumination, display technology, medical treatment, transportation, communications, agriculture, and biotechnology. Moreover, continuous advances in solid-state lighting technology will add new functionalities and launch a wide range of applications.

In a broad sense, solid-state lighting covers both light-emitting diodes (LEDs) and organic light-emitting diodes (OLEDs). LEDs are based on inorganic material including III-V group semiconductor materials or inorganic luminescent materials, which are called *semiconductor devices*, that produce noncoherent, narrow-spectrum light when an electrical current flows through it. The first LED was patented by Holonyak and Bevacqua (1962), and it was an infrared (IR)-emitting device. Nowadays LEDs can emit in wavelengths from the UV band to IR and are commercially available in packages ranging from milliwatts to more than 10 W. On the other hand, OLEDs rely on organic materials (polymers and small molecules) that give off light when tweaked with an electrical current. In contrast to LEDs, which are small point sources, OLEDs are made in sheets, which provide a diffuse-area light source. OLED technology is developing rapidly and is increasingly being used in display applications such as cell phones and personal digital assistant (PDA) screens. However, OLEDs are still some years away from becoming a practical general illumination source because they have some problems with light output, color, efficiency, cost, and lifetime. Because it is beyond the scope of this book to discuss the issues related to OLEDs, only LEDs and related luminescent materials will be covered.

1.1.2 Light-Emitting Diodes (LEDs)

As mentioned above, LED is a semiconductor device in which the light emission originates from a very thin crystalline layer composed of semiconductor compounds, for example, in the Al-Ga-In-N alloy system, when an electric current is flown through it. This thin layer, sandwiched between *p* type GaN (positively charged particles called *holes*) and *n* type GaN (negatively charged particles called *electrons*) thick layers, is called a *quantum well* (see Figure 1.1). Basically, the boundary interface where *p* type and *n* type semiconductors meet

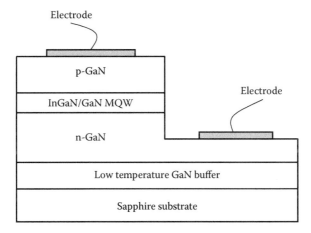

FIGURE 1.1
Schematic LED structure consisting of InGaN/GaN quantum well.

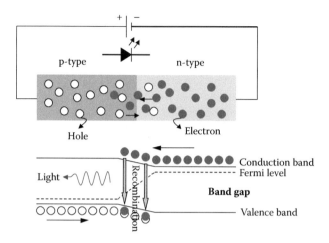

FIGURE 1.2
(See color insert.) Schematic diagram of a light-emitting diode. The characteristics of the emitted light are related to the energy gap, E_g, between the conduction and valence band in the semiconductor. E_F is the Fermi energy.

is called a *p-n* junction, so it is quantum-well-based junction when a quantum well exists. The *p-n* junction is characterized by a band gap E_g (see Figure 1.2), the quantity of which determines the minimum energy required to excite an electron from the valence band to the conduction band. The band gap, of course, also determines the photon energy released as an electron in the conduction band recombines with a hole in the valence band.

When the *p-n* junction is connected to an electric power source, current flows from the *p* side to the *n* side, with electrons injected into the InGaN/GaN quantum well from the *n* type GaN and holes injected from the *p* type GaN. The electrons and holes are located in the quantum well at different energy levels separated by E_g. As shown in Figure 1.2, when the electrons meet holes and recombine subsequently, the energy is released in the form of a photon (light) with energy equivalent to the band gap energy (hv ~ E_g). The specific wavelength or color emitted by the LED depends on the materials used to make the diode, which means that the emission color of LEDs can be tuned by varying either the composition or the thickness of the quantum well (Humphreys 2008). For example, red LEDs are based on aluminum gallium arsenide (AlGaAs), blue LEDs are made from indium gallium nitride (InGaN), and green from aluminum gallium phosphide (AlGaP) (Madelung 1996; Schubert 2003). These are also called monochromatic LEDs. Figure 1.3 schematically shows the structure of a LED lamp.

1.1.3 Semiconductor Materials for LEDs

The performance of LEDs has dramatically improved, typically in the last decade, since the first commercial LED was developed 50 years ago. The first

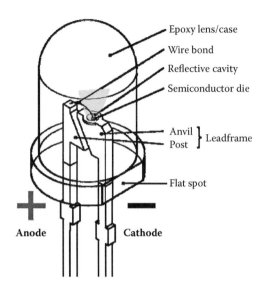

Epoxy lens/case
Wire bond
Reflective cavity
Semiconductor die

Anvil ⎫
Post ⎬ Leadframe

Flat spot

Anode **Cathode**

FIGURE 1.3
Schematic image of a packaged round (the normal type) LED.

LED was based on GaAsP semiconductor material and emitted red light with a very low luminous efficiency of ~0.1 lm/W (Craford 1992). Today LEDs cover a very broad emission wavelength, ranging from *ultraviolet* (including deep UV), to *visible* (blue to red), to *infrared* regions of the electromagnetic spectrum. Moreover, LEDs are getting more and more powerful in emission, as well as more and more efficient in conversion. Certainly, the remarkable progress made in LED technologies is strongly dependent on the advances in semiconductor materials that fundamentally determine the emission color, luminous efficiency, quantum efficiency, and durability of LED devices.

LEDs are semiconductor devices that are made of semiconductor materials that convert electricity into light through a *p-n* junction. Most of the technologically important semiconductor materials for LEDs are known as III-V compound semiconductors (see Table 1.1) (Madelung 1996; Steranka 1997; Schubert 2003; Gessmann and Schubert 2004). Among these semiconductors, III-nitride wide-band-gap semiconductors, such as GaN, AlN, and (In, Al, Ga) N, are recognized as a very important material system for the fabrication of short-wavelength-emitting diodes (UV-to-blue LEDs) (Strite and Morkoc 1992; Morkoc et al. 1994; Morkoc and Mohammad 1995; Ponce and Bour 1997; Hirayama 2005; Wu 2009).

As addressed previously, the performance of LEDs is primarily decided by the band gap or band structure of semiconductor materials. Therefore, the size of the band gap or the structure of the band is of great importance. The engineering of the band structure of semiconductor materials is a common and necessary method to achieve a desired band gap. The simplest band structure engineering involves the formation of semiconductor alloys.

TABLE 1.1

Emission Wavelengths of LED Semiconductor Materials

Semiconductor Material	Wavelength (nm)	Emission Color
GaAs	840	Infrared
AlGaAs	780, 880	Red, infrared
GaP	555, 565, 700	Green, yellow, red
GaAsP	590–620	Yellow, orange, red
AlGaInP	590–620	Amber, yellow, orange
InGaN	390, 420, 515	Ultraviolet, blue, green
AlGaInN	450, 570	Blue, green
AlGaN	220–360	Ultraviolet
AlN	210	Ultraviolet

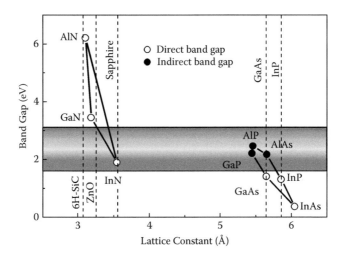

FIGURE 1.4

(See color insert.) Band gaps of selected III-V semiconductor materials against their lattice constant. The lattice constant data of commonly used substrate materials (6H-SiC, ZnO, sapphire, GaAs, and InP) are also included.

Figure 1.4 plots the band gap of several III-V semiconductor materials used for LEDs vs. their lattice constant. As seen, the materials can be divided into two distinct groups. On the right, one can find arenic and phosphorous compounds having the cubic zinc blend crystal structure, whereas on the left one can see the nitride compounds with the hexagonal wurtzite crystal structure. For both groups, alloyed semiconductor compounds can be formed in a broad composition range.

AlxGa1–xAs is formed by alloying GaAs with AlAs. Within the entire range of Al mole fraction x, the ternary compounds have excellent lattice matching due to almost identical lattice constants of GaAs and AlAs,

TABLE 1.2

Fundamental Physical Properties of Wurtzite GaN, AlN, and InN

Materials	Lattice Parameters	Thermal Conductivity	Thermal Expansion Coefficients	Band Gap Energy
GaN	a = 3.189 Å c = 5.185 Å	1.3 W/cm.K	5.6×10^{-6}/K 3.2×10^{-6}/K	3.40 eV
AlN	a = 3.112 Å c = 4.982 Å	2.0 W/cm.K	4.2×10^{-6}/K 5.3×10^{-6}/K	6.20 eV
InN	a = 3.548 Å c = 5.760 Å	0.8 W/cm.K	4.0×10^{-6}/K 3.0×10^{-6}/K	1.89 eV

enabling substrates with very few defects to be grown epitaxially on GaAs. $Al_xGa_{1-x}As$ has a direct band gap at x < 0.45, and an indirect band gap at x > 0.45. $Al_xGa_{1-x}As$ emits in the range of 630–870 nm, which is widely used as high-brightness red LEDs (Monemar et al. 1976).

(AlxGa1–x)yIn1–yP is a quaternary alloy containing four elements. Although GaP and AlP have an indirect band gap, by alloying with the direct-band-gap InP, the quaternary compounds with a direct band gap can be produced. Nearly perfect lattice matching to GaAs substrates can be achieved at y ~ 0.5, and to InP substrates at y ~ 0.47 x. $(Al_xGa_{1-x})_yIn_{1-y}P$ emitters can be used as amber-to-red LEDs (Schubert et al. 1999; Schubert 2003; Kovac et al. 2003; Kangude 2005).

AlxInyGa1–x–yN quaternary compounds can be formed by alloying GaN with AlN and InN (Nakamura 2000; Jiang and Lin 2002; Hirayama 2005). A great breakthrough was made in the growth of high-quantum-efficiency $Al_xIn_yGa_{1-x-y}N$ thin films on lattice-mismatched substrates, such as sapphire, by S. Nakamura and coworkers (1993, 1997, 1999, 2000). Other important substrate materials are 6H-SiC (Koga and Yamaguchi 1991; Strite and Morkoc 1992; Morkoc et al. 1994) and ZnO (Willander et al. 2009). Table 1.2 summarizes some physical properties of end members of III-nitrides. As seen, the lattice mismatch among the ternary nitrides is so small that it permits great range and flexibility in band structure (heterojunction) design. A continuous alloy system $Al_xIn_yGa_{1-x-y}N$ can be formed by just varying the Al(In) mole fraction x and y, and the band gaps in the entire compistions are direct, which vary from 1.89 eV for InN, to 3.4 eV for GaN, to 6.2 eV for AlN. $Al_xIn_yGa_{1-x-y}N$ semiconductor alloys have band gap energy that can be tuned in the blue-green region of the visible spectrum and UV region, which allows their use as high-brightness UV, blue, and green LEDs (Nakamura and Fasol 1997; Nakamura and Chichibu 2000; Hirayama 2005).

High-quality semiconductor thin layers are usually grown on a suitable single-crystal substrate, known as epitaxial growth method (Strite and Morkoc 1992; Morkoc et al. 1994; Fox 2001). Two techniques, molecular

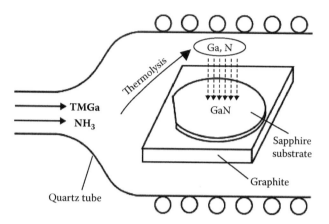

FIGURE 1.5
Schematic drawing of growth of GaN by MOVPE.

beam epitaxy (MBE) and metal-organic vapor phase epitaxy (MOVPE), are commonly used to produce epitaxial layers of semiconductors. MOVPE is the method of choice to grow III-V nitride thin films commercially (see Figure 1.5). In the process, individual components of the semiconductor layer to be grown are formed as gaseous elements by thermolysis of gas source precursors, such as trimethyl gallium ($Ga(CH_3)_3$), trimethyl indium ($In(CH_3)_3$), trimethyl aluminum ($Al(CH_3)_3$), and ammonia (NH_3). Then, the gaseous components, carried by the flowing stream of NH_3, travel to the substrate. Close to the substrate, which is heated by a radiofrequency coil, the thermolyzed elements react, depositing the semiconductor product on the substrate. A typical chemical reaction is

$$Ga(CH_3)_3 + NH_3 \rightarrow GaN + 3CH_4 \qquad (1.1)$$

1.2 Basics of White-Light-Emitting Diodes

1.2.1 Generation of White Light with LEDs

The ultimate goal of solid-state lighting is to replace traditional incandescent and fluorescent lamps for general illumination. Therefore, white LEDs with properties superior to those of traditional light sources, such as high luminous efficiency, tunable color temperatures, high color rendering properties, and long lifetime, are continuously pursued. Although the first commercial red LED was developed in 1968, the first white LED was on sale in 1996 after S. Nakamura invented highly efficient GaN-based blue LEDs in

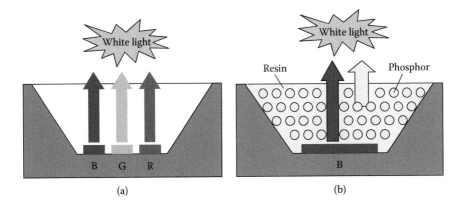

FIGURE 1.6
(See color insert.) White light produced by (a) multi-LED chip and (b) phosphor conversion approaches. RGB stands for red, green, and blue primary LEDs, respectively.

1993. This white LED was manufactured by combining a blue LED chip and a yellow-emitting cerium-doped yttrium aluminum garnet (YAG:Ce) phosphor.

In general, white light can be created with LEDs in two primary methods, as shown in Figure 1.6:

1. *Multi-LED chip approach*, in which light coming from three mono-chromatic red, green, and blue (RGB) LEDs is mixed, resulting in white light to human eyes.
2. *Phosphor conversion*, in which a GaN-based blue or UV LED chip is coated with a yellow or multichromatic phosphor. The mixing of light from the phosphor and the LED chip appear white to human eyes.

When considering the performance (luminous efficacy, color quality, lifetime, etc.) and cost of white LEDs produced by the approaches described above, each approach definitely has certain advantages and disadvantages, as outlined in Table 1.3.

Among phosphor-converted white LEDs, the most common and simplest one consists of a yellow-emitting phosphor and a blue LED chip, which is called a *one-phosphor-converted* (1-pc) white LED. Although the 1-pc white LED exhibits high luminous efficacy, the color rendering index is so low (Ra < 80) that it cannot be accepted for general illumination. This is caused by the relatively narrow emission spectrum of the phosphor. Moreover, the 1-pc white LED usually does not allow for great tunability in color temperature, which is due to the deficiency in red color of the phosphor, which only results in high color temperatures (cold/blue light). To solve these problems, multiple phosphors (e.g., red and green phosphors) are coupled to a blue LED to "fill in" the gap between the blue and yellow light and strengthen the red part of the spectrum, resulting in a multiband white LED with high color

TABLE 1.3

Comparison between Phosphor-Free and Phosphor-Converted White LEDs

LED Type	Advantages	Disadvantages
Multi-LED chip	High color rendering Highest potential efficiency Color flexibility Wide color gamut	Each monochromatic LED responds differently to driven current, temperature, time, and dimming Each LED has its own controlling circuit, making the system more complicated Controls needed for color consistency add expense The performance is limited by green efficiency (green gap) High cost is expected
Phosphor conversion	High luminous flux High luminous efficacy Great color stability over a wide range of temperatures Low cost Mass production process	Does not allow for extensive tunability in terms of spectral modulation May have color variability in beam Low color rendition Technical challenges such as Stokes loss

rendition and tunable color temperature. Alternatively, high-color-rendering white LEDs can be achieved by pumping an RGB phosphor blend with a UV or near-UV chip. For phosphor-converted white LEDs, phosphors play a key role in downconverting the wavelength of the LED chip into suitable visible light, and finally in controlling the light quality of white LEDs. Consequently, downconversion phosphors are one of the key materials in white LED technologies, and will be discussed in later chapters.

1.2.2 Characteristics of White LEDs

White LEDs are proposed to take the place of conventional light sources for general illumination, due to their promises of huge energy savings and reducing carbon emission. Thus, when considering the spectra of light sources for general illumination, the most important characteristics of white LEDs are luminous efficacy and color rendering, the same as those of the commonly used bulb and tube lamps. Other important quantities for white LEDs are chromaticity coordinates and color temperature. In this section, these parameters will be introduced.

Luminous efficacy is the most commonly used measure of the energy efficiency of a light source. It is the ratio of the emitted *luminous flux* from a light source to the amount of the absorbed energy to transmit it, which is expressed in lumens/watt (lm/W). The luminous efficacy of a light source (or luminous efficiency), η_v, describes how well the source provides visible light from a given amount of electricity, which is determined by the following equation (Zukauskas et al. 2002a):

$$\eta_v = \eta_e \times K \tag{1.2}$$

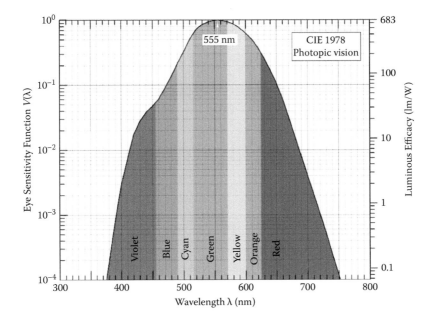

FIGURE 1.7
(See color insert.) Eye-sensitive function, V(λ) (left ordinate), and luminous efficacy measured in lumens per Watt of optical power (right ordinate). V(λ) is highest at 555 nm.

where η_e is the radian efficiency of the source, designating the ability of the light source to convert the consumed power P into radian flux (Φ_e), or the conversion efficiency from the electrical energy to electromagnetic radiation:

$$\eta_e = \Phi_e / P \tag{1.3}$$

K is the *luminous efficacy of radiation*, a measure of the ability of the radiation to produce a visual sensation, which is a characteristic of a given spectrum that describes how sensitive the human eye is to the mix of wavelength involved. It is the ratio of luminous flux (Φ_v) to radiant flux, given as below (Zukauskas et al. 2002a, 2002b):

$$\eta = \frac{\Phi_v}{\Phi_e} = 683 lm / W \times \frac{\displaystyle\int_{380}^{780} V(\lambda) S(\lambda) d\lambda}{\displaystyle\int_{0}^{\infty} S(\lambda) d\lambda} \tag{1.4}$$

In Equation 1.4, V(λ) is the 1924 CIE relative luminous efficiency function for photopic vision that is defined in the visible range of 380–780 nm, as shown in Figure 1.7. S(λ) is the spectral power distribution (SPD), which describes

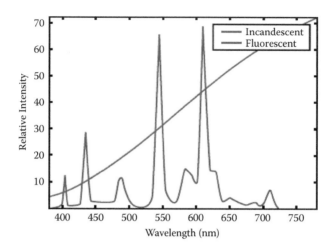

FIGURE 1.8
(See color insert.) Spectral power distribution of 3,000 K lamps.

the power per unit area per unit wavelength of an illumination, or more generally, the per-wavelength contribution to radiant flux:

$$S(\lambda) = \frac{d\Phi_e}{d\lambda} \tag{1.5}$$

To help the reader better understand Equation 1.5, Figure 1.8 presents the spectral power distribution of incandescent and fluorescent lamps with a color temperature of 3,000 K. Chhajed et al. (2005) and Mirhosseini et al. (2009) proposed the calculations of the spectral power distribution of multichip white LEDs as well as phosphor-converted white LEDs. The SPD of an LED can be described by using the Gaussian function:

$$S(\lambda) = P \times \frac{1}{\sigma\sqrt{2\pi}} \times \exp\left[-\frac{1}{2}\left(\frac{\lambda - \lambda_{peak}}{\sigma}\right)^2\right] \tag{1.6}$$

where

$$\sigma = \frac{\lambda_{peak}^2}{2hc\sqrt{2\ln 2}}\, E \tag{1.7}$$

where ΔE is the full width at half maximum (FWHM) of the emission spectrum and P is the optical power of the LED.

For trichromatic white LEDs the SPD can be given as

$$S_{white}(\lambda) = S_{blue}(\lambda) + S_{green}(\lambda) + S_{red}(\lambda) \tag{1.8}$$

For phosphor-converted bichromatic white LEDs, by considering the overlap between the absorption spectrum of the phosphor and the emission spectrum of a blue LED, the number of photons absorbed by the phosphor per second is determined as follows:

$$photon_{abs} = \int \frac{S_{blue}(\lambda)}{hc/\lambda} \left[1 - \exp(-S_{abs}(\lambda)t)d\lambda \right] \tag{1.9}$$

where t denotes the thickness of phosphor, and $Sabs(\lambda)$ is the absorption spectrum of phosphor. Conversion of pumped LED power into yellow power is

$$S_{yellow}(\lambda) = photon_{abs} \times \eta \times S_{ems}(\lambda) \tag{1.10}$$

where η is the quantum efficiency of the phosphor, and $Sems(\lambda)$ is the emission spectrum of the phosphor. Therefore, the SPD of the phosphor-converted bichromatic white LEDs is given as

$$S_{white}(\lambda) = S_{emblue}(\lambda) + S_{yellow}(\lambda) \tag{1.11}$$

where $Semblue(\lambda)$ is the blue power transmitted through the phosphor.

Color rendering is a figure of merit of a light source, which indicates how a light source renders the colors of illuminated people and objects. In most cases, a general *color rendering index* (CRI) R_a is used to evaluate color rendering of a light source, which is obtained by averaging the values of the special color rendering indices of eight standard color samples (Ohno 2004):

$$R_a = \frac{1}{8} \sum_{i=1}^{8} R_i \tag{1.12}$$

In Equation 1.12, R_i is the special color rendering indices for each sample, and is given by

$$R_i = 100 - 4.6\Delta E_i \ (i=1,...,14) \tag{1.13}$$

where ΔE_i is the color differences of 14 selected Munsell samples when illuminated by a reference illuminant and when illuminated by a given illumination. The reference illuminant is the Planckian radiation for test sources with a correlated color temperature (CCT) of <5,000 K, or a phase of daylight for test sources with CCT ≥ 5,000 K.

The color rendering of a light source is determined by its spectral power distribution. A high-color-rendering index generally requires that the light source have a broad emission spectrum covering the entire visible spectrum. The sun and blackbody radiation have a maximum CRI value of 100. A lower CRI value means that some colors may appear unnatural when illuminated by the light. Basically, for illumination purposes, CRI values in the 70s are considered acceptable, and values greater than 80 are regarded as good. Incandescent lamps have a CRI of 95, the cool white fluorescent lamp has a value of 62, and fluorescent lamps using tricolor rare earth phosphors have a value of >80.

There is a trade-off relationship between luminous efficacy and the color rendering index, because the luminous efficacy of radiation is highest for monochromatic radiation at 555 nm, where the human eye is most sensitive, whereas the color rendering index is best achieved by broad band spectra distributed throughout the visible region. Zukauskas et al. (2002b) proposed a function to optimize the trade-off between *Ra* and *K* in multichip LED lamps, as below:

$$F_\sigma \left(\lambda_1, ..., \lambda_n, I_1, ..., I_n \right) = \sigma K + \left(1 - \sigma \right) R_a \qquad (1.14)$$

where σ is the weight that controls the trade-off between the efficacy and color rendering ($0 \le \sigma \le 1$), *n* is the number of primary LEDs, and λi, *Ii* are peak wavelengths and luminous fluxes of the primary LED sources. Although two-band white LEDs are most efficient theoretically, having an efficacy of >400 lm/W, they suffer from a low CRI. Mirhosseini et al. (2009) demonstrated through simulation that both the color rendering index and luminous efficacy could be enhanced significantly over a broad range of correlated color temperatures by using dual-blue LED chips in phosphor-converted white LEDs. Alternatively, an excellent balance between color rendering and luminous efficacy is attained in trichromatic or tetrachromatic white LEDs (Zukauskas et al. 2002b; Ohno 2004). White LEDs with CRI > 85 and K > 300 lm/W can be achieved by optimizing the number and wavelength of primary LEDs.

Chromaticity coordinates are one of basic concepts of *colorimetry* that quantify and describe physically the human color perception. The 1931 International Commission on Illumination (CIE) chromaticity coordinates (x,y) are computed from the spectral power distribution of the light source and the color-matching functions (see Figure 1.9). Tristimulus values X, Y, and Z are obtained by integrating the spectrum with the standard color-matching functions $\bar{x}(\lambda)$, $\bar{y}(\lambda)$, and $\bar{z}(\lambda)$ through the following equations (Zukauskas et al. 2002a):

$$X = \int \bar{x}(\lambda) S(\lambda) d\lambda \qquad (1.15a)$$

$$Y = \int \bar{y}(\lambda) S(\lambda) d\lambda \qquad (1.15b)$$

$$Z = \int \bar{z}(\lambda) S(\lambda) d\lambda \qquad (1.15c)$$

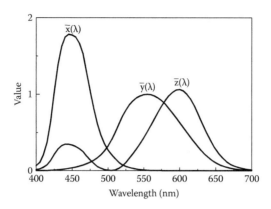

FIGURE 1.9
CIE 1931 XYZ color-matching functions.

Tristimulus values X, Y, and Z specify a color's:

- Lightness—light or dark
- Hue—red, orange, yellow, green, blue, purple
- Saturation—pink-red, pastel-fluorescent, baby blue–deep blue

Therefore, the 1931 CIE chromaticity coordinates (x,y) of a light source with the spectrum $S(\lambda)$ are calculated as follows:

$$x = \frac{X}{X+Y+Z} \tag{1.16a}$$

$$y = \frac{Y}{X+Y+Z} \tag{1.16b}$$

$$z = \frac{Z}{X+Y+Z} \equiv 1-x-y \tag{1.16c}$$

Using x,y as the coordinates, a two-dimensional chromaticity diagram (the CIE 1931 color space diagram) can be plotted as shown in Figure 1.10. As seen, the spectral locus, the purple boundary (a straight bottom line), and the blackbody locus comprise the chromaticity diagram, which is horseshoe shaped. The area covered by the horseshoe-shaped curve comprises the chromaticity coordinates of all real colors. The blackbody locus represents the chromaticities of blackbodies having various (color) temperatures. In the chromaticity diagram, we can see that:

- Linear combinations of colors form straight lines.
- Any color in the interior (i.e., convex hull) of the horseshoe can be achieved through the linear combination of two pure spectral colors.

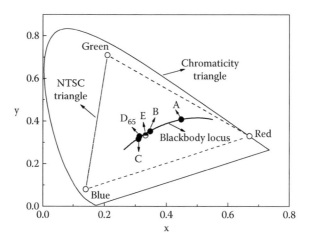

FIGURE 1.10
1931 CIE chromaticity diagram.

- The straight line connecting red and blue is referred to as the line of purples.
- RGB primaries form a triangular color gamut.
- The color white falls in the vicinity of the blackbody locus.

Four standard white points are also included in the chromaticity diagram: A (tungsten lamp, 2,865 K, x = 0.448, y = 0.408), B (direct sunlight, 4,870 K, x = 0.348, y = 0.352), C (overcast sunlight, 6,770 K, x = 0.310, y = 0.316), and D65 (daylight, 6,504 K, x = 0.313, y = 0.329). Point E denotes the equal energy white (5,400 K, x = 0.333, y = 0.333).

In colorimetry, the *CIE 1976 (L', u', v') color space*, also known as the *CIELUV color space*, is a color space adopted by the International Commission on Illumination (CIE) in 1976, as a simple mathematical transformation of the 1931 CIE XYZ color space, but which attempted perceptual uniformity (i.e., the same distance between any two points are presumed to be perceptually equal). It is extensively used for applications such as computer graphics, which deal with colored lights.

$$u' = \frac{4x}{12y - 2x + 3} \qquad (1.17a)$$

$$v' = \frac{9y}{12y - 2x + 3} \qquad (1.17b)$$

Color temperature is a term used to describe the relationship of a white light source with a Planck's blackbody radiator. At increasing temperatures it emits

TABLE 1.4

Performance Characteristics of Some Light Sources

Light Source	CCT (K)	CRI	Efficacy (lm/W)
Equal energy spectrum	5,457	95	—
Daylight, 5,500 K	5,000	100	—
Daylight, 11,000 K	11,000	100	—
Xenon, 1,000 W	5,900	96	22
Incandescent, 60–100 W A-lamp	2,800	100	17
Compact fluorescent			
15–20 W, 2,700–3,000 K	2,800	82	61
15–20 W, 5,000 K	5000	85	47
T8 fluorescent			
32 W, 3,500 K	3,300	84	92
32 W, 5,000K	4,800	87	88
Metal halide			
320–400 W/warm color	3,600	68	92
175–400 W/cool color	4,300	61	90
Ceramic metal halide			
35–70 W/warm color	2,900	84	79
100 W/warm color	3,100	81	93
100–150 W/cool color	4,100	93	90
High-pressure sodium, 200–400 W	2,000	12	120
Low-pressure sodium, 180 W	1,800	52	180
Light-emitting diode (LED)			
White LED	5,000	78	13
RGB (615/525/470 nm) LED mix	4,400	65	22

visible radiation in the red, orange, yellow, white, and ultimately bluish white. If a light source generates a different spectrum from a blackbody radiator, a *correlated color temperature* (CCT) is used, which is an extrapolation of the color of the light source to the color of a blackbody radiator of a given color temperature such that they appear the same color to the human eye. CCT is given in Kelvin (unit of absolute temperature). The CCT for any white light can be determined by drawing a perpendicular line from the measured 1931 CIE chromaticity coordinates to the blackbody locus to obtain the (x,y) intercept. Then the CCT of the white light would correspond to the surface temperature of a blackbody radiator with the same (x,y) chromaticity coordinates.

The correlated color temperature, color rendering index, and luminous efficacy of some popular light sources are listed in Table 1.4.

The most popular white LEDs, made of a blue LED and a YAG:Ce phosphor, usually have high correlated color temperatures, often above 5,000 K, producing a "cold" bluish light. Today, thanks to advances in phosphor technologies, tunable correlated color temperatures (from 2,700 to 6,500 K) can be achieved through the combination of suitable phosphors. Furthermore, the luminous efficacy and color rendering of warm white LEDs (2,600 to 3,500 K, typically for house illumination) have recently been improved remarkably, now approaching or exceeding those of compact fluorescent lamps (CFLs).

1.2.3 Phosphors for White LEDs

Phosphor is one of the key materials in LED-based solid-state lighting technologies, and it absorbs the light emitted from ultraviolet, violet, or blue LEDs and converts it into visible light. Therefore, phosphor plays an important role in determining the performance of white LEDs, such as luminous efficacy, color rendering, correlated color temperature, color gamut, lifetime, and so on.

Luminescent materials or phosphors have a history of about 100 years, and today are a mature technology, finding their applications in lighting, display, and imaging. There are tremendous reports on phosphors in cathode-ray tubes (CRTs), fluorescent lamps, x-ray films, and plasma display panels (PDPs), but most of these traditional phosphors cannot be accepted for use in white LEDs due to their quite low efficiency in converting LED radiation. Furthermore, these traditional phosphors usually exhibit large thermal quenching, resulting in luminescence quenching at the temperature of LED packages (>150°C). Therefore, the search for novel phosphors for white LEDs is continuously carried out, and the development of phosphors with suitable emission color, high conversion efficiency, and small thermal quenching is an urgent mission in order to keep pace with advances in solid-state lighting technology and the expanding applications of white LEDs.

The application field of white LEDs spreads considerably annually, varying from backlights and flashlights for electronic devices to general illumination. For certain applications, the specifications of white LEDs, such as luminous efficacy, color rendition, and color temperature, will vary from case to case. Accordingly, the requirements of phosphors are basically dependent on the end use of white LEDs. For example, the white LED backlight for liquid crystal displays (LCDs) should have a color gamut as wide as possible, which requires that the phosphors have a narrow emission band and suitable emission colors to match well with LCD color filters. When white LED is used for general lighting or medical applications, its color rendering index should be as high as possible; phosphors in such lamps thus should have very broad emission spectra to cover the whole visible spectral region. Of course, the brightness (i.e., luminous efficacy) of white LEDs for various applications is generally required to be high, which indicates that the phosphors should have a high luminescence intensity as well. In summary, the phosphors for white LEDs should have the following common specifications:

- Strong absorption of emitted light from ultraviolet, violet, or blue LED chips
- High conversion efficiency of LED chips with desired wavelengths

- Useful emission colors
- Small thermal quenching of luminescence intensity
- Chemically or compositionally stable against temperature and humidity
- Fine and uniform particle size
- Ease of synthesis and mass production
- No hazardous elements contained

To find phosphors that meet those requirements, the host lattice of phosphors and the activator ions should be carefully selected. The selection rules for the host lattice and activator ions are summarized below:

- Host lattice exhibiting large crystal field splitting to lower the 5d energy levels
- Host lattice having strong covalent chemical bonds (nephelauxetic effect)
- Host lattice being chemically stable
- Activator ions having a small coordination number with ligands
- Activator ions having short chemical bonding to ligands
- Activator ions being spin allowed
- Activator ions showing broad emission bands

To date, a number of new and promising LED phosphors have been invented, developed, or modified that have demonstrated their suitability in white LEDs. These phosphors include oxides, sulfides, phosphates, and nitrides. Among those, nitride phosphors are considered a novel class of luminescent materials, and are getting much more attention in recent years. Nitride phosphors show significantly red-shifted excitation and emission spectra, high chemical stability, and very small thermal degradation; thus they are very suitable for use in white LEDs, typically warm white LEDs, and high-power LEDs. This book focuses on the crystal chemistry, synthesis, photoluminescent properties, and applications of nitride phosphors, and these will be introduced in Chapters 4–7. Of course, other types of LED, phosphors, mainly oxide phosphors, will also be presented in Chapter 3.

1.3 Applications of Solid-State Lighting

As a type of solid-state lighting, light-emitting diodes (LEDs) offer the lighting market a new and revolutionary light source that saves energy, improves quality, reliability, performance, and service, and reduces CO_2 emission. Today, LEDs are competing or challenging traditional light sources: incandescent, halogen, fluorescent, neon, and high-intensity discharge (HID). With great progress made in enhancing the extraction efficiency of LED chips, packaging design, packaging materials, and phosphor technologies, LEDs are now penetrating steadily into the light market and replacing conventional light sources in a very broad range. Generally, the applications of LEDs fall into the following major categories:

- Visual signal applications to convey a message or meaning, including indicators, traffic signals, decorative holiday lights, exit signs, electric signage, and rear lamps of autos
- Illumination where LED light is reflected from objects to give the visual response of these objects, including backlighting for LCDs and lightweight laptop displays, flashlights for cameras and videos, street lighting, and general lighting
- Light generation for measuring and interacting with processes that do not involve the human visual system, such as communication, sensors, plant growth, biological imaging, medical treatment, water purification, and machine vision systems (Goins et al. 1997; Dietz et al. 2003; Pan et al. 2003; Hu et al. 2008)

1.3.1 Visual Signals (Colored Light Applications)

LEDs are competing successfully with conventional, incandescent light sources that use color filters to generate the desired colored light emission, such as those found in traffic signals and exit signs (see Figure 1.11). In such applications, LEDs are chosen because they offer more cost-effective performance than incandescent lamps. Compared with incandescent signals, LED signals have advantages, as mentioned below:

- *Very low power consumed.* LED signals consume only 10–20 W, compared with 135 W for incandescent lamps, saving about 93% energy.
- *Very long lifetime expected.* Traffic signals operate 24 h per day year-round, amounting to an annual operating cycle of 8,760 h. The lifetime of incandescent lamps is usually about 1,000 h, whereas it is up to 50,000 h for LED signals. This indicates low maintenance costs and is environment friendly.
- *Catastrophic failures avoided.* An incandescent lamp has only one filament; when it fails, the lamp loses functions and needs to be replaced immediately. Although an LED signal is made out of LED arrays that contain dozens of LEDs, it continues to work even if several LEDs darken.

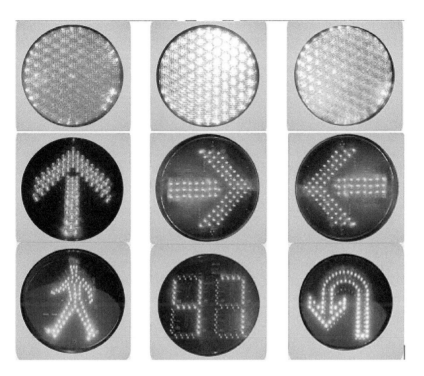

FIGURE 1.11
(See color insert.) LED traffic signals. (From http://www.szslr.com/proshow.asp?id=1431.)

- *Intersection safety enhanced.* LED signals are brighter than incandescent ones, and their brightness improves steadily.

- *Phantom effect eliminated.* Incandescent traffic signals use reflectors behind the bulbs. For signals on east-west approaches during morning and evening hours, all colors seem to light up when the sunrays fall directly on these signals. This problem is eliminated when LED signals are used because there are no reflectors.

- *Battery-driven possibility.* Because LED signals require very low power to operate, it is feasible to run the signals with battery backup during power failures.

The revolutionary triumph of LEDs in automotive lighting started at the end of the 1980s and the beginning of the 1990s with the first central high-mounted stop lamps (CHMSLs). The fact that today more than 95% of these lamp functions are already being generated with the aid of LEDs underlines the dynamic advanced development and penetration of this technology. The implementation of LEDs in rear combination lamps (RCLs) began a decade later. At first, the individual tail lamp, stop lamp, and direction indicator functions were developed on an LED basis to supplement conventional filament bulb technology. In 2005 the first full-LED rear combination lamp was launched on the market. All the functionalities, including the reverse lamp, are realized on the basis of LEDs. They clearly demonstrate the unique combination of advantages of a complete LED lamp solution: vehicle service life, reduced design space thanks to extremely low-profile design, and top-level energy savings.

In addition, monochromatic LEDs with low power consumption, a long lifetime, and small size are well suited for status indicators, exit signs, decorative holiday lights, and message and destination displays.

1.3.2 Lighting and Illumination

With the development of high-efficiency and high-power LEDs it has become possible to incorporate LEDs in lighting and illumination. For white light applications, the competition between high-power white LEDs and traditional light sources is currently quite intense. LEDs are chosen because they offer superior benefits to customers in terms of energy savings, an extended operational life span, high reliability, and are environmentally friendly, when compared with conventional white light sources, such as incandescent bulbs, fluorescent tubes, halogen lamps, and HIDs. These advantages are summarized below:

> *Low energy consumption.* LEDs consume up to 90% less energy. Currently the energy efficiency of white LEDs is ~50%, indicating white LEDs are ten times more efficient than incandescent bulbs, three times more efficient than compact fluorescent lamps, two times more efficient than fluorescent tubes, and even more efficient than a sodium lamp. This means that substantial amounts of electricity/energy can be saved in white light applications where LED light sources are adopted to substitute for conventional lamps.
>
> *Long service life.* Usually a filament or halogen bulb has a typical lifetime of a few hundred hours, and fluorescent tubes have a service life of 6,000–30,000 h, whereas LEDs have much longer lifetimes, up to 100,000 h. The long service life indicates low maintenance costs and less disposal waste issues.
>
> *Highly directional light source.* This nature is very suitable for street lighting or task lighting, highlighting particular architectural or decorative features, or providing local spot lighting.
>
> *Less heat generation.* White LEDs promise higher energy efficiency and less Stokes losses than conventional light sources, producing less forward infrared heat or UV radiation and projecting less heat into the illuminated objects.

FIGURE 1.12
(See color insert.) LED lighting in an aircraft cabin of a Virgin America A320. (From http://flickr.com/photos/72398575@N00/975198348.)

High reliability. LEDs are solid-state devices with no moving parts, glass, or filaments; and they have reduced maintenance costs, particularly in hard-to-access or remote locations, such as bridges, tall buildings, helicopter landing pads, and even the tops of wind generators.

Better color quality. The color temperature and color rendering of white LEDs are tunable, by controlling the wavelength spectra through appropriate selection of phosphors.

Free of mercury. Unlike fluorescent lamp tubes, which use gaseous mercury as the excitation source for phosphors (each tube contains 3–5 mg mercury), white LEDs are fabricated by combining InGaN-based LED chips and phosphors, both of which are free of mercury.

The drastic increase in the performance of white LEDs in particular in the past few years has made LED technology interesting for indoor illuminations (see Figure 1.12), such as recessed downlights, retail displays, refrigerated display cases, task lights, office undershelf lights, and kitchen undercabinet lights, as well as outdoor illuminations, such as street and area lights, and step, path, and porch lights.

In addition to energy savings and long lifetime, increases in reliability, compact design (space savings), and high efficiency have led to the fact that white LEDs are fast becoming a viable alternative to halogen and high-intensity discharge headlamps for automotive designers. As early as

FIGURE 1.13
(See color insert.) LED headlamps of an Audi A8. (From http://www.motortrend.com/roadtests/sedans/2011_audi_a8_first_look_and_photos/photo_26.html.)

2002, the first position light to use white high-power LEDs and be integrated in the headlamp was presented to the market. Although main-beam headlights using LEDs are currently at the prototype stage of development, some new models, such as Audi's A8 luxury saloon (see Figure 1.13) and Lexus's LS600, are using LED arrays for daytime running lights, which can improve safety.

A field where white LEDs show great promise is in LCD backlight applications. The present backlight units are cold-cathode fluorescent lamps (CCFLs) that are actually made out of very fine fluorescent tubes. Compared with CCFL backlights, white LEDs are able to (1) produce an image with greater dynamic contrast, (2) offer a wider color gamut, (3) make thinner displays possible, and (4) reduce environmental pollution on disposal. The first commercial LCD TV using LED backlights was the Sony Qualia 005, which was introduced in 2004 (Kakinuma 2006). Today, a variety of LCD TV sets, manufactured by Samsung Electronics, Sharp, Sony, Panasonic, Philips, and LG Electronics, are adopting LED backlights, which are also named LED-backlit LCD televisions (LED-TVs).

1.3.3 Nonvisual Applications

LEDs have many applications other than seeing or lighting. These applications can be categorized into three groups: *communication, sensors,* and *light matter interaction.*

LEDs are used extensively in optical fiber and free space optics communications because the solid-state light can be modulated very quickly. This

FIGURE 1.14
(See color insert.) A panel of red LEDs used for illumination for a plant growth experiment with possible future application to food growing in space. (From http://www.msfc.nasa.gov/news/news/photos/2000/photos00-336.htm.)

includes remote controls, such as TVs and VCRs, where infrared LEDs are often used. Opto-isolators use an LED combined with a photodiode or phototransistor to provide a signal path with electrical isolation between two circuits. It is especially useful in medical equipment where the signals from a low-voltage sensor circuit in contact with a living organism must be electrically isolated from any possible electrical failure in a recording or monitoring device operating at potentially dangerous voltages.

LEDs can be used as a photodiode for both light detection and emission. This capability has been demonstrated and used in a variety of applications, including ambient light detection, bidirectional communications, and touch-sensing screens. In addition, RGB LEDs are challenging CCFLs in some flat scanners, which have independent control of three illuminated colors, which allows the scanners to calibrate themselves for more accurate color balance, and no warm-up is required.

Lots of materials and biological systems are sensitive to or rely on light. The use of LEDs is particularly interesting to plant cultivators, because they are more energy efficient, produce less heat (which can damage plants), and provide the optimum light frequency for plant growth and bloom periods (see Figure 1.14). LEDs are also developed for healthcare applications: UV LEDs for sterilization or purification to remove bacteria and viruses from water or air; for use in surface skin treatments, cancer cell destruction, and endoscopes; specialized white light sources for changing moods and encouraging patients to sleep or wake; white light sources for tiny cameras to take pictures inside the body; etc.

References

Chhajed, S., Xi, Y., Li, Y.-L., Gessmann, Th., and Schubert, E. F. 2005. Influence of junction temperature on chromaticity and color rendering properties of trichromatic white-light sources based on light-emitting diodes. *J. Appl. Phys.* 97:054506-1–054506-8.

Craford, M. G. 1992. LEDs challenge the incandescents. *Circuits and Devices*, 24–29.

Dietz, P., Yerazunis, W., and Leigh, D. 2003. Very low-cost sensing and communication using bidirectional LEDs. Mitsubishi Electric Research Laboratories. http://www.merl.com/reports/docs/TR2003-35.pdf (accessed August 27, 2010).

Fox, M. 2001. *Optical properties of solids*. New York: Oxford University Press.

Gessmann, Th., and Schubert, E. F. 2004. High-efficiency AlGaInP light-emitting diodes for solid-state lighting applications. *J. Appl. Phys.* 95:2203–2216.

Goins, G. D., Yorio, N. C., Sanwo, M. M., and Brown, C. S. 1997. Photomorphogenesis, photosynthesis, and seed yield of wheat plants grown under red light-emitting diodes (LEDs) with and without supplemental blue lighting. *J. Exp. Botany* 48:1407–1413.

Hirayama, H. 2005. Quaternary InAlGaN-based high-efficiency ultraviolet light-emitting diodes. *J. Appl. Phys.* 97:091101.

Holonyak, N., Jr., and Bevacqua, S. F. 1962. Coherent (visible) light emission from Ga(As1-xPx) junctions. *Appl. Phys. Lett.* 1:82–83.

Hu, H., Zhang, J., Davitt., K. M., Song, Y.-K., and Nurmiko, A. V. 2008. Application of blue-green and ultraviolet micro-LEDs to biological imaging and detection. *J. Phys. D Appl. Phys.* 41:094013.

Humphreys, C. J. 2008. Solid-state lighting. *MRS. Bull.* 33:459–470.

Jiang, H. X., and Lin, J. Y. 2002. AlGaN and InAlGaN alloys—Epitaxial growth, optical and electrical properties, and applications. *Opto-electronics Rev.* 10:271–286.

Kakinuma, K. 2006. Technology of wide color gamut backlight with light-emitting diode for liquid crystal display television. *Jpn. J. Appl. Phys.* 45:4330–4334.

Kangude, Y. 2005. Red emitting photonic devices using InGaP/InGaAlP material system. Master diss., Massachusetts Institute of Technology.

Kim, J. K., and Schubert, E. F. 2008. Transcending the replacement paradigm of solid-state lighting. *Opt. Express.* 16:21835–21842.

Kitai, A. 2008. *Luminescent materials and application*. West Sussex, UK: John Wiley & Sons Ltd.

Koga, K., and Yamaguchi, T. 1991. Single crystals of SiC and their application to blue LEDs. *Prog. Crystal Growth Charact.* 23:127–151.

Kovac, J., Peternai, L., and Lengyel, O. 2003. Advanced light emitting diodes structures for optoelectronic applications. *Thin Solid Film* 433:22–26.

Madelung, O. 1996. *Semiconductors, basic data*. 2nd ed. Berlin: Springer-Verlag.

Mirhosseini, R., Schubert, M. F., Chhajed, S., Cho, J., Kim, J. K., and Schubert, E. F. 2009. Improved color rendering and luminous efficacy in phosphor-converted white light-emitting diodes by use of dual-blue emitting active regions. *Opt. Express.* 17:10805–10813.

Monemar, B., Shih, K. K., and Pettit, G. D. 1976. Some optical properties of the AlxGa1-xAs alloy system. *J. Appl. Phys.* 47:2604–2613.

Morkoc, H., and Mohammad, S. N. 1995. High-luminosity blue and blue-green gallium nitride light-emitting diodes. *Science* 267: 51–55.

Morkoc, H., Strite, S., Gao, G. B., Lin, M. E., Sverdlov, B., and Burns, M. 1994. Large-band-gap SiC, III-V nitride, and II-VI ZnSe-based semiconductor device technology. *J. Appl. Phys.* 76:1363–1398.

Nakamura, S. 1999. InGaN-based blue light-emitting diodes and laser diodes. *J. Cryst. Growth* 201–202:290–295.

Nakamura, S., and Chichibu, S. F. 2000. *Introduction to nitride semiconductor blue lasers and light emitting diodes*. London: Taylor & Francis.

Nakamura, S., and Fasol, G. 1997. *The blue laser diode*. New York: Spinger.

Nakamura, S., Mukai, T., Senoh, M., Nagahama, S., and Iwasa, N. 1993. $In_xGa_{(1-x)}/In_yGa_{(1-y)}N$ superlattices grown in GaN films. *J. Appl. Phys.* 74:3911–3915.

Ohno, Y. 2004. Color rendering and luminous efficacy of white LED spectra. In *Proceedings of SPIE*, ed. I. T. Ferguson, N. Narendran, S. P. Denbaars, and J. C. Carrano, 88–98. Vol. 5530. Bellingham, WA: SPIE.

Pan, Y. L., Boutou, V., Chang, R. K., Ozden, I., Davitt, K., and Nurmikko, A. V. 2003. Application of light-emitting diodes for aerosol fluorescence detection. *Opt. Lett.* 28:1707–1709.

Ponce, F. A., and Bour, D. P. 1997. Nitride-based semiconductors for blue and green light-emitting devices. *Nature* 386:351–359.

Schubert, E. F. 2003. *Light emitting diodes*. Cambridge: Cambridge University Press.

Schubert, E. F., and Kim, J. K. 2005. Solid-state light sources getting smart. *Science* 308:1274–1278.

Schubert, E. F., Kim, J. K., Luo, H., and Xi, J. Q. 2006. Solid state lighting—A benevolent technology. *Rep. Prog. Phys.* 69:3069–3099.

Schubert, M., Woollam, J. A., Leibiger, G., Rheinlander, B., Pietzonka, I., Saβ, T., and Gottschalch, V. 1999. Isotropic dielectric functions of highly disordered $Al_xGa_{1-x}InP$ ($0 \leq x \leq 1$) lattice matched to GaAs. *J. Appl. Phys.* 86:2025–2033.

Steranka, F. M. 1997. High brightness light emitting iodes. In *Semiconductors and semimetals*, ed. G. B. Stringfellow and M. G. Craford, 65–95. Vol. 48. San Diego: Academic Press.

Strite, S., and Morkoc, H. 1992. GaN, AlN, and InN: A review. *J. Vac. Sci. Technol. B* 10:1237–1266.

Wu, J. 2009. When group-III nitrides go infrared: New properties and perspectives. *J. Appl. Phys.* 106:011101.

Willander, M., Nur, O., Zhao, Q. X., Yang, L. L., Lorenz, M., Vao, B. Q., Perez, J. Z., Czekalla, C., Zimmermann, G., Grundmann, M., Bakin, A., Behrends, A., Al-Suleiman, M., El-Shaer, A., Mofor, A. C., Postel, B., Waag, A., Boukos, N., Travlos, A., Kwack, H. S., Guinard, J., and Dang, D. L. S. 2009. Zinc oxide nanorod based photonic devices: Recent progress in growth, light emitting diodes and laser. *Nanotechnology* 20:332001-1–332001-40.

Zukauskas, A., Shur, M. S., and Gaska, R. 2002a. *Introduction to solid state lighting*. New York: John Wiley & Sons.

Zukauskas, A., Vaicekauskas, R., Ivanauskas, F., Gaska, R., and Shur, M. S. 2002b. Optimization of white polychromatic semiconductor lamps. *Appl. Phys. Lett.* 80:234–236.

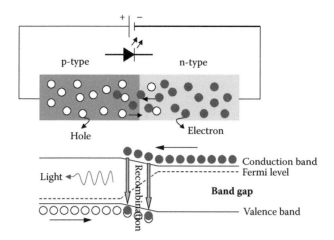

FIGURE 1.2
Schematic diagram of a light-emitting diode. The characteristics of the emitted light are related to the energy gap, E_g, between the conduction and valence band in the semiconductor. E_F is the Fermi energy.

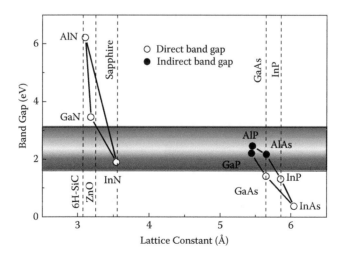

FIGURE 1.4
Band gaps of selected III-V semiconductor materials against their lattice constant. The lattice constant data of commonly used substrate materials (6H-SiC, ZnO, sapphire, GaAs, and InP) are also included.

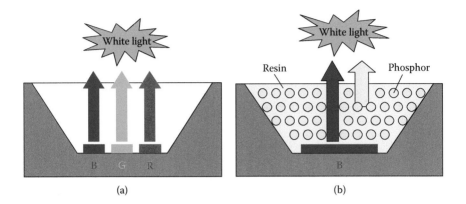

FIGURE 1.6
White light produced by (a) multi-LED chip and (b) phosphor conversion approaches. RGB
stands for red, green, and blue primary LEDs, respectively.

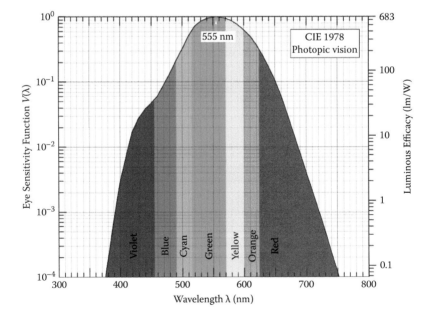

FIGURE 1.7
Eye-sensitive function, V(λ) (left ordinate), and luminous efficacy measured in lumens per Watt
of optical power (right ordinate). V(λ) is highest at 555 nm.

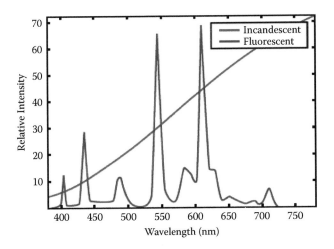

FIGURE 1.8
Spectral power distribution of 3,000 K lamps.

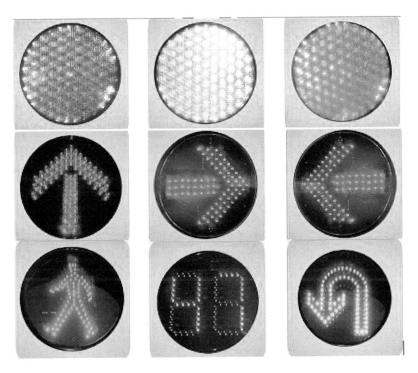

FIGURE 1.11
LED traffic signals. (From http://www.szslr.com/proshow.asp?id=1431.)

FIGURE 1.12
LED lighting in an aircraft cabin of a Virgin America A320. (From http://flickr.com/photos/72398575@N00/975198348.)

FIGURE 1.13
LED headlamps of an Audi A8. (From http://www.motortrend.com/roadtests/sedans/2011_audi_a8_first_look_and_photos/photo_26.html.)

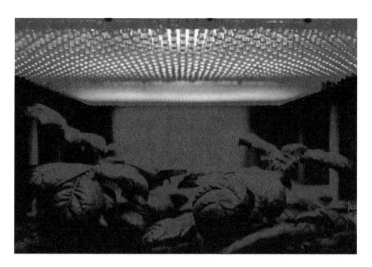

FIGURE 1.14
A panel of red LEDs used for illumination for a plant growth experiment with possible future application to food growing in space. (From http://www.msfc.nasa.gov/news/news/photos/2000/photos00-336.htm.)

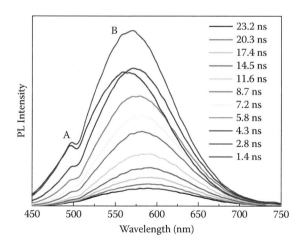

FIGURE 2.30
Time-resolved luminescence of $(Sr_{0.99}Eu_{0.01})AlSi_4N_7$. (Reprinted from Ruan, J., Xie, R.-J., Hirosaki, N., and Takeda, T., *J. Am. Ceram. Soc.*, DOI: 10.1111/j.1551-2916.2010.04104.x, 2010. With permission.)

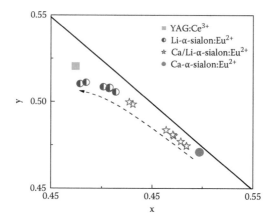

FIGURE 4.70
Chromaticity coordinates of Ca-, Li-, and (Ca,Li)-α-sialon:Eu²⁺.

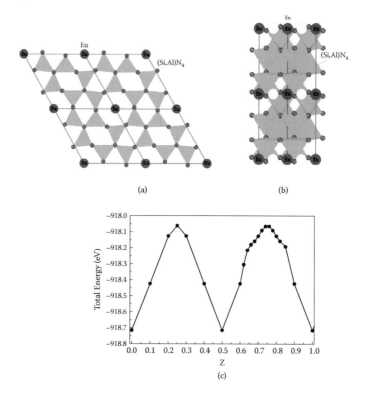

FIGURE 5.10
Crystal structure of β-SiAlON:Eu²⁺, viewed along (a) (001) and (b) (100). (Reprinted from Li, Y. Q., Hirosaki, N., Xie, R.-J., Takeda, T., and Mitomo, M., *J. Solid State Chem.*, 181, 3200–3210, 2008b. With permission.) (c) The total energy of β-SiAlON:Eu²⁺ as a function of the fractional position of the Eu atom along (0, 0, Z).

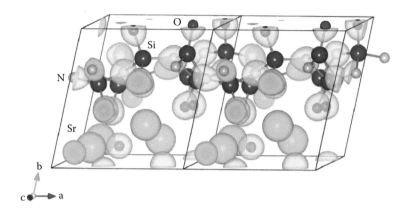

FIGURE 5.31
Three-dimensional valence electronic density distribution image of $SrSi_2O_2N_2$, viewed along (001).

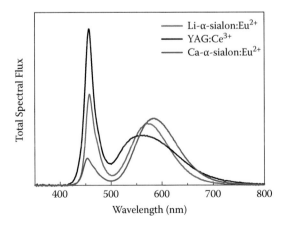

FIGURE 7.5
Emission spectra of white LEDs using Li-α-sialon:Eu^{2+}, YAG:Ce^{3+}, and Ca-α-sialon:Eu^{2+}.

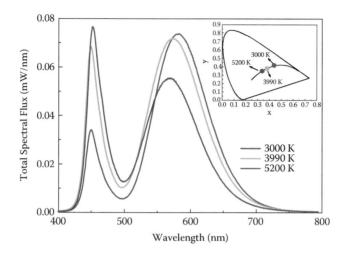

FIGURE 7.6
Emission spectra and chromaticity coordinates of white LEDs using Li-α-sialon:Eu²⁺ with different emission wavelengths. (Reprinted from Xie, R.-J., Hirosaki, N., Mitomo, M., Sakuma, K., and Kimura, N., *Appl. Phys. Lett.*, 89, 241103, 2009. With permission.)

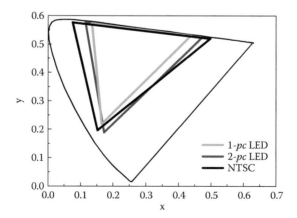

FIGURE 7.19
CIE 1976 chromaticity coordinates of white LEDs. (Reprinted from Xie, R.-J., Hirosaki, N., and Takeda, T., *Appl. Phys. Express*, 2, 022401, 2009. With permission.)

2

Introduction to Luminescence

Luminescence is a general term given to the phenomenon that a substance releases extra energy in the form of electromagnetic radiation either immediately or over a long period, in excess of thermal radiation, after it is excited by a certain type of external energy, such as electromagnetic radiation (light), an electric field, a high-energy electron, or an electron beam. Obviously, the luminescence process basically involves the excitation, energy transfer, and subsequent relaxation of the valence electrons from their excited states (emission). According to different excitation sources, luminescence can be classified into photoluminescence, electroluminescence, cathodoluminescence, radioluminescence, chemiluminescence, thermoluminescence, triboluminescence, etc. (García Solé et al. 2005; Yen et al. 2006), as seen in Table 2.1. In this book, we pay attention to the photoluminescence of a material.

Luminescent materials, also called *phosphors,* are usually inorganic crystalline materials (host lattices) that are intentionally doped with a small amount of impurities or activators (i.e., rare earth ions or transition metal ions). Therefore, when regarding the luminescence mechanism or luminescence process, the influences of both host lattices and activators should be considered. In the chapters that follow, we will look at the luminescent properties and applications of a variety of phosphors. Before that, we must first understand the fundamental physical processes of luminescence. This is the subject of the present chapter.

The structure of this chapter is as follows. Section 2.1 briefly introduces optical processes when light interacts with a substance, such as refraction, reflection, absorption, scattering, and luminescence. Section 2.2 describes the

TABLE 2.1

Classification of Luminescence

Type	Excitation Source	Application
Photoluminescence	Light	Fluorescent lamps, white LEDs, PDPs, afterglow
Electroluminescence /Electric current	Electric field	LED, flat-panel displays, screens
Cathodoluminescecne	Electron beam	FED, CRT
Radioluminescence	X-rays, α-, β-, γ-rays	X-ray detectors, scintillators
Chemiluminescence	Chemical reaction	Gas sensors, labeling
Thermoluminescence	Heating	Detectors, dosimeters
Triboluminescence	Mechanical energy	Remote sensing

luminescence process (i.e., excitation, emission, and nonradiative transitions) when an activator is promoted to its excited state by absorption of enough energy, and then returns to its ground state by emitting photons. Section 2.3 discusses the luminescence center, such as lanthanide ions, transition metal ions, and complex ions, which are commonly used in practical phosphors.

2.1 Classification of Optical Processes

When light interacts with a medium, some of the light is reflected from the front surface, whereas the rest of the light penetrates into the medium and propagates through it, as seen in Figure 2.1. Furthermore, the following general phenomena take place when light propagates through an optical medium such as a phosphor: refraction, absorption, scattering, and luminescence. In this section, we will briefly introduce these basic concepts. Following this, the luminescence process—excitation and emission and nonradiative energy transfer—will be discussed.

2.1.1 Refraction

Refraction causes the light to pass through the medium with a smaller velocity than in free space, but it does not affect the intensity of the light wave. Different mediums slow down the speed of passing light at different rates. This property of matter is called the *refractive index* (symbol *n*). It is defined as the ratio of the velocity of light in free space *c* to the velocity of light in the medium v (Fox 2001):

$$n = \frac{c}{v} \tag{2.1}$$

The refractive index depends on the frequency of the light.

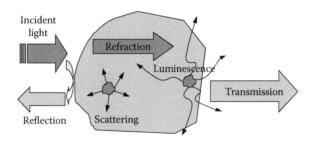

FIGURE 2.1
Optical processes of the incident light interacting with an optical medium.

2.1.2 Reflection

Reflection causes the light to bounce off the surface of a medium. It bounces away at the same angle that it hits the material. The reflection at the surfaces is described by *reflectivity* (symbol R), which is defined as the ratio of a reflected intensity to the intensity incident on the surface:

$$R = \frac{I_R}{I_0} \qquad (2.2)$$

where I_R is the reflected optical intensity, and I_0 the incident intensity (García Solé et al. 2005).

There are two types of visible light reflection: specular reflection and diffuse reflection. Specular reflection occurs on a smooth surface where all the light hits the surface at the same angle, and reflects at the same angle. Although diffuse reflection occurs on a rough or textured surface where the light hits the material at the same angle, the light reflects in different directions because there are microscopic variations in the surface. Correspondingly, reflectivity spectra have two different modes: direct reflectivity and diffuse reflectivity. Direct reflectivity is measured with well-polished samples at normal incidence, while diffuse reflectivity is generally applied to unpolished or powdered samples. As for phosphors, diffuse reflectivity spectra are usually measured, which will be discussed in the next section.

2.1.3 Absorption

Absorption occurs if the frequency of the traveling light is resonant with the transition frequencies (for example, a ground- to excited-state transition) of the atoms in the medium. In this case, the light beam will be attenuated as it propagates. That is, a fraction of the intensity is generally emitted, usually at a lower frequency than that of an incident light, giving rise to *luminescence*. The other fraction of the absorbed intensity is dissipated by a nonradiative process. Selective absorption is responsible for the coloration of phosphors.

The process of absorption for a two-level system is shown in Figure 2.2. The atom is shifted from the ground state (with energy of E1) to the excited state (with energy of E2) by absorbing the required energy from a photon. For this to be possible, the energy difference between E_2 and E_1 must satisfy the following equation:

$$h\nu = E_2 - E_1 \qquad (2.3)$$

where $h = 6.62 \times 10^{-34}$ J is the Planck's constant, and ν is the frequency of the photon. Clearly, absorption is not a spontaneous process. The electron cannot be promoted to the excited state without being stimulated by an incoming photon.

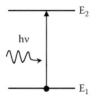

FIGURE 2.2
Absorption between two levels of an atom.

The absorption of light by an optical medium is quantified by the absorption coefficient (symbol α). This coefficient is defined as the fraction of the power absorbed in a unit length of the medium. If the light beam is traveling in the z direction, and the intensity (i.e., optical power per unit area) at location z is $I(z)$, then the light intensity attenuation dI after propagating a differential thickness dz can be written as (Fox 2001; García Solé et al. 2005)

$$dI = -\alpha I(z) dz \qquad (2.4)$$

Upon integration of Equation 2.4, we obtain the Lambert–Beer law:

$$I(z) = I_0 e^{-\alpha z} \qquad (2.5)$$

where I_0 is the optical intensity at z = 0.

The *transmission* of the medium clearly has relations with the absorption, because only unabsorbed light can be transmitted. The coefficient of transmission or transmissivity (T) is defined as the ratio of the transmitted intensity to the incident intensity of light. If absorption or scattering does not occur, then we obtain

$$R + T = 1 \qquad (2.6)$$

As for an absorbing medium with thickness z, if the front and back surfaces of the medium have equal reflectivities, then the transmissivity is given by (Fox 2001)

$$T = (1-R)^2 e^{-\alpha z} \qquad (2.7)$$

In addition, the absorption of an optical medium is also sometimes described in terms of the optical density (*O.D.*) or absorbance. It is defined as

$$O.D. = \log_{10}\left(\frac{I_0}{I(z)}\right) \qquad (2.8)$$

Obviously, the optical density is directly connected to the absorption coefficient α according to Equation 2.5 as follows:

$$O.D. = \log_{10}\left(e^{\alpha z}\right) = \alpha z \log_{10}(e) = 0.434\alpha z \qquad (2.9)$$

It is apparent that the optical density is dependent on the sample thickness, but the absorption coefficient is not. Furthermore, the optical density can be easily related to other optical magnitudes that can be directly measured by spectrophotometers, such as transmissivity, according to Equation 2.10:

$$T = 10^{-O.D.} \qquad (2.10)$$

2.1.4 Scattering

Scattering is the phenomenon in which the light changes its traveling direction, and possibly also its frequency, when it propagates through an optical medium. The total number of photons remains unchanged, but it is spatially distributed. In this sense, scattering has the same attenuating effect as absorption. There are two types of scattering processes: elastic scattering and inelastic scattering (Raman scattering), depending on the frequency of the scattered light. If the frequency of the scattered light is the same as that of incident light, it is called *elastic scattering*; if the frequency of the scattered light is different from that of incident light, it is called *inelastic scattering*.

Scattering occurs due to variations of the refractive index of the medium on a length scale shorter than the wavelength of the light. It could result from the presence of impurities, defects, or inhomogeneities in the medium. Analogous to absorption, scattering causes the attenuation of a light beam according to (Fox 2001)

$$I(z) = I_0 e^{-N\delta_s Z} \qquad (2.11)$$

where N is the number of scattering centers per unit volume, and δ_s is the *scattering cross section* of the scattering center (Fox 2001). This equation is the same in form as the Lambert–Beer law given in Equation 2.5, with $\alpha \equiv N\delta_s$.

If the size of the scattering center is much smaller than that of the light wavelength, then the scattering is described as Rayleigh scattering. In this case, the scattering cross section will change with the wavelength λ according to

$$\delta_s(\lambda) \propto \frac{1}{\lambda^4} \qquad (2.12)$$

It is clearly seen in Equation 2.12 that inhomogeneous materials tend to scatter short wavelengths more strongly than longer wavelengths.

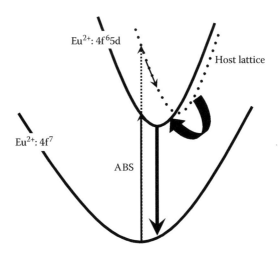

FIGURE 2.3
A configurational coordinate model illustrating the luminescence process (ABS = absorption).

2.1.5 Luminescence

Luminescence is the phenomenon in which the atoms return to the ground state by spontaneous emission of photons after they are promoted into excited states by absorbing photons of appropriate frequency. It is, in some ways, the inverse process to absorption. Luminescence, however, does not always accompany absorption. It usually takes a certain period of time for the excited atoms to go back to the ground state by spontaneous emission. This implies that the radiative de-excitation process occurs after the excited atoms release the excitation energy as heat (nonradiative process). Therefore, the luminescence efficiency is closely related to the dynamics of the de-excitation mechanism in the atoms (Fox 2001).

The physical processes involved in luminescence are more complicated than those in absorption. This is due to the fact that (1) the emission of light by luminescence is intimately tied up with the energy relaxation mechanisms in the solid, and (2) the shape of the emission spectrum is affected by the thermal distribution of the electrons and holes within their bands. Therefore, to understand the luminescence efficiency and the luminescence spectrum, we need to think about the emission rates and the thermal spread of carriers.

Figure 2.3 schematically shows the luminescence process in a solid. When atoms in a solid absorb enough excitation energies, they will be lifted up to the excited state by forming valence electrons in the excited state and leaving behind holes in the ground state. The photon is emitted when an electron in an excited state drops down to recombine holes in the ground state.

The emission rate for radiative transition between two levels (i.e., excited state and ground state) is determined by the Einstein A coefficient. If the upper level has a population N at time t, the radiative emission rate is given by (Fox 2001)

$$\left(\frac{dN}{dt}\right)_{radiative} = -AN \tag{2.13}$$

This indicates that the number of photons emitted in a given time is closely related to both the *A* coefficient of the transition and the population of the upper level. By solving the rate equation, we obtain

$$N(t) = N(0)e^{-At} = N(0)e^{-\frac{t}{\tau_R}} \tag{2.14}$$

where $\tau_R = 1/A$ is the radiative lifetime of the transition.

Figure 2.3 also indicates that, besides radiative emission, it is possible for nonradiative relaxation to take place during the process of electrons in an excited state dropping down to recombine with holes in the ground state. For instance, the electrons lose their excitation energy in terms of heat by emitting photons, or they may transfer the energy to traps, such as impurities or defects. Very little emission will be observed if the nonradiative relaxation dominates the transition process.

The luminescent efficiency η_R can be calculated by considering the nonradiative relaxation processes in the rate equation for the population of the excited state (Fox 2001):

$$\left(\frac{dN}{dt}\right)_{total} = -\frac{N}{\tau_R} - \frac{N}{\tau_{NR}} = -N\left(\frac{1}{\tau_R} + \frac{1}{\tau_{NR}}\right) \tag{2.15}$$

where τ_{NR} is the nonradiative lifetime. η_R is defined by the ratio of the radiative emission rate to the total de-excitation rate. By dividing Equation 2.15 by Equation 2.13, we obtain (Fox 2001)

$$\eta_R = \frac{AN}{N\left(\dfrac{1}{\tau_R} + \dfrac{1}{\tau_{NR}}\right)} = \frac{1}{1 + \dfrac{\tau_R}{\tau_{NR}}} \tag{2.16}$$

This indicates that if $\tau_R \ll \tau_{NR}$, then η_R approaches one, and the maximum possible amount of photons is emitted. On the other hand, if $\tau_R \gg \tau_{NR}$, then η_R is ultimately small, and the light emission is extremely weak. Therefore, for gaining high luminescent efficiency, the radiative lifetime should be much shorter than the nonradiative lifetime.

2.2 Fundamentals of Luminescence

In this section we consider the physical processes that are involved in promoting atoms to an excited state by absorbing the excitation energy, and dropping them down to the ground state by emitting photons. These processes include excitation, emission, energy transfer, and nonradiative transitions.

2.2.1 Excitation

2.2.1.1 General Consideration

A luminescent material does not emit light unless enough photons with required frequency are absorbed. This implies that, unlike emission, excitation is not a spontaneous process, as mentioned previously. When we consider that luminescent materials are usually composed of host lattices and activators, the excitation process has two types: direct excitation and indirect excitation. Direct excitation means that the activators directly absorb the excitation energy, promoting atoms to the excited state of the activator ions. Indirect excitation means that of the host lattice first absorbs the excitation energy, and then transfers it to the excited state of the activator ions. Generally, the energy levels of activator ions are located in between the conduction band and valence band of the host material; the required energy for direct excitation is therefore smaller than that of indirect excitation. Moreover, the efficiency of direct excitation is generally larger than that of indirect excitation. These are typically important for phosphors in white LEDs, because the excitation sources in these devices are usually near ultraviolet or blue light that has lower energy than that in fluorescent lamps or cathode ray tubes (CRTs).

The excitation spectrum of a phosphor is measured by a spectrophotometer. Let us look at an excitation spectrum of a well-known garnet phosphor for white LEDs: cerium-doped yttrium aluminum garnet ($Y_3Al_5O_{12}:Ce^{3+}$). Figure 2.4 illustrates the typical room temperature excitation spectrum of YAG:Ce^{3+}, monitored at 460 nm. The spectrum has the following features:

1. Two strong and well-shaped broad bands centered at 340 and 455 nm
2. A weak broad band in the range of 200–300 nm
3. No narrow lines observed

It is reported that the undoped YAG only shows an absorption edge located at ~190 nm. Therefore, the excitation spectrum of YAG:Ce^{3+} in Figure 2.4 mainly stands for the absorption of Ce^{3+}. Two main excitation bands at 340 and 455 nm are ascribed to the electronic transitions from the ground state ($^2F_{5/2}$) to different energy levels of the 5d excited state of Ce^{3+} (Blasse and Grabmaier 1994). Although the 5d orbit of Ce^{3+} can be split into a maximum of five energy levels, the third weak band in the range of 200–300 nm is perhaps due to the Ce^{3+} absorption or the impurity/defect absorptions. The more intense peak located at 455 nm matches very well the blue emission of GaN-based LED, indicating that the YAG:Ce^{3+} phosphor can absorb the blue light efficiently and convert it into visible light at a longer-wavelength range.

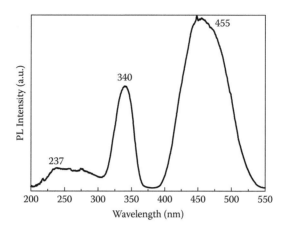

FIGURE 2.4
Excitation spectrum of $Y_3Al_5O_{12}$:Ce^{3+}. The broad bands are electronic transitions between 4f and 5d of Ce^{3+}.

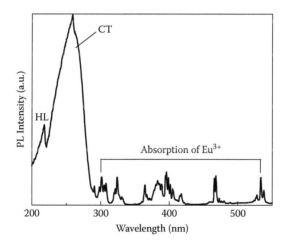

FIGURE 2.5
Excitation spectrum of Y_2O_3:Eu^{3+}. The narrow lines are transitions with the $4f^6$ configuration of Eu^{3+}, CT is the charge transfer between O^{2-} and Eu^{3+}, and HL represents absorption of the host lattice.

We can see different shapes and information from the excitation spectrum of the second example. Figure 2.5 presents the excitation spectrum of the Eu^{3+}-doped Y_2O_3 nanophosphor. It has the following features:

1. A very intense broad band with a maximum at ~260 nm

2. An absorption peak at ~220 nm

3. A number of weak sharp lines at 300, 323, 364, 382, 395, 468, and 532 nm

The undoped Y_2O_3 only shows the host lattice absorption, which is located at ~220 nm. So the intense band and the narrow spectral lines are associated with Eu^{3+}. The very strong broad band in the excitation spectrum is ascribed to a charge-transfer (CT) transition in which an electron is promoted from the highest-filled oxygen orbital to Eu^{3+}, and the groups of narrow line spectra are due to the Eu^{3+} absorption, resulting from the electronic transitions of $^7F_j \rightarrow {}^5D_0$ of Eu^{3+} (Blasse and Grabmaier 1994). From the excitation spectrum of $Y_2O_3:Eu^{3+}$, we can see that Eu^{3+} can be excited both directly (CT band and narrow lines) and indirectly (HL absorption). For indirect excitation, the energy is absorbed by the Y_2O_3 host lattice first, followed by transferring absorbed energy to Eu^{3+}.

From these observations, we can understand that the excitation spectrum of a luminescent material is greatly dependent on the activator species and host lattice. Moreover, the excitation spectrum tells us that the emission of a phosphor can be observed if the activator ions (such as Ce^{3+}, Eu^{3+}, Eu^{2+}, Mn^{2+}, etc.) are directly excited, or if the energy absorbed by either the host lattice or some ligands is transferred to activator ions through the nonradiative transition process. On the other hand, there are still some questions about the excitation spectrum that need to be clarified, for example:

- Why are some spectra so broad and others so narrow? Or what factors affect the shape of the excitation spectrum?
- Why are some spectra so intense and others so weak?
- How does the host lattice affect the spectral positions and shapes?

In the following section, we will answer these questions in a very simple form by using the configurational coordinate diagram.

2.2.1.2 The Width of an Excitation Band

The width of an excitation band, i.e., broad or narrow, can be interpreted by the configurational coordinate diagram (Klick and Schulman 1957; Curie 1963; Dibartolo 1968; Kamimura et al. 1969). Such a diagram is often used to explain optical properties, particularly the effect of lattice vibrations of a localized center on the basis of potential curves, representing the total energy of the absorbing center in its ground or excited state as a function of the configurational coordinate.

Figure 2.6 shows the configurational coordinate diagram, plotting the energy E vs. the metal-ligand distance R (Blasse and Grabmaier 1994). The configurational coordinate represents the arrangement of nuclei in the vicinity of an absorbing center. The energy levels for the harmonic oscillator are quantized. Within the ground state, represented by the parabola U, the wave functions of these vibrational levels are labeled $v_0, v_1, v_2, \ldots, v_n$. For the lowest vibration level (v_0), the system has the highest probability of being located in the center

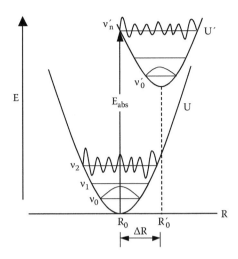

FIGURE 2.6
Configurational coordinate diagram. The ground state (U) has the equilibrium distance R_0, and its vibrational levels of $v = 0, 1$, and 2 are shown. The excited state (U') has the equilibrium distance R_0', and its vibrational levels of $v' = 0, 1, 2$, and n are illustrated.

of the parabola at R_0. For higher vibration levels, the probability is greater for finding the system at the turning points than in the center of the parabola.

The description of the ground state in the *E-R* diagram also holds true for the excited states. Similar to the ground state, the excited states of the absorbing center also have a parabolic shape (U') with a minimum at R_0'. As the chemical bond in the excited state is different from that in the ground state, the parabolas are shifted relative to each other over a distance of ΔR.

Absorption of energy causes transition from the lowest vibrational level (v_0) of the ground state (U) to some high vibrational level (e.g., v'_n) in the excited energy state (U'). The probability is therefore greatest for transitions initiating from R_0, where the vibrational wave function has its maximum value. The transition will terminate on the edge of the excited state parabola where the vibrational levels of the excited state find their highest amplitude. The absorption band (given by the bold arrow labeled E_{abs}) will have a maximum energy corresponding to that transition. Of course, it is also possible for transitions to initiate at R values smaller ($R < R_0$) or larger ($R > R_0$) than R_0. This distribution of probabilities leads to the width of the absorption band, because the energy difference of the transition for $R < R_0$ will be greater than that for $R = R_0$, and it is smaller for $R > R_0$.

In the configurational coordinate diagram, the value of $\Delta R = R'_0 - R_0$ represents how strong the interaction between the absorbing center and the vibrations of its surroundings is. It is said that the probability of a transition between the ground state and the excited state is proportional to both electronic wave functions and vibrational wave functions of the ground state and the excited state. The former determines the intensity of the transition, and the latter gives the width of the absorption band.

If $\Delta R = 0$, the ground-state and excited-state parabolas lie exactly above each other, and the vibrational overlap has its maximum for levels of $v = 0$ and $v' = 0$ since the vibrational wave functions of the ground state and excited state have their maxima at R_0. This finally will lead to a sharp-line excitation spectrum, corresponding to the transition from the vibrational level of $v = 0$ to that of $v' = 0$. This transition is also called a *zero-phonon line*, as it occurs without the participation of a phonon. On the other hand, if $\Delta R \neq 0$, then the $v = 0$ vibrational level will find its maximal vibrational overlap with several levels of $v' > 0$. This will result in a broad excitation band at the end. With these in mind, we can therefore understand that the value of ΔR determines the width of an excitation spectrum, and the larger the value of ΔR, the broader the excitation spectrum.

2.2.1.3 Selection Rules and the Intensity of an Excitation Band

The variation in intensity of an optical transition is very pronounced because not all transitions between different energy levels occur with equal probability. The probability of electronic transitions from the ground state to the excited state is governed by selection rules that are derived from quantum mechanics. The selection rules determine which transitions are allowed or which are forbidden. Two important selection rules controlling the probability of electronic transitions are shown below (Weckhuysen et al. 2000):

1. *The spin selection rule.* Only electronic transitions are allowed between states of the same spin ($\Delta S = 0$). In other words, only those transitions are allowed that result in a conservation of the number of unpaired electrons in the d orbitals.

2. *The orbital selection rule.* Transitions involving only a redistribution of electrons within the same set of atomic orbitals are forbidden, or orbitally allowed transitions obey $\Delta l = \pm 1$ (l is the orbital angular momentum quantum number).

In fact, however, there are various factors that weaken these rules, so that forbidden transitions still appear, but with very reduced intensity. These factors are mainly due to interactions between atoms that produce hybrid electron states. Taking Eu^{3+} as an example in Figure 2.5, optical absorption transitions of Eu^{3+} occur within the f-shells ($4f^6$), so that they are strongly forbidden by the parity selection rule, but odd components of the crystal field make them partially allowed due to the admixture of sites with the opposite parity. The net effect of these selection rules is that forbidden transitions are weak, even for those partially allowed transitions.

On the other hand, the charge-transfer transitions are allowed. The intensity of charge-transfer transitions arises from the movement of electronic charge across a typical interatomic distance, producing a large transition dipole moment and a concomitant large oscillator strength for the absorption process (Henderson and Imbusch 1989). Therefore, the charge-transfer band has a higher intensity than the 4f-4f transition lines, as shown in Figure 2.5.

The absorption of Ce^{3+} is assigned to allowed electronic transitions between 4f and 5d, so that the intensity of Ce^{3+} absorptions is very strong, as seen in Figure 2.4.

2.2.1.4 Influence of Host Lattice

Figure 2.7 illustrates a scheme of an absorbing center consisting of a dopant ion O coordinated to six ligand ions L. This center, OL_6, is also called an *optically active center*. For a given dopant ion, its absorption band not only is dependent on its energy level, but also is significantly affected by the host lattice. This is because the energy levels of a free ion O will vary with the presence of the nearest neighboring ions L in the lattice. That is, the absorption band of a dopant ion is affected by the nature of chemical bonds between O and L (i.e., covalency, bond length, coordination number, symmetry, etc.). The two important factors responsible for the variations in excitation spectra of a given ion are *nephelauxetic effect* and *crystal field splitting*.

The nephelauxetic effect, also called the *cloud-expanding effect*, is regarded as a measure of covalency of the bond between the absorbing ion and ligand. As covalency increases, the interaction between the electrons is reduced due to the expansion of electron clouds. This therefore leads to the shift of electronic transitions to lower energy with increasing covalency. According to Reisfeld (1973; Gaft et al. 2005), the degree of the nephelauxetic effect is expressed by Equation 2.17 using a parameter β:

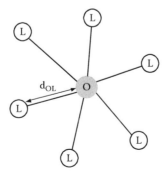

FIGURE 2.7
An optically active center, OL_6, containing a doping ion O that is coordinated to six ligand ions L. (Reproduced from García Solé, J., Bausa, L. E., and Jaque, D., *An introduction to the optical spectroscopy of inorganic solids*, John Wiley & Sons, West Sussex, UK, 2005. With permission.)

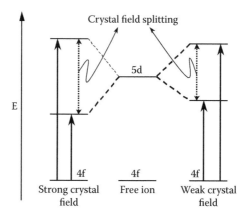

FIGURE 2.8
Schematic image of the crystal field splitting of Ce^{3+}.

$$\beta = \frac{\left(\sigma_f - \sigma\right)}{\sigma} \tag{2.17}$$

where σ_f is the energy of a free ion (in wavenumber), and σ is the absorption maximum, which can vary depending on host materials.

The charge-transfer absorption band is usually observed in the UV region. Jørgensen (1975) and Reisfeld (1976) defined optical electronegativities χ_{opt}, expressed by

$$\sigma_{abs} = \frac{\chi_{opt}(X)}{\chi_{uncorre}(M)} \times 30000 \tag{2.18}$$

where σ_{obs} is the wavenumber for the absorption peak of the charge-transfer states, $\chi_{uncorre}(M)$ is the optical negativity without correction for subshell energy differences, M is the RE ion, and X is the host lattice. The dependence of σ_{obs} on host lattices is attributable to the nephelauxetic effect, which results from the expansion of the partially filled shell due to the electronic transition from the ligands to the core of the central RE ion. Therefore, as the covalency for the chemical bond between an RE ion and the ligand increases, the electronegativity difference between the RE ion and ligand reduces, resulting in the shift of the charge-transfer transitions to lower energy. For example, in Eu^{3+}-doped oxide phosphors, the charge-transfer band varies with the host lattice: 45,000 cm^{-1} in YPO_4, 37,000 cm^{-1} in $LaPO_4$, and 30,000 cm^{-1} in Y_2O_2S (Blasse and Grabmaier 1994).

Crystal field splitting is another key factor that affects the absorption band of an absorbing ion. As for the 4f-4f interconfigurational transitions, the absorption band does not vary with the host lattice because the 4f electrons are well shielded by the outer electronic layer. On the other hand, the

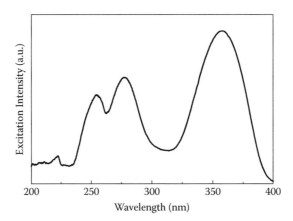

FIGURE 2.9
Excitation spectrum of Li_2SrSiO_4:Ce^{3+}.

absorption band due to the 4f-5d electronic transitions is greatly influenced by the circumstances surrounding the rare earth ions. With strong crystal field splitting, the energy level of 5d is significantly lowered, leading to the red-shift of the absorption band (see Figure 2.8). This is why YAG:Ce^{3+} can absorb blue light, as shown in Figure 2.4, but Li2SrSiO4:Ce3+ only absorbs UV light, as seen in Figure 2.9.

2.2.2 Emission

2.2.2.1 The Emission Process

Figure 2.10 presents the configurational coordinate diagram for explaining the emission process. As described in Section 2.1.1, the absorption of the excitation energy promotes the absorbing center to a high vibrational level of the excited state (U'). Then the excited ion releases some of its energy to its surroundings and returns to the lowest vibrational level of the excited state. This process is also called *relaxation*, which brings the excited ions to their equilibrium state, with their interatomic distances corresponding to an equilibrium position. The relaxation process is nonradiative, and therefore no emission is observed.

From the lowest vibrational level of the excited state, the excited ions can go back to the ground state (U) spontaneously through the radiative emission of photons. By this process, the ions then reach a high vibrational level of the ground state, and finally relax to the lowest vibrational levels of the ground state nonradiatively.

2.2.2.2 Stokes Shift

Again, there is a distribution of probabilities for the starting points of the emission, with the greatest probability being at R'_0. Because the relaxation

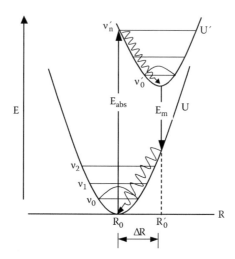

FIGURE 2.10
Configurational coordinate diagram showing the emission and relaxation processes.

process dissipates some of the energy of the excited ion, radiative decay to the ground state occurs at a lower energy than the absorption process (i.e., $E_{abs} > E_{em}$). This energy difference between positions of the band maxima of the excitation and emission spectra of the same electronic transition is known as *Stokes shift* (in wavenumber or frequency unit). Assuming that the parabolas of the excited state and ground state have the same shape, the energy lost during relaxation is equal to $Sh\nu$ per parabola. Then, the Stokes shift (SS) is expressed by

$$SS = 2Sh\nu \qquad (2.19)$$

where $h\nu$ is the energy difference between two vibrational levels, and S is the Huang-Rhys coupling factor that measures the strength of the electron-lattice coupling. S is proportional to $(\Delta R)^2$ (Henderson and Imbusch 1989).

The Stokes shift is an importance feature of luminescence. It can be seen from Figure 2.10 and Equation 2.19 that a larger value for ΔR will result in a larger Stokes shift. An appropriate small Stokes shift is desired, generally for achieving high luminescence, because (1) it avoids strong overlap between the absorption and emission bands (otherwise, the emitted light would be reabsorbed by the emitting center), and (2) it results in small thermal quenching.

2.2.2.3 Temperature Dependence of Emission

Temperature affects the emission with the results of *band broadening* and *thermal quenching*.

By reinspecting Figures 2.6 and 2.10, we can easily understand that the increased temperature will lead to a broader absorption or emission band.

This is due to the fact that the vibrational levels of the ground state ($v = 0, 1, 2, 3, \ldots$) or those of the excited state ($v' = 0, 1, 2, 3, \ldots$) are populated as a result of temperature increase, so that they also participate in the absorption or emission process. By considering this thermalization effect, Henderson and Imbush (1989) proposed the following equation to relate the bandwidth (ΔE) of the absorption or emission band to temperature:

$$\Delta E(T) \approx \Delta E(0) \sqrt{\coth\left(h\nu/2KT\right)} \tag{2.20}$$

where $\Delta E(0)$ is the bandwidth at 0K, and $h\nu$ is the energy of the coupling phonon. It is seen from Equation 2.20 that the excitation or emission band broadens with temperature.

Thermal quenching of emission is the phenomenon in which the emission efficiency decreases as the temperature increases. The luminescence intensity at temperature T, $I(T)$, can be related to temperature according to (Chen and McKeever 1997)

$$I(T) = \frac{I(0)}{1 + \dfrac{\Gamma_0}{\Gamma_v}\exp(\dfrac{-\Delta E_{tq}}{K_B T})} \tag{2.21}$$

where Γ_v is the probability of a radiative transition, Γ_0 is a preexponential constant, K_B is Boltzmann's constant, and ΔE_{tq} is the energy barrier for thermal quenching. Then, assuming that $K_B T \gg \Delta E_{tq}$, the relation between the emission efficiency and temperature can be given as follows:

$$\eta \approx c\exp\left(\frac{-\Delta E_{tq}}{K_B T}\right) \tag{2.22}$$

where c is a constant.

2.2.2.4 Concentration Quenching of Luminescence

In principle, the emission intensity usually increases with an increase in the concentration of a luminescent center in a given host lattice. This is due to the increase of the absorption efficiency that contributes to the increase in the luminescence intensity. However, such behavior is only observed when the concentration of a luminescent center reaches a critical value. Above this value, the luminescence intensity drops. This phenomenon is called *concentration quenching of luminescence*.

An example of $La_2Si_3O_6N_8:Ce^{3+}$ is given in Figure 2.11, where concentration quenching occurs at ~6 mol% Ce^{3+}. Above this critical concentration, the emission of Ce^{3+} quenches dramatically. Let's look at another sample: Tb^{3+}-doped Ca-α-sialon in Figure 2.12. It is clearly seen that with increasing Tb^{3+}

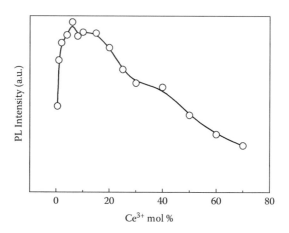

FIGURE 2.11
Emission intensity of $La_2Si_3O_6N_8$:Ce^{3+} as a function of Ce^{3+} concentration.

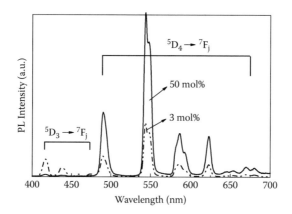

FIGURE 2.12
Emission spectra of Tb^{3+}-doped Ca-α-sialon with varying Tb^{3+} concentrations.

concentration, the emission peaks assigned to $^5D_4 \rightarrow {}^7F_j$ intensify, whereas those assigned to $^5D_3 \rightarrow {}^7F_j$ weaken correspondingly. Figure 2.13 illustrates the changes in the luminescence intensity of $^5D_4 \rightarrow {}^7F_5$ and $^5D_3 \rightarrow {}^7F_5$ transitions with increasing the Tb^{3+} concentration. It indicates that the excitation energy of $^5D_3$ is effectively transferred to the emission of $^5D_4 \rightarrow {}^7F_j$.

The origin of concentration quenching is closely related to energy transfer among the luminescence centers. The quenching initiates at a critical concentration, where the distance between luminescent centers becomes so short that the efficient energy transfer cannot occur any more. It is generally accepted that *donor-killer energy migration* and *cross-relaxation* are two main mechanisms that are responsible for the concentration quenching of luminescence (García Solé et al. 2005).

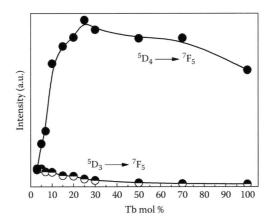

FIGURE 2.13
Emission intensity of Tb^{3+}-doped Ca-α-sialon as a function of Tb^{3+} concentration.

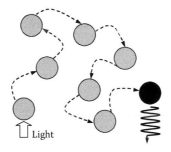

FIGURE 2.14
Donor-killer energy migration mechanism for concentration quenching of luminescence. The excitation energy is transferred to a killer center (black) along a chain of donors (grey). (Reproduced from García Solé, J., Bausa, L. E., and Jaque, D., *An introduction to the optical spectroscopy of inorganic solids*, John Wiley & Sons, West Sussex, UK, 2005. With permission.)

2.2.2.4.1 Donor-Killer Energy Migration Mechanism

The excitation energy can transfer to a large number of centers before being released radiatively, due to very efficient energy transfer. On the other hand, a phosphor material contains more or less a certain amount of defects or impurity ions that can act as acceptors. Therefore, the excitation energy can also be transferred to these sites in the end, if the distance between the donors and acceptors is suitable for efficient energy transfer with increasing the concentration of donors. These acceptors relax the absorbed energy to their ground state via a nonradiative process, so that they are called *killers* that act as an energy sink within the transfer chain to quench the luminescence, as seen in Figure 2.14.

2.2.2.4.2 Cross-Relaxation Mechanism

Not always does all of the excitation energy migrate to luminescent centers; the excitation energy can also be lost via a *cross-relaxation mechanism*. It occurs

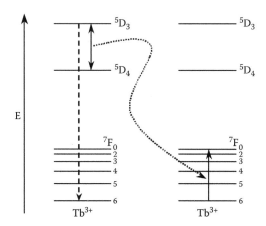

FIGURE 2.15
Quenching of higher levels of emission ($^5D_3$-^7F_j) by the cross-relaxation mechanism in Tb^{3+}-doped phosphors.

by resonant energy transfer between two identical adjacent centers, due to the special energy-level structure of these centers. The mechanism accounts for the concentration quenching of Tb^{3+}-doped phosphors (Van Uitert 1971), as seen in Figures 2.12 and 2.13. A possible energy-level scheme involving the $^5D_3$-$^5D_4$ cross relaxation of Tb^{3+} is schematically shown in Figure 2.15. At a low concentration of Tb^{3+}, the radiative emissions from $^5D_3$ to ^7F_j dominate. However, with increasing the Tb^{3+} concentration, a resonant energy transfer occurs between two neighboring Tb^{3+} ions, transferring part of the energy of $^5D_3$-$^5D_4$ from excited ions (donors) to those ions in the ground state (acceptors), as described by ($^5D_3$, $^7F_6$)–($^5D_4$, $^7F_0$). As a result of cross relaxation, the donor ions will be in the $^5D_4$ energy level, while the acceptor ions will be promoted to the ground state of $^7F_0$. Correspondingly, the luminescence from the transition of $^5D_3 \rightarrow {}^7F_j$ is quenched, whereas the emission arising from the transition of $^5D_4 \rightarrow {}^7F_j$ is enhanced.

2.2.3 Nonradiative Transitions

We have discussed the radiative de-excitation process (i.e., emission) when the excited ions go back to the ground state by releasing the excitation energy. On the other hand, in addition to luminescence, there is the possibility of nonradiative de-excitation, which means that the luminescent center returns to its ground state without emitting photons. In fact, some nonradiative mechanisms have already been used to explain the concentration quenching of luminescence in the above section. In this part, we will discuss this nonradiative transition process that competes with the radiative de-excitation from the excited state.

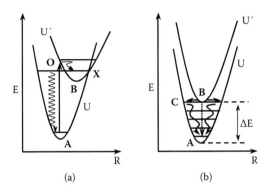

FIGURE 2.16
Configuration coordinate diagram illustrating the nonradiative de-excitation process of multiphonon emission: (a) strong electron-lattice coupling and (b) weak electron-lattice coupling ($S = 0$).

2.2.3.1 Multiphonon Emission

Multiphonon emission is one of the important nonradiative de-excitation processes (Weber 1973; García Solé et al. 2005). A configurational coordinate diagram is again used to qualitatively explain this process. Figure 2.16 illustrates two configurational coordinate models, corresponding to cases of strong and weak electron-lattice coupling, respectively (Blasse and Grabmaier 1994). In the strong electron-lattice coupling case, there is a crossover point X between the ground state U and excited state U′ parabolas, while in the weak coupling case, two parabolas are parallel and will never cross with each other ($S = 0$).

As seen in Figure 2.16a, when a luminescent center is promoted to the excited state, the maximum of absorption will occur along the line AO at absolute temperature. If it is in a normal case, as given in Figure 2.10, the center will then be relaxed to the lowest vibrational level corresponding to point B, and finally return to the ground state along the line BA with the emission of photons. However, this is not the case when the crossover point X is located at an energy level lower than point O. The luminescent center is downrelaxed to the vibrational level corresponding to point X, other than to point B. This level belongs to both the ground state and excited state, so that the de-excitation probability from this level is much higher through the phonon states of parabola U. This finally leads to the luminescent center going back to the ground state by means of a nonradiative multiphonon emission mechanism. No emission is thus observed in this case.

A different situation is given in Figure 2.16b. Nonradiative de-excitation is impossible since there is no crossover point between the parabolas of the ground state and excited state. However, it is possible for the luminescent center to return to the ground state nonradiatively if the energy gap

ΔE is not four to five times larger than the higher vibrational frequency of the surroundings. Therefore, the de-excitation process will occur at point B, followed by the instantaneous transfer to point C, and finally nonradiative relaxation to point A (Yen et al. 2006). This multiphonon emission mechanism is often observed in rare earth ions (RE^{3+}). The nonradiative rate for this mechanism is closely related to the energy gap, and decreases exponentially with the corresponding energy gap. Therefore, this multiphonon emission mechanism is also known as the *energy gap law* (Riseberg and Weber 1975; Weber 1973; García Solé et al. 2005).

2.2.3.2 Thermally Activated Nonradiative De-excitation

As discussed in Section 2.2.2.3, luminescence can be quenched with increasing temperature, which is called *thermal quenching*. We can make use of the configurational coordinate diagram to explain this mechanism, as illustrated in Figure 2.17. A crossing point X is indicated in Figure 2.17, but it is at a higher energy level than point O. In this case, after being promoted to the excited state corresponding to point O, the luminescent center is relaxed nonradiatively to the lowest vibrational level corresponding to point B. At this time, when temperature increases, the center will be thermally activated from point B to the crossover point X, where the parabolas of the ground state and excited state cross, so that it easily reaches the ground state without the emission of phonons through the phonon states of the excited state corresponding to line XC. Therefore, the luminescence due to the transition from point B to point C is quenched.

The probability of a thermally activated de-excitation process is strongly dependent on the energy separation between point X and point B, as given by Equation 2.23 (Yen et al. 2006):

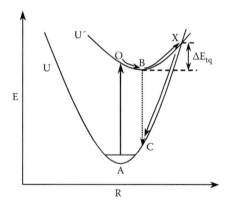

FIGURE 2.17
Configurational coordinate diagram of thermally activated nonradiative transition.

$$W_{NR} = s \exp\left(\frac{-\Delta E_{tq}}{K_B T}\right) \tag{2.23}$$

where ΔE_{tq} is the activation energy for thermal quenching, and s is the frequency factor. This equation has the same meaning as Equation 2.22.

2.2.3.3 Energy Transfer

An excited luminescent center can also return to its ground state by *nonradiative energy transfer* to a neighboring center. Such an energy transfer process is illustrated in Figure 2.18. Let us consider two different luminescent centers, donor D and acceptor A, with a distance of R. First, the donor absorbs the excitation energy $h\nu_D$, and is promoted from the ground state D to an excited state D*. Then, the donor is relaxed to its ground state by transferring its absorbed energy to an acceptor, leading to the shift of the acceptor from its ground state A to an excited state A*. In this process, no donor emission is observed as the donor returns to its ground state (dashed line). Finally, the acceptor decays to its ground state by emitting its own photons with the energy of $h\nu_A$.

For energy transfer to take place, some conditions, for example, the interaction between the excited donor D* and the acceptor center A, as well as the energy differences between the excited and ground states of the donor and acceptor centers, should be fulfilled. Foster (1948) and Dexter (1953) proposed the following equation to describe the probability of energy transfer from the donor centers to the acceptor centers:

$$P_{D \to A} = \frac{2\pi}{\hbar} |\varphi_D \varphi_{A^*}| H_{int} |\varphi_{D^*} \varphi_A|^2 \int g_D(E) g_A(E) dE \tag{2.24}$$

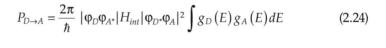

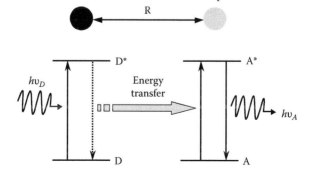

FIGURE 2.18

Nonradiative energy transfer between the donor D and acceptor A. The distance between D and A is R. The energy schemes of D and A are illustrated.

where φ_D and φ_{D^*} are the wavefunctions of the donor center in the ground and excited states, respectively, φ_A and φ_{A^*} are the wavefunctions of the acceptor center in the ground and excited states, respectively, and H_{int} represents the donor-acceptor Hamiltonian interaction. The integral in Equation 2.24 denotes the spectral overlap between the normalized donor emission spectrum ($g_D(E)$) and the normalized acceptor excitation spectrum ($g_A(E)$).

As seen, a high-energy transfer probability requires (1) strong spectral overlap between the donor emission and acceptor absorption, and (2) large Hamiltonian interaction.

The strongest spectral overlap occurs when the donor and acceptor centers have coincident energy levels; i.e., the energy differences between the excited and ground states of the donor and acceptor centers are equal. This case is also known as *resonant energy transfer* (García Solé et al. 2005). On the other hand, when the donor and acceptor centers have different energy levels, an energy mismatch between the transitions of the donor and acceptor centers exists. To allow energy transfer in this case, a lattice phonon with appropriate energies is required. This energy transfer mechanism is often called *phonon-assisted energy transfer* (García Solé et al. 2005).

The donor-acceptor interaction strongly depends on the distance between the donor and acceptor centers and the nature of their wavefunctions. There are generally three types of interaction mechanisms that are involved in the Hamiltonian interaction: multipolar electric interactions, multipolar magnetic interactions, and exchange interactions (García Solé et al. 2005).

Multipolar electric interactions can be divided into *electric dipole–dipole, electric dipole–quadrupole*, and *quadrupole–quadrupole* interaction mechanisms, depending on the nature of the involved transitions between the donor and acceptor centers. The transfer probability varies with R^{-n}, with $n = 6, 8$, and 10 for electric dipole–dipole, electric dipole–quadrupole, and quadrupole–quadrupole interactions, respectively.

Multipolar magnetic interactions are analogous to multipolar electric interactions, having a similar dependence on the distance of the energy transfer probability. In any case, the multipolar magnetic interactions are less important than the s ones.

Exchange interaction occurs only if there is direct overlap between the electronic wavefunctions of the donor and acceptor ions. The exchange interaction probability is closely tied to the distance between the donor and acceptor ions: $P_{ei} \propto e^{-2R/L}$, where L is an average of the radii of the donor and acceptor ions. This indicates that for exchange interactions, the distance dependence is exponential.

For all energy transfer mechanisms discussed above, the luminescence decay time of the donor center is usually shortened as a result of a nonradiative energy transfer process.

Nonradiative energy transfer is often used to improve the efficiency or tune the emission color of phosphors in practical applications. For example, Yang et al. (2005) reported the Eu^{2+}-Mn^{2+} energy transfer in $CaAl_2Si_2O_8$, and

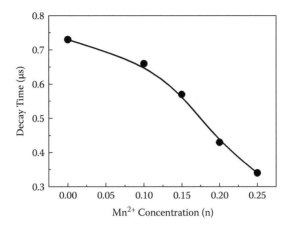

FIGURE 2.19
Decay time of the Eu^{2+} luminescence in $(Ca_{0.99-n}Eu_{0.01}Mn_n)Al_2Si_2O_8$ phosphors.

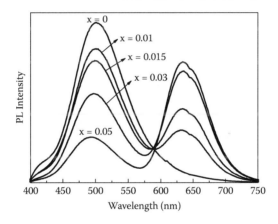

FIGURE 2.20
Emission spectra of $(Ca_{0.995-x}Eu_{0.005}Mn_x)_9La(PO_4)_7$ with varying Mn^{2+} concentrations, showing the energy transfer between Eu^{2+} and Mn^{2+}.

observed the increase of the green emission of Mn^{2+} and the decrease of the blue emission of Eu^{2+} as the Mn^{2+} concentration increases. This also shifted the emission color from blue in Eu^{2+} solely doped material to white in Eu^{2+}-Mn^{2+} codoped samples. Figure 2.19 presents the decay time of the Eu^{2+} emission as a function of the Mn^{2+} concentration in $CaAl_2Si_2O_8$. It is clearly seen that the Eu^{2+}-Mn^{2+} energy transfer shortens the decay time of the Eu^{2+} emission. Figure 2.20 shows the typical emission spectra of Mn^{2+} and Eu^{2+} codoped $Ca_9La(PO_4)_7$ phosphors, in which the energy transfer of $Eu^{2+} \rightarrow Mn^{2+}$ contributes to the enhancement of the Mn^{2+} red emission. Such an energy transfer can also be found in many other phosphors, such as $BaMg_2Si_2O_7$:Eu^{2+}, Mn^{2+} (Yao et al. 1998) and $Ba_3MgSi_2O_8$:Eu^{2+}, Mn^{2+} (Kim et al. 2004).

2.3 Luminescent Centers

In the previous sections we discussed the luminescence process of optical centers in a solid material. One of the key roles of these luminescent centers is to introduce totally new energy levels in between the energy gap of the solid crystals, creating new absorption and emission bands that are not characteristic of the perfect crystals.

Not all of the elements of the periodic table can be considered optically active centers, but only a number of them can be accommodated into the lattice structure of crystals and produce suitable energy levels within the band gap of the crystals. The optically active centers that are of technological and practical importance in luminescent materials are usually based on rare earth ions and transition metal ions. In this section, we will concentrate on these luminescent centers and discuss their energy levels.

2.3.1 Lanthanide Ions

2.3.1.1 Electronic Configuration

Lanthanide ions refer to those ions after lanthanum in the periodic table: from La^{3+} (atomic number 57) to Lu^{3+} (atomic number 71). Lanthanide ions are now commonly used as luminescent centers in phosphors, and they are usually doped in the crystals in the trivalent or divalent state. The electronic configuration is shown in Table 2.2 for both trivalent and divalent lanthanide

TABLE 2.2

Electronic Configurations and Ground State of Lanthanide Ions

Ions	4f Electrons	Ground State	Ions	4f Electrons	Ground State
La^{3+}	0	$^1S_0$			
Ce^{3+}	1	$^2F_{5/2}$			
Pr^{3+}	2	$^3H_4$			
Nd^{3+}	3	$^4I_{9/2}$			
Pm^{3+}	4	$^5I_4$			
Sm^{3+}	5	$^6H_{5/2}$	Sm^{2+}	6	$^7F_0$
Eu^{3+}	6	$^7F_0$	Eu^{2+}	7	$^8S_{7/2}$
Gd^{3+}	7	$^8S_{7/2}$			
Tb^{3+}	8	$^7F_6$			
Dy^{3+}	9	$^6H_{15/2}$			
Ho^{3+}	10	$^5I_8$			
Er^{3+}	11	$^4I_{15/2}$			
Tm^{3+}	12	$^3H_6$			
Yb^{3+}	13	$^2F_{7/2}$	Yb^{2+}	14	$^1S_0$
Lu^{3+}	14	$^1S_0$			

ions (García Solé et al. 2005; Yen et al. 2006). As seen, the lanthanide ions from Ce^{3+} to Yb^{3+} have partially filled 4f orbitals, so that they exhibit their characteristic energy levels, which can induce luminescence processes. On the other hand, the 4f orbit is empty for La^{3+} and fully filled for Lu^{3+}, leading to no energy levels required for luminescence. There are only three divalent lanthanide ions that are stable: Sm^{2+}, Eu^{2+}, and Yb^{2+}. The electronic configurations of these divalent ions are the same as those of Eu^{3+}, Gd^{3+}, and Lu^{3+}, respectively.

The electronic state is denoted as $^{2S+1}L_J$, where S, L, and J are the spin angular momentum, orbital angular momentum, and total angular momentum, respectively. L represents S, P, D, F, G, H, I, K, L, M, ..., when L is equal to 0, 1, 2, 3, 4, 5, 6, 7, 8, 9, ..., respectively. Table 2.2 also shows the ground state of trivalent and divalent lanthanide ions (García Solé et al. 2005; Yen et al. 2006).

The optical transition of trivalent lanthanide ions is different from that of divalent ones, due to their differences in electronic configurations. Trivalent ions show a 4f-4f interconfigurational optical transition, whereas divalent ions show a 4f-5d interconfigurational optical transition. This leads to variations in luminescent properties of lanthanide ions, as will be discussed in the following sections.

2.3.1.2 Energy Levels and Luminescence of Trivalent Rare Earth Ions

As shown in Table 2.2, the outer electronic configuration of trivalent lanthanide ions only differs in the number of 4f electrons, n, having a general form of $5s^2 5p^6 4f^n$. It is the 4f electrons that are involved in the optical transitions of trivalent ions. The 4f electrons, however, are well shielded by the outer $5s^2$ and $5p^6$ electrons, so that the energy levels of 4f electrons are not much affected by the ligand ions in the host crystals.

The characteristic energy levels of trivalent lanthanide ions are illustrated in Figure 2.21. This energy-level diagram was obtained by experimentally measuring the optical spectra of lanthanide ions in lanthanum chloride ($LaCl_3$) by Dieke (1968) and Carnall et al. (1989), and is also known as a *Dieke diagram*. It presents the energy of electronic states $^{2S+1}L_J$ for trivalent ions in $LaCl_3$ and almost any other crystals. The maximum number of split sublevels for each $^{2S+1}L_J$ multiplet is $(2J + 1)$ for integer J, or $(J + 1/2)$ for half-integer J, respectively. The number of levels is determined by the symmetry of the crystal field surrounding the lanthanide ions, and the width of each level is a measure of the crystal field splitting.

By taking advantage of the Dieke diagram, one can explain the absorption and emission spectra and assign each spectral peak of rare earth ions in luminescent materials. Taking Tb^{3+} as an example, Figure 2.22 shows the emission spectrum of Tb^{3+}-doped Ca-α-sialon. Tb^{3+}, an activator mostly used in green lamp phosphors (for example, $LaPO_4{:}Tb^{3+}$), shows a number of sharp lines in the visible light region. With the help of the Dieke diagram, one can understand that these lines correspond to electronic transitions from two excited

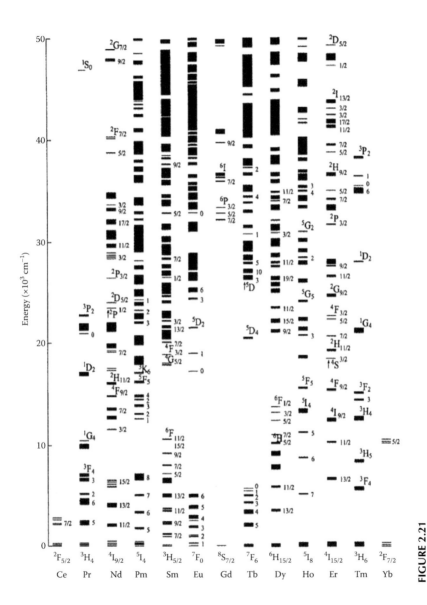

FIGURE 2.21
Characteristic energy levels of trivalent lanthanide ions, proposed by Dieke and coworkers. (Reproduced from Carnall W. T., Goodman, G. L., Rajnak, K., and Rana, R. S., *J. Chem. Phys.*, 90, 3443–3457, 1989. With permission.)

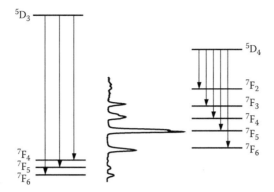

FIGURE 2.22
The emission spectrum of Tb^{3+} in Ca-α-sialon. The energy levels of Tb^{3+} are also schematically illustrated.

states, $^5D_3$ and $^5D_4$, to sublevels of the ground state ^7F_J (J = 0, 1, 2, 3, 4, 5, 6). The $^5D_3 \rightarrow {}^7F_J$ (J = 0, 1, 2, 3) and $^5D_4 \rightarrow {}^7F_J$ (J = 0, 1) transitions are so weak that their corresponding emission lines are not clearly observed in Figure 2.22.

Photoluminescence spectra of various trivalent rare earth ions in Ca-α-sialon are given in Figure 2.23. Depending on the rare earth ions, the luminescence spectra can be divided into two types: broad band and sharp lines. The luminescence of Ce^{3+} is characteristic of broad absorption and emission bands, which are due to the 4f-5d electronic transitions, whereas the luminescence spectra of other trivalent lanthanide ions consist of a group of sharp lines, which are attributable to the 4f-4f electronic transitions. Each group of sharp lines corresponds to an electronic transition between an excited state and a ground state designated by the total angular momentum, J, and it can be properly assigned by employing the Dieke diagram, as discussed before.

2.3.1.2.1 Ce^{3+} ($4f^1$)

The Ce^{3+} ion has the simplest electron configuration among the rare earth ions. The $4f^1$ ground-state configuration is divided into two sublevels, $^2F_{5/2}$ and $^2F_{7/2}$, and these two sublevels are separated by about 2,000 cm^{-1} as a result of spin-orbit coupling. This is the reason for the double structure usually observed in the Ce^{3+} emission band. The $5d^1$ excited state configuration is split into two to five components by the crystal field, with the splitting number depending on the crystal field symmetry. The Ce^{3+} emission is strongly affected by the host lattice through the crystal field splitting of the 5d orbital and the nephelauxetic effect, and usually varies from the ultraviolet to the blue spectral region. But in covalent and strong crystal field surroundings, the 5d orbital significantly shifts to lower energies, resulting in yellow and even red emission colors of Ce^{3+}. Typical examples are green YAG:Ce^{3+} (Holloway and Kestigian 1969),

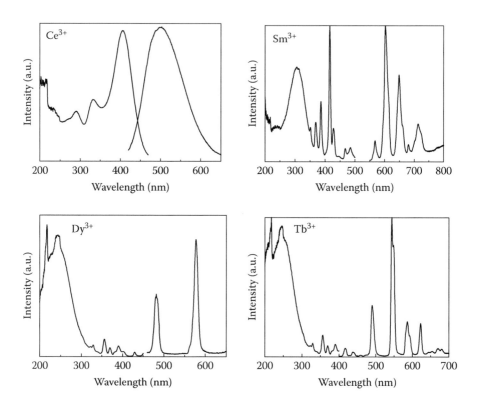

FIGURE 2.23
Excitation and emission spectra of trivalent lanthanide ions in Ca-α-sialon.

yellow CaAlSiN$_3$:Ce^{3+} (Li et al. 2008), and red CaSiN$_2$:Ce^{3+}(Le Toquin and Cheetham 2006).

Ce^{3+}-doped garnet and nitride phosphors are of technical importance for white LEDs.

2.3.1.2.2 Pr^{3+} (4f^2)

The luminescence of Pr^{3+} is greatly dependent on the host lattice, as it consists of many multiplets, such as $^3P_0 \rightarrow {}^3H_4$ (green), $^3P_0 \rightarrow {}^3F_2$ (red), $^3P_0 \rightarrow {}^3H_6$ (red), and $^1D_2 \rightarrow {}^3H_6$ (red). The luminescence spectrum of Pr^{3+} in Ca-α-sialon is composed of five bands: $^3P_0 \rightarrow {}^3H_4$ (~506 nm), $^3P_0 \rightarrow {}^3H_5$ (~545 nm), $^1D_2$ 45$^3H_4$ (~615 nm), $^3P_0 \rightarrow {}^3H_6$ (~630 nm), and $^3P_0 \rightarrow {}^3F_2$ (~669 nm), which show a red emission color. The decay time of the $^3P_0$-^3H_J or ^3F_J emission is very short for a lanthanide ion (tens of microseconds), due to the spin-allowed 4f-4f transitions.

2.3.1.2.3 Sm^{3+} (4f^5)

The emission color of Sm^{3+} is orange-red, originating from the transitions of $^4G_{5/2} \rightarrow {}^6H_{7/2}$ (~610 nm) and $^4G_{5/2} \rightarrow {}^6H_{9/2}$ (~650 nm).

2.3.1.2.4 *Eu^{3+} ($4f^6$)*

The emission of Eu^{3+} is often situated in the red region with sharp spectral lines, due to the $^5D_0 \rightarrow {}^7F_J$ transitions. The red emission at ~600 nm originating from the magnetic dipole transition of $^5D_0 \rightarrow {}^7F_1$ dominates when the Eu^{3+} site has inversion symmetry (in this case, the electric-dipole transition is strictly forbidden due to the parity selection rule). On the other hand, the red emission at 610–630 nm from the electric-dipole transition of $^5D_0 \rightarrow {}^7F_2$ dominates if it is a noninversion symmetry site.

Eu^{3+}-doped vandate, molybdate, and tungstate phosphors are interesting red phosphors for white LEDs (Neeraj et al. 2004; Wang et al. 2005; Sivakumar and Varadaraju 2005; Wang et al. 2007; Shimomura et al. 2008; Choi et al. 2009; Rao et al. 2010).

2.3.1.2.5 *Tb^{3+} ($4f^8$)*

The emission color of Tb^{3+} is mainly in the green spectral region due to the $^5D_4 \rightarrow {}^7F_J$ transitions. It also can be in the blue spectral region as a result of the $^5D_3 \rightarrow {}^7F_J$ transitions, typically when the Tb^{3+} concentration is low. A cross-relaxation transition occurs between $^5D_3$ and $^5D_4$ with increasing Tb^{3+} concentration, leading to a decrease of the $^5D_3$ emission.

2.3.1.2.6 *Dy^{3+} ($4f^9$)*

The emission color of Dy^{3+} is close to white due to the $^4F_{9/2} \rightarrow {}^6H_{15/2}$ transition (~480 nm) and the $^4F_{9/2} \rightarrow {}^6H_{13/2}$ (~577 nm) transition.

2.3.1.3 Luminescence of Divalent Rare Earth Ions

There are mainly three divalent rare earth ions: Eu^{2+}, Sm^{2+}, and Yb^{2+}. Their 4f electronic configuration is equivalent to that of Gd^{3+}, Eu^{3+}, and Lu^{3+}, respectively. Unlike their corresponding trivalent ions, the luminescence of these divalent ions is due to the $4f^n \rightarrow 4f^{n-1}5d$ transition that is parity allowed, leading to intense, broad absorption and emission spectra. In addition, the absorption and emission bands of divalent rare earth ions depend largely on the host lattice, resulting in great changes in emission color from one host to another.

2.3.1.3.1 *Eu^{2+} ($4f^7$)*

The Eu^{2+} ion has the ground state of $4f^7$ ($^8S_{7/2}$) and the excited state of $4f^65d^1$. The luminescence is strongly dependent on the host lattice, with the emission color varying in a very broad range, from ultraviolet to red. The abundant emission colors are due to the changes in covalency (nephelauxetic effect) and crystal field splitting from one host to another. With increasing the covalency or crystal field splitting, the 5d energy level of Eu^{2+} is lowered greatly, resulting in the red-shift of the absorption and emission bands. Figure 2.24 gives two examples. As seen, the emission color of Eu^{2+} is blue in $SrSiAl_2O_3N_2$ (Xie et al. 2005a) and yellow in Ca-α-sialon (Xie et al. 2002; van Krevel et al. 2002),

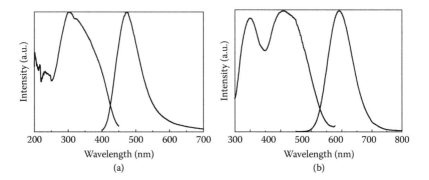

FIGURE 2.24
Excitation and emission spectra of Eu^{2+}-doped $SrAlSi_2O_3N_2$ (a) and $CaSi_9Al_3ON_{15}$ (b). (Reprinted from Xie, R.-J., Mitomo, M., Uheda, K., Xu, F. F., and Akimune, Y., *J. Am. Ceram. Soc.*, 85, 1229–1234, 2002; Xie, R.-J., Hirosaki, N., Yamamoto, Y., Suehiro, T., Mitomo, M., and Sakuma, K., *J. Ceram. Soc. Jpn.* 113, 462–465, 2005a. With permission.)

and both excitation and emission bands are significantly shifted to a longer wavelength in Ca-α-sialon. This is due to the very large crystal field splitting in Ca-α-sialon (Xie et al. 2002; van Krevel et al. 2002).

Eu^{2+} is also a very important rare earth ion in phosphors for white LEDs, which will be shown and discussed in the following chapters.

2.3.1.3.2 Sm^{2+} ($4f^6$)

If the $4f^55d^1$ energy level is located below the 4f levels, the luminescence of Sm^{2+} shows a broad band and is red (~715 nm) in color, as seen in Figure 2.23. On the other hand, a line emission spectrum due to the intraconfigurational transition of $4f \rightarrow 4f5D_0 \rightarrow {}^7F_1$ is observed if the lowest $4f^55d^1$ level is at a higher energy than its 4f levels.

2.3.1.3.3 Yb^{2+} ($4f^{14}$)

The luminescence of Yb^{2+} is due to the electronic transition of $4f^{13}5d^1 \rightarrow 4f^{14}$, with the emission color varying from ultraviolet to red. For example, it is blue (432 nm) in $Sr_3(PO_4)_2$, yellow (560 nm) in $Sr_5(PO_4)_3Cl$, and red (624 nm) in $Ba_5(PO_4)_3Cl$ (Palilla et al. 1970). The luminescence of Yb^{2+} is also observed in nitride phosphors, such as green emission (λ_{em} = 550 nm) in Ca-α-sialon (Xie et al. 2005b) and red emission (λ_{em} = 615 nm) in $SrSi_2O_2N_2$ (Bachmann et al. 2006). The decay time of Yb^{2+} (several tens of microseconds) is very long for a $5d \rightarrow 4f$ transition, due to the spin selection rule.

2.3.2 Transition Metal Ions

2.3.2.1 Introduction

Transition metal ions are frequently used as optically active centers in commercial phosphors. They have an electronic configuration of $1s^22s^22p^63s^23p^63d^n4s^2$,

TABLE 2.3

Commonly Used Transition Metal Ions
and Their Corresponding Numbers of
3d Valence Electrons

Ions	Number of 3d Electrons
Ti^{3+}, V^{4+}	1
V^{3+}, Cr^{4+}, Mn^{5+}	2
V^{2+}, Cr^{3+}, Mn^{4+}	3
Cr^{2+}, Mn^{3+}	4
Mn^{2+}, Fe^{3+}	5
Fe^{2+}, Co^{3+}	6
Fe^+, Co^{2+}, Ni^{3+}	7
Co^+, Ni^{2+}	8
Ni^+, Cu^{2+}	9

where n means the number of 3d electrons (1 < n < 10). These 3d electrons are valence electrons, and thus are responsible for the optical transitions. In addition, the 3d electron orbitals are not well shielded from the outer shells, leading to either broad spectral bands ($S > 0$) or sharp spectral lines ($S \sim 0$) due to the strong electron-lattice coupling. Table 2.3 lists the most commonly used transition metal ions in phosphors.

The luminescence of transition metal ions is strongly dependent on the crystal field of host crystals, because the splitting of the energy levels of 3d orbitals varies in different site symmetries. Taking the ions with the simplest $3d^1$ electron configuration as an example, Figure 2.25 shows the splitting of 3d orbitals in different symmetries.

As seen, the fivefold degenerate 3d orbitals split into t_{2g} and e_g energy levels, respectively. The energy separation between e_g and t_{2g} is equal to $10Dq$, which is expressed as (Yen et al. 2006)

$$E_{e_g} - E_{t_{2g}} = 10Dq$$

$$D = \frac{35Ze}{4d_{OL}^5} \quad (2.25)$$

$$q = \frac{2e}{105} \int |d_{OL}(r)|^2 r^4 \, dr$$

where D is a factor related to the ligand ions, Ze is the charge of each ligand ion, r is the radial position of the electron, and d_{OL} is the distance between the activator and the ligand ions (see Figure 2.7).

For those transition metal ions with $n > 1$, the calculation of crystal field slitting is much more complicated. In this case, electrostatic interactions between 3d electrons, as well as interactions between 3d electrons and the

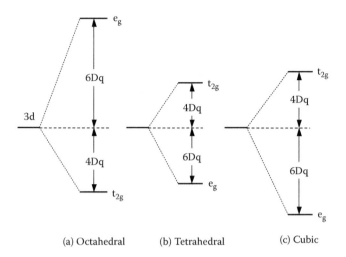

(a) Octahedral   (b) Tetrahedral   (c) Cubic

FIGURE 2.25
Splitting of 3d^1 energy level caused by crystal field in (a)–(c). (Reproduced from García Solé, J., Bausa, L. E., and Jaque, D., *An introduction to the optical spectroscopy of inorganic solids*, John Wiley & Sons, West Sussex, UK, 2005. With permission.)

crystal field, should be considered. For 3dn ($n > 1$) ions, the *Sugano-Tanabe diagram*, which represents the 3d energy levels as a function of the octahedral crystal field strength, is very useful for understanding the optical spectra of transition metals ions in most crystals (Kamimura et al. 1969).

2.3.2.2 Luminescence of Transition Metal Ions

Transition metal ions are an important class of dopants in luminescent materials and solid-state laser materials, and their emission colors are very abundant. Among the commonly used transition metal ions listed in Table 2.3, Mn^{2+} and Mn^{4+} are two of the most important and useful luminescent centers in LED phosphors, which will be discussed below. The luminescence of other transition metal ions is covered in several books (Glasse and Grabmaier 1994; Henderson and Imbusch 1989; Weckhuysen et al. 2000; Gaft et al. 2005; García Solé et al. 2005; Yen et al. 2006).

2.3.2.2.1 Mn^{2+} (3d^5)

The emission color of Mn^{2+} varies from green to orange-red, depending on the crystal field strength of the host crystal. The emission spectrum of Mn^{2+} displays a broad band, and the emission corresponds to the $^4T_1$ (^4G) → $^6A_1$ (^6S) transition. Figure 2.26 presents the excitation and emission spectra of Mn^{2+}-doped Υ-ALON. It shows a green emission at ~512 nm. A number of absorption bands are observed, resulting from $^6A_1$ → $^4A_1$, ^4E (^4G), $^4T_1$ (^4G), and $^4T_2$ (^4G) transitions. In general, the emission color is green when Mn^{2+} is

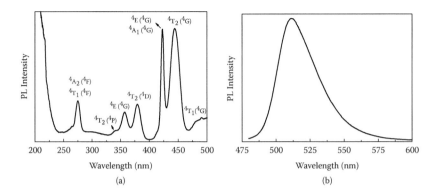

FIGURE 2.26
Excitation (a) and emission (b) spectra of Y-alon:Mn^{2+}. (Reprinted from Xie, R.-J., Hirosaki, N., Liu, X.-J., Takeda, T., and Li, H.-L., *Appl. Phys. Lett.*, 92, 201905, 2008. With permission.)

tetrahedrally coordinated, and it is orange-red when Mn^{2+} is octahedrally coordinated. In the case of Y-ALON, Mn^{2+} substitutes Al^{3+} in a tetrahedral site, thus giving the green emission (Xie et al. 2008).

The absorption of Mn^{2+} is usually low due to the spin-forbidden transition. To improve the absorption of Mn^{2+}, energy transfer mechanisms are employed to sensitize Mn^{2+}. Commonly used sensitizing ions are Sb^{3+}, Pb^{2+}, Sn^{2+}, Ce^{3+}, and Eu^{2+}, which absorb the UV light efficiently and then transfer the energy to Mn^{2+} (Yao et al. 1998; Kim et al. 2004; Yen et al. 2006). A typical example is shown in Eu^{2+}-Mn^{2+} codoped $Ca_9La(PO_4)_7$ in Figure 2.20.

2.3.2.2.2 Mn^{4+} $(3d^3)$

The emission of Mn^{4+} is usually in the red spectral region, due to the $^2E \rightarrow {}^4A_2$ transition. In $CaAl_{12}O_{19}$, the emission peak is located at ~656 nm (Murata et al. 2005), and it is positioned at ~715 nm in $YAlO_3$ (Loutts et al. 1998). Mn^{4+}-doped K_2TiF_6 can absorb blue or violet light efficiently, and emits at ~632 nm, enabling it to be used in high-color-rendering white LEDs (Setlur et al. 2010).

2.3.3 Complex Ions

Complex ions, such as molybdate (MoO_4^{2-}), vanadate (VO_4^{3-}), and tungstate (WO_4^{2-}), are widely used luminescence centers in practical phosphors. The electronic configurations of the outer shell are $[Xe]4f^{14}$, $[Kr]$, and $[Ar]$ for WO_4^{2-}, MoO_4^{2-}, and VO_4^{3-}, respectively. The materials containing these complex ions are also named *scheelite compounds* (Yen et al. 2006). The scheelite phosphors have the general compositions of ABO_4 and $A'Ln(BO_4)_2$, where $A = Ca^{2+}$, Sr^{2+}, Ba^{2+}, and Pb^{2+}; $A' = Li^+$, Na^+, K^+, and Ag^+; Ln = rare earth ions; and $B = W^{6+}$ and Mo^{6+}.

The intrinsic luminescence of the complex ions, which is often observed in the blue to green spectral regions due to the spin-forbidden $^3T_1 \rightarrow {}^1A_1$ transition, is not interesting for LED phosphors. On the other hand, Eu^{3+}-doped scheelite phosphors have efficient absorption of blue or near-ultraviolet light, and show bright and sharp-line red emissions arising from the Eu^{3+} ions (~610 nm), enabling them to be applied in white LED. In addition, a broad charge-transfer band usually observed in the absorption spectra of Eu^{3+}-doped scheelite phosphors suggests that energy transfer from the complex ions to Eu^{3+} occurs.

Typical red-emitting scheelite phosphors are $NaM(WO_4)_{2-x}(MoO_4)_x:Eu^{3+}$ (M = Gd, Y, Bi) (Neeraj et al. 2004), $AgGd(WO_4)_{2-x}(MoO_4)_x:Eu^{3+}$ (Sivakumar and Varadaraju 2005), $YVO_4:Eu^{3+}$, Bi^{3+} (Nguyen 2009; Xia et al. 2010), $Ca_3La(VO_4)_3:Eu^{3+}$ (Rao et al. 2010), $Ca_{1-2x}Eu_xLi_xMoO_4$ (Wang et al. 2005), $M_5Eu(MoO_4)_4$ (M = Na, K) (Pan et al. 1988), and $CaMoO_4:Eu^{3+}$ (Hu et al. 2005; Liu et al. 2007).

2.4 Measurement of Luminescence

2.4.1 Excitation and Emission Spectra

The excitation and emission spectra are usually measured by a spectro-fluorometer, as shown in Figure 2.27. The system basically consists of a light source, excitation and emission monochromators, sample holders, a photo-detector, and a computer. There are several types of *light sources* that may be used as excitation sources, including lasers, photodiodes, and lamps (e.g., xenon arcs and mercury vapor lamps). A laser emits light of high

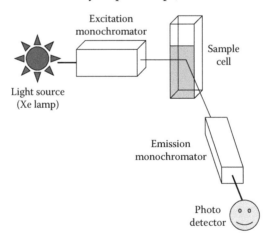

FIGURE 2.27
Optical layout of a spectrofluorometer. A light beam produced by a Xe lamp is passed through a monochromator to select a particular excitation wavelength that is then focused onto a sample. The ensuring emission is observed at right angles, analyzed spectrally with the emission monochromator, and recorded by an appropriate photodetector.

irradiance at a very narrow wavelength interval, which makes an excitation monochromator unnecessary. But the emission wavelength of a laser cannot be varied too much. A xenon lamp, however, has a continuous emission spectrum with a nearly constant intensity in the range of 300–800 nm and a sufficient irradiance for measurements down to just above 200 nm, which is often used as an excitation source in many types of spectrofluorometers. The selection of radiation of a required wavelength from the excitation source may be achieved with filters or with an *excitation monochromator* using entrance and exit slits to give the required spectral bandwidth. Luminescence radiation of the required wavelength is selected from the sample by an *emission monochromator*. For allowing anisotropy measurements, the addition of two polarization filters are necessary: one after the excitation monochromator, and the other before the emission monochromator. For luminescence spectroscopy, the photodetectors used include thermopiles, bolometers, pyroelectric detectors, quantum counters, or silicon photodiodes. The *photodetector* can be either single channeled or multichanneled. The single-channeled detector can only detect the intensity of one wavelength at a time, whereas the multichanneled detector can detect intensity at all wavelengths simultaneously.

The spectrofluorometer works with the following scheme: the light from an excitation source passes through a filter or the excitation monochromator, and strikes the sample in a sample cell. A part of the incident light is absorbed by the sample, and the sample emits light when the absorbed energy returns to the ground state by a radiative transition process. Some of the emitted light passes through a second filter or emission monochromator, and finally is recorded by a photodetector, which is usually placed 90° to the incident light beam to minimize the risk of transmitted or reflected incident light.

To measure the excitation spectrum of a phosphor, the luminescence intensity is recorded at a fixed emission wavelength. The monochromator is used to change the excitation wavelength. So the excitation spectrum can be obtained by plotting the luminescence intensity vs. the excitation wavelength. It is used to determine the best excitation wavelength for analysis and is related to the absorption spectrum of the phosphor. Since the excitation light source has different light intensities at different wavelengths, and the optical system has different transmissions for each wavelength, a number of errors must be corrected. The correction is done by measuring the responses of the system simultaneously with a photodiode with a known response. The ratio between the measured photodiode response and known response then gives the correction factor for the excitation spectrum.

The emission spectrum is recorded by exciting the sample at a fixed wavelength and scanning the emission monochromator. The recorded emission intensity is then plotted as a function of the emission wavelength. The obtained spectrum also must be corrected for different sensitivities of photomultipliers (PMs) to different wavelengths, as well as for the varying transmission behaviors of the optics. To correct the emission spectrum, light from the lamp that passes through exactly the same optical path as the sample is measured. The relative light intensities for each wavelength are known (color temperature) since the lamp has a current stabilized power supply. Then, the ratio between the measured and theoretical color temperatures is computed for each wavelength and used as the correction factor for the measured emission spectrum. Another correction method is to measure the emission spectrum of a cuvette with a rhodamine solution. The ratio of this emission to the theoretical rhodamine spectrum can then be used to correct the emission spectra of phosphors.

2.4.2 Quantum Efficiency

The quantum efficiency of a phosphor is how much light (photons) absorbed by the phosphor is converted into luminescence, which is in the range of 0–1. This is also called *internal quantum efficiency*. The external quantum efficiency denotes the ratio of the number of photons of the emitted light to that of the incident light on a phosphor. Ohkubo and Shigeta (1999) proposed a method to calculate the absolute (external) quantum efficiency of phosphor materials. The internal (η_i) and external (η_0) quantum efficiencies can be computed by the following equations:

$$\eta_0 = \frac{\int \lambda \bullet P(\lambda)d\lambda}{\int \lambda \bullet E(\lambda)d\lambda}$$

$$\eta_1 = \frac{\int \lambda \bullet P(\lambda)d\lambda}{\int \lambda \{E(\lambda) - R(\lambda)\}d\lambda} \tag{2.26}$$

where $E(\lambda)/h\nu$, $R(\lambda)/h\nu$, and $P(\lambda)/h\nu$ are the number of photons in the spectrum of excitation, reflectance, and emission of the phosphor, respectively.

Figure 2.28 illustrates the measurement system for the quantum efficiency of phosphors. It basically includes the light source, monochromator, integrated sphere, multichannel photodetector (MCPD) spectrophotometer, and a computer. According to Equation 2.26, the quantum efficiency can be obtained by measuring the following spectra:

- Absorption spectrum of the $BaSO_4$ white reflectance standard sample, $R(\lambda)$
- Absorption spectrum of the phosphor sample, $E(\lambda)$
- Emission spectrum of the phosphor sample, $P(\lambda)$

By using this method, the absorption and quantum efficiency of a yellow Li-α-sialon:Eu^{2+} are measured as a function of the excitation wavelength

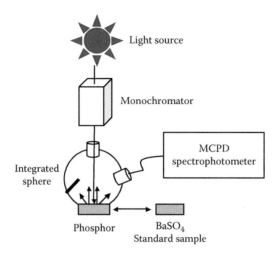

FIGURE 2.28
Experimental setup for measuring the quantum efficiency of a phosphor.

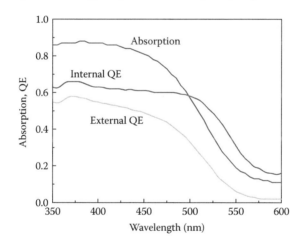

FIGURE 2.29
Absorption and quantum efficiency of Li-α-sialon:Eu^{2+} as a function of excitation wavelength.

and given in Figure 2.29. It shows that upon the 450 nm excitation, Li-α-sialon:Eu^{2+} has the absorption and internal and external quantum efficiencies of 82, 61, and 50%, respectively.

2.4.3 Time-Resolved Luminescence

Time-resolved emission spectroscopy provides information on the kinetic behavior of luminescent excited states and intermediates. The experimental setup used for time-resolved luminescence is similar to that of Figure 2.27,

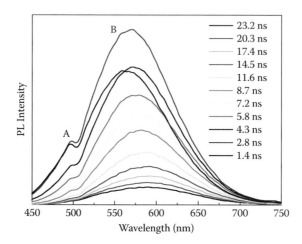

FIGURE 2.30
(See color insert.) Time-resolved luminescence of $(Sr_{0.99}Eu_{0.01})AlSi_4N_7$. (Reprinted from Ruan, J., Xie, R.-J., Hirosaki, N., and Takeda, T., *J. Am. Ceram. Soc.*, DOI: 10.1111/j.1551-2916.2010.04104.x, 2010. With permission.)

but a light source must be pulsed (pulsed or modulated lamps, light-emitting diodes, or lasers), and the detector must be connected to a time-sensitive system, such as an oscilloscope, a multichannel analyzer, or a boxcar integrator. The basic idea behind this spectroscopy is to record the emission spectrum at a certain *delay time*, *t*, with respect to the excitation pulse. Therefore, for using different delay times, different spectral shapes of phosphors are obtained.

Figure 2.30 shows the time-resolved luminescence of Eu^{2+}-doped $SrAlSi_4N_7$ measured at varying delay times. As seen, two emission bands, A and B, can be clearly observed as the delay time decreases. These two bands are ascribed to Eu^{2+} ions occupying two different crystallographic sites of Sr in $SrAlSi_4N_7$ (Ruan et al. 2010). Sohn et al. (2009) measured the time-resolved luminescence of $Sr_2Si_5N_8:Eu^{2+}$ (see Figure 2.31) and found that the emission band of $Sr_2Si_5N_8:Eu^{2+}$ was deconvoluted into two Gaussian peaks irrespective of the Eu^{2+} concentration. Sohn et al. (2009) addressed that the two-peak emission was closely associated with the energy transfer taking place between Eu^{2+} activators located at two different crystallographic sites in the $Sr_2Si_5N_8$ structure.

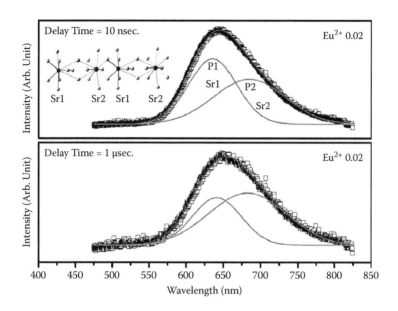

FIGURE 2.31

Time-resolved luminescence of $(Sr_{0.98}Eu_{0.02})Si_5N_8$. (Reprinted from Sohn, K.-S., Lee, S., Xie, R.-J., and Hirosaki, N., *Appl. Phys. Lett.*, 95, 121903, 2009. With permission.)

References

Bachmann, V., Justel, T., Meijerink, A., Ronda, C., and Schmidt, P. J. 2006. Luminescence properties of $SrSi_2O_2N_2$ doped with divalent rare earth ions. *J. Lumin.* 121:441–449.

Blasse, G., and Grabmaier, B. C. 1994. *Luminescent materials*. Berlin: Springer-Verlag.

Carnall W. T., Goodman, G. L., Rajnak, K., and Rana, R. S. 1989. A systematic analysis of the spectra of the lanthanides doped into single crystal lanthanum fluoride (LaF_3). *J. Chem. Phys.* 90:3443–3457.

Chen, R., and McKeever, S. W. S. 1997. *Theory of thermoluminescence and related phenomena*. Singapore: World Scientific Press.

Choi, S., Moon, Y. M., Kim, K., Jung, H. K., and Nahm, S. 2009. Luminescent properties of a novel red-emitting phosphor: Eu^{3+}-activated $Ca_3Sr_3(VO_4)_4$. *J. Lumin.* 129:988–990.

Curie, D. 1963. *Luminescence in crystals*. London: Methuen and Co.

Dexter, D. L. 1953. A theory of sensitized luminescence in solids. *J. Chem. Phys.* 21:836–850.

Dibartolo, B. 1968. *Optical interactions in solids*. New York: John Wiley & Sons.

Dieke, G. H. 1968. *Spectra and energy levels of rare earth ions in crystals*. New York: Interscience.

Foster, T. 1948. Zwischenmolekulare Energiewanderung und Fluoreszenz. *Ann. Phys. (Leipzig)* 2:55–75.

Fox, M. 2001. *Optical properties of solids*. New York: Oxford University Press.

Gaft, M., Reisfeld, R., and Panczer, G. 2005. *Luminescence spectroscopy of minerals and materials.* Berlin: Springer-Verlag.

García Solé, J. Bausa, L. E., and Jaque, D. 2005. *An introduction to the optical spectroscopy of inorganic solids.* West Sussex, UK: John Wiley & Sons.

Henderson, B., and Imbusch, G. F. 1989. *Optical spectroscopy of inorganic solids.* Oxford: Oxford Science Publications.

Holloway, W. W., and Kestigian, M. 1969. Optical properties of cerium-activated garnet crystals. *J. Opt. Soc. Am.* 59:60–63.

Hu, Y. S., Zhuang, W. D., Ye, H. Q., Wang, D. H., Zhang, S. S., and Huang, X. W. 2005. A novel red phosphor for white light emitting diodes. *J. Alloys Compd.* 390:226–229.

Jørgensen, C. K. 1975. Partly filled shells constituting anti-bonding orbitals with higher ionization energy than their bonding counterparts. In *Structure and bonding*, ed. J. D. Dunitzin, P. Hemmerich, and J. Ibers, 49–81. Vol. 22. Berlin: Springer-Verlag.

Kamimura, H., Sugano, S., and Tanabe, Y. 1969. *Ligand field theory and its applications.* Tokyo: Shokabo.

Kim, J. S., Jeon, P. E., Choi, J. C., Park, H. L., Mho, S. I., and Kim, G. C. 2004. Warm-white-light emitting diode utilizing a single-phase full-color $Ba_3MgSi_2O_8:Eu^{2+}$, Mn^{2+} phosphor. *Appl. Phys. Lett.* 84:2931–2933.

Klick, C. C., and Schulman, J. H. 1957. Luminescence in solids. In *Solid state physics*, ed. F. Seitz and D. Turnbull, 97–116. Vol. 5. New York: Academic Press.

Le Toquin, R., and Cheetham, A. K. 2006. Red-emitting cerium-based phosphor materials for solid state lighting applications. *Chem. Phys. Lett.* 423:352–356.

Li, Y. Q., Hirosaki, N., Xie, R.-J., Takeda, T., and Mitomo, M. 2008. Yellow-orange-emitting $CaAlSiN_3:Ce^{3+}$ phosphor: Structure, photoluminescence, and application in white LEDs. *Chem. Mater.* 20:6704–6714.

Liu, J., Lian, H. Z., and Shi, C. S. 2007. Improved optical photoluminescence by charge compensation in the phosphor system $CaMoO_4:Eu^{3+}$. *Opt. Mater.* 29:1591–1594.

Loutts, G. B., Warren, M., Taylor, L., Rakhimov, R. R., Ries, H. R., Miller III, G., Noginov, M. A., Curley, M., Noginova, N., Kukhtarev, N., Caulfield, H. J., and Venkateswarlu, P. 1998. Manganese-doped yttrium orthoaluminate: A potential material for holographic recording and data storage. *Phys. Rev. B.* 57:3706–3709.

Murata, T., Tanoue, T., Iwasaki, M., Morinaga, K., and Hase, T. 2005. Fluorescence properties of Mn^{4+} in $CaAl_{12}O_{19}$ compounds as red-emitting phosphor for white LED. *J. Lumin.* 114:207–212.

Neeraj, S., Kijima, N., and Cheetham, A. K. 2004. Novel red phosphors for solid-state lighting: The system $NaM(WO_4)_{2-x}(MoO_4)_x:Eu^{3+}$ (M = Gd, Y, Bi). *Chem. Phys. Lett.* 387:2–6.

Nguyen, H. D., Mho, S., and Yeo, I. H. 2009. Preparation and characterization of nanosized (Y,Bi)VO4:Eu3+ and Y(V,P)O4:Eu3+ red phosphors. *J. Lumin.* 129:1754–1758.

Ohkubo, K., and Shigeta, T. 1999. Absolute fluorescent quantum efficiency of NBS phosphor standard samples. *J. Illum. Eng. Inst. Jpn.* 83:87–93.

Palilla, F. C., O'Reilly, B. E., and Abbruscato, V. J. 1970. Fluorescence properties of alkaline earth oxyanions activated by divalent ytterbium. *J. Electrochem. Soc.* 117:87–91.

Pan, J., Liu, L. Z., Chen, L. G., Zhao, G. W., Zhou, G. E., and Guo, C. X. 1988. Studies on spectra properties of $Na_5Eu(WO_4)_4$ luminescent crystal. *J. Lumin.* 40–41:856–857.

Rao, B. V., Jang, K., Lee, H. S., Yi, S. S., and Jeong, J. H. 2010. Synthesis and photo-luminescence characterization of RE^{3+} (=Eu^{3+}, Dy^{3+})-activated $Ca_3La(VO_4)_3$ phosphors for white light-emitting diodes. *J. Alloys Compd.* 496:251–255.

Reisfeld, R. 1973. Spectra and energy transfer of rare earths in inorganic glasses. In *Structure and bonding*, ed. J. D. Dunitzin, P. Hemmerich, and J. Ibers, 53–98. Vol. 13. Berlin: Springer-Verlag.

Reisfeld, R. 1976. Excited states and energy transfer from donor cations to rare earths in the condensed phase. In *Structure and bonding*, ed. J. D. Dunitz, J. A. Ibers, and B. D. Reinin, 65–97. Vol. 30. Berlin: Springer-Verlag.

Riseberg, L. A., and Weber, M. J. 1975. Relaxation phenomena in rare earth lumines-cence. In *Progress in optics*, ed. E. Wolf, 89–159. Vol. 14. New York: Elsevier.

Ruan, J., Xie, R.-J., Hirosaki, N., and Takeda, T. 2010. Nitrogen gas pressure sintering and photoluminescence properties of orange-red $SrAlSi_4N_7$:Eu^{2+} phosphors for white light-emitting diode. *J. Am. Ceram. Soc.*, DOI: 10.1111/j.1551-2916.2010.04104.x.

Setlur, A. A., Radkov, E. V., Henderson, C. S., Her, J. H., Srivastava, A. M., Karkada, N., Kishore, M. S., Kumar, N. P., Aesram, D., Deshpande, A., Kolodin, B., Grigorov, L. S., and Happek, U. 2010. Energy-efficient, high-color-rendering LED lamps using oxyfluoride and fluoride phosphors. *Chem. Mater.* 22:4076–4082.

Shimomura, G., Kijima, N., and Cheetman, A. K. 2008. Novel red phosphors based on vanadate garnets for solid state lighting applications. *Chem. Phys. Lett.* 455:279–283.

Sivakumar, V., and Varadaraju, U. V. 2005. Intense red-emitting phosphors for white light emitting diodes. *J. Electrochem. Soc.* 152:H168–H171.

Sohn, K.-S., Lee, S., Xie, R.-J., and Hirosaki, N. 2009. Time resolved photoluminescence analysis of two-peak emission behavior in $Sr_2Si_5N_8$:Eu^{2+}. *Appl. Phys. Lett.* 95:121903.

van Krevel, J. W. H., van Rutten, J. W. T., Mandal, H., Hintzen, H. T., and Metselaar, R. 2002. Luminescence properties of terbium-, cerium-, or europium-doped α-sialon materials. *J. Solid State Chem.* 165:19–24.

Van Uitert, L. G. 1971. Energy transfer between rare earth ions in tungstates. *J. Lumin.* 4:1–7.

Wang, J., Jing, X., Yan, C., and Lin, J. 2005. $Ca_{1Cag}Eu_xLi_xMoO_4$: A novel red phosphor for solid state lighting based on a GaN LED. *J. Electrochem. Soc.* 152:G186–G188.

Wang, X. X., Xian, Y. L., Shi, J. X., Su, Q., and Gong, M. L. 2007. The potential red emitting $Gd_{2-y}Eu_y(WO_4)_{3OO}(MoO_4)_x$ phosphors for UV InGaN-based light-emitting diodes. *Mater. Sci, Eng. B* 140:69–72.

Weber, M. J. 1973. Multiphonon relaxation of rare-earth ions in yttrium ortho-aluminate. *Phys. Rev. B* 8:54–64.

Weckhuysen, B. M., Voort, P., and Catana, G. 2000. *Spectroscopy of transition metal ions on surfaces*. Leuven, Belgium: Leuven University Press.

Xia, Z. G., Chen, D. M., Yang, M., and Ying, T. 2010. Synthesis and luminescence properties of YVO_4:Eu^{3+}, Bi^{3+} phosphor with enhanced photoluminescence by Bi^{3+} doping. *J. Phys. Chem. Solids* 71:175–180.

Xie, R.-J., Hirosaki, N., Liu, X.-J., Takeda, T., and Li, H.-L. 2008. Crystal structure and photoluminescence of Mn^{2+}-Mg^{2+} codoped gamma aluminum oxynitride (γ-alon): A promising green phosphor for white light-emitting diodes. *Appl. Phys. Lett.* 92:201905.

Xie, R,-J., Hirosaki, N., Mitomo, M., Uheda, K., Suehiro, T., Xu, X., Yamamoto, Y., and Sekiguchi, T. 2005b. Strong green emission from α-sialon activated by divalent ytterbium under blue light irradiation. *J. Phys. Chem. B* 109:9490–9494.

Xie, R.-J., Hirosaki, N., Yamamoto, Y., Suehiro, T., Mitomo, M., and Sakuma, K. 2005a. Fluorescence of Eu^{2+} in strontium oxonitridoaluminosilicates (SiAlONS). *J. Ceram. Soc. Jpn.* 113:462–465.

Xie, R.-J., Mitomo, M., Uheda, K., Xu, F. F., and Akimune, Y. 2002. Preparation and luminescence spectra of calcium- and rare-earth (R = Eu, Tb, and Pr)-codoped α-SiAlON ceramics. *J. Am. Ceram. Soc.* 85:1229–1234.

Yang, W. J., Luo, L., Chen, T.-M., and Wang, N.-S. 2005. Luminescence and energy transfer of Eu- and Mn-coactivated $CaAl_2Si_2O_8$ as a potential phosphor for white-Light UVLED. *Chem. Mater.* 17:3883–3888.

Yao, G. Q., Lin, J. H., Zhang, L., Lu, G. X., Gong, M. L., and Su, M. Z. 1998. Luminescent properties of $BaMg_2Si_2O_7$:Eu^{2+}, Mn^{2+}. *J. Mater. Chem.* 8:585–588.

Yen, W. M., Shionoya, S., and Yamamoto, H. 2006. *Phosphor handbook*. 2nd ed. Boca Raton, FL: CRC Press.

3

Traditional Phosphors in White LEDs

Phosphor has a history of more than 100 years; now it is almost everywhere in our daily life. For example, we turn on fluorescent lamps, switch on the computer in our office, turn on the TV to watch programs, and even see x-ray images of our chest, brain, or legs at the hospital. The light or the colorful images we see are generated by means of some phosphors, such as lamp, cathode-ray tube (CRT), or x-ray phosphors. These phosphors have been extensively studied, and notable progress has been made in their efficiency and quality over the past 40 years. On the other hand, as a definite type of phosphor, LED phosphors are quite new to the family of luminescent materials, which appeared after white LEDs came onto the stage several years ago. As stated in Chapter 1, it is necessary to combine suitable phosphors with a LED chip to produce white light, as phosphors play key roles in determining optical properties of white LED lamps, such as luminous efficiency, chromaticity coordinate, color temperature, and lifetime. LED phosphors differ greatly from other types of phosphors in the excitation source. The excitation source is ultraviolet (350–410 nm) or blue (440–470 nm) LEDs for LED phosphors, whereas it is 254 nm, electron beam, and x-rays for lamp, CRT, and x-ray phosphors, respectively. This difference leads to different requirements for materials design and luminescent properties of LED phosphors, and also leads to the conclusion that other types of phosphors cannot be directly used as LED phosphors without any modifications. This means that it is necessary to search for and develop novel and suitable host crystals for LED phosphors, or to modify the currently available lamp phosphors.

Extensive investigations on LED phosphors have been conducted in recent years, which include the discovery of new host crystals (e.g., nitride compounds), the modification of luminescent properties (e.g., tailoring the excitation or emission band by elemental substitution), the development of new synthetic approaches (e.g., gas reduction and nitridation for nitride phosphors), and the evaluation of their behaviors in LEDs (e.g., tuning of color rendering or color temperature by means of phosphor blends). With these great efforts, a number of LED phosphors have been developed, and some of them have been commercialized. These phosphors are composed of a broad range of materials with different shapes and sizes, varying from oxides, sulfides, and phosphates to nitrides; from aluminates and silicates to scheelites; and from polycrystalline powders to quantum dots.

The main content of this book treats the photoluminescence properties and applications of phosphors for LEDs, with emphasis on nitride luminescent materials. Before we present detailed descriptions of nitride phosphors, we need to first understand useful nitrogen-free phosphors, which we name "traditional phosphors." Actually, they are not traditional phosphors; some of them are even quite new. We just use this term to distinguish nitride phosphors from those phosphors that have been investigated and known about for a long time, based on the common view of nitride phosphors as a new class of luminescent material.

The structure of this chapter is as follows: Section 3.1 treats the general requirements for LED phosphors. Section 3.2 classifies the traditional phosphors and presents their photoluminescence properties. Section 3.3 gives some examples of white LEDs using traditional phosphors.

3.1 Requirements for Phosphors in White LEDs

The principle of white LEDs has been discussed in Chapter 1. In the case of phosphor-converted white LEDs, phosphor is the key material to down-convert the light of UV or blue LEDs into visible light, so that phosphor has a great impact on the performance of white LEDs. To achieve highly efficient and reliable white LEDs, phosphor itself should exhibit excellent luminescent properties together with great enhancement in overall efficiencies of LED chips and the structure of LED products. In addition, although there are a huge number of phosphors, only a few of them are suitable for use in LEDs. This depends on the requirements or selection rules for LED phosphors. Obviously, only those phosphors that meet the requirements can be potentially applied in white LEDs. These selection rules or requirements help phosphor researchers or engineers design, develop, and utilize phosphors correctly.

The general requirements for LED phosphors are listed below:

Strong absorption. The phosphor should have strong absorption of the light emitted by the LED chips, i.e., UV (350–410 nm) or blue (440–480 nm) light. In general, the stronger the absorption, the higher the efficiency. Strong absorption means a larger number of absorbed photons are involved in the luminescence process. Direct absorption by activators themselves, as a result of allowed 4f-5d or 4f-4f electronic transitions, is usually desired. In the case of activators with low absorption, such as Mn^{2+} due to the spin-forbidden transition of 3d, sensitizing ions (Eu^{2+}, Ce^{3+}, etc.) are often utilized to enhance the absorption of activators by means of nonradiative energy transfer.

Broad excitation spectrum. The excitation spectrum of the phosphor should match well with the emission spectrum of LED chips.

Typically in the case of using blue LED chips, which is the main-stream method to create white light, the excitation spectrum needs to be significantly red-shifted, to cover the whole blue spectral region. This implies that the selected host crystals should have large crystal field splitting or a strong nephelauxetic effect. In addition, the excitation spectrum in the blue spectral region should be as flat as possible, to accept the deviation in emission peaks of LED chips from one to another.

Useful emission spectrum. The emission spectrum of the phosphor should be as broad as possible in the case of white LEDs used for general illumination, and should be as narrow as possible in the case of white LEDs used for liquid crystal display (LCD) backlight. The former is for obtaining high color rendition, and the latter for achieving wide color gamut (or high color purity). The position of the emission spectrum, or the emission color, should be useful. Generally, phosphors with blue (λem = 420–500 nm), green (λ_{em} = 500–560 nm), yellow (λ_{em} = 560–590 nm), and red (λ_{em} = 590–650 nm) emission colors are required.

High quantum/conversion efficiency. Higher quantum or conversion efficiency means less energy loss during the luminescence process. Therefore, the phosphor should have high quantum or conversion efficiency under the UV or blue light excitation. In other words, the phosphor should convert the light emitted by LED chips into desired light very efficiently. The quantum efficiency is closely related to the processing conditions that affect the crystallinity, particle morphology, particle size, and particle size distribution of the phosphor.

Small thermal quenching. Thermal quenching is a measure of thermal stability of phosphors, which is one of the most important technical parameters in their practical uses. Thermal quenching is one of the reasons for the degradation of LED lamps, having a great influence in LED's lifetime and chromaticity. To attain high efficiency and long lifetime of LED lamps, it is therefore required that the thermal quenching of a phosphor be as small as possible. Thermal quenching relies significantly on the crystal structure and chemical composition of the host lattice.

High chemical stability. Chemical stability refers to the stability in chemical composition and crystal structure of a phosphor, as well as the valence of doped rare earth ions upon chemical attacks or upon UV or blue light irradiation. Low chemical stability of a phosphor will not only make the production process more complex and costly, but also reduce the luminous efficiency and shorten the lifetime of LED products dramatically. Consequently, the phosphor should be stable under the ambient atmosphere, and will not chemically react with CO, CO_2, H_2O, and air.

Suitable particle size and morphology. The LED phosphor will be dispersed into an epoxy or a silicone resin in practical use, so that the phosphor particles should be fine and uniform to avoid sediment. Usually, the particle size is in the range of 5–20 µm, and the particle size distribution will be as narrow as possible. Moreover, spherical particles are generally preferred, and the density of phosphors will not differ too much from each other for the phosphor blending.

Low cost. Usually, users or producers will welcome the low cost of a phosphor, although the phosphor is much cheaper than LED chips, and the amount of a phosphor used in a white LED is only several tenths of a gram. The cost of a phosphor is dependent on raw materials, synthetic approaches, processing conditions, etc.

3.2 Classification of Phosphors

Phosphors can be divided into different groups, depending on their chemical composition, the emission color, or the LED type. There are a large number of compounds that are suitable host lattices for LED phosphors. According to their chemical compositions, the corresponding phosphors can be grouped into

- Garnet
- Aluminate
- Silicate
- Sulfide and oxysulfide
- Phosphate
- Scheelite
- Nitride

The phosphors can also be grouped into *UV and blue LED phosphors*, depending on the type of LED chips that they combine with, and into *red, green, and blue* (RGB) *phosphors*, according to their emission colors. The former is closely related to the excitation spectrum of the phosphor, and the latter to the emission spectrum of the phosphor. For users or customers, these classifications are more convenient.

In the following, we will discuss the photoluminescent properties of traditional phosphors grouped by their chemical compositions. Of course, the reader can easily classify these phosphors according to their photoluminescence spectra.

3.3 Photoluminescent Properties of Traditional Phosphors

Photoluminescence spectra, such as excitation and emission spectra, are a first and important feature to allow one to judge the value of the phosphors. Other important photoluminescent properties include absorption, temperature-dependent luminescence, quantum efficiency, decay time, etc. In the next two sections, we will present photoluminescence spectra of some typical phosphors for white LEDs, including aluminates, silicates, sulfide, phosphates, and nitrides. We first make a brief introduction of the photoluminescent properties of so-called traditional phosphors.

3.3.1 Garnet Phosphors

Materials with a garnet structure are promising host lattices for LED phosphors, because they usually have such a strong crystal field surrounding the activator ions that the excitation and emission bands of activator ions are significantly red-shifted, due to the large crystal field splitting of their 5d energy levels. This makes garnet phosphors very suitable for combinations with blue LED chips.

The general formula of the garnet structure can be written as $X_3Y_2Z_3O_{12}$, where X refers to eightfold-coordinated cations (mostly divalent Ca, Mn, Fe, and Mg); Y refers to sixfold-coordinated cations such as trivalent Al, Fe, and Cr; and Z refers to the tetrahedral site of Si (Pavese et al. 1995; Meagher 1975; Chenavas 1978). The cubic garnet structure belongs to space group Ia-3d with eight molecules per unit cell, so that there are ion total 24 dodecahedral sites, 16 octahedral sites, and 24 tetrahedral sites in the garnet structure. The three different cations X, Y, and Z are surrounded by different oxygen polyhedra (i.e., XO_8, YO_6, and ZO_4).

Several phosphors with the garnet structure have been developed for white LEDs; the most famous one is cerium-activated yttrium-aluminum garnet $Y_3Al_5O_{12}:Ce^{3+}$ (YAG:Ce). Other interesting silicate garnet phosphors are $Ca_3Sc_2Si_3O_{12}:Ce^{3+}$ (Shimomura et al. 2007), $Lu_2CaMg_2(Si,Ge)_3O_{12}:Ce^{3+}$ (Seltur et al. 2006), and $Mg_3Y_2Ge_3O_{12}:Ce^{3+}$ (Jiang et al. 2010).

3.3.1.1 YAG:Ce³⁺

3.3.1.1.1 Crystal Structure

Yttrium-aluminum garnet $Y_3Al_5O_{12}$ can be rewritten as $Y_3Al_2Al_3O_{12}$ according to the general structural formula of garnet. The structural data are summarized in Table 3.1 (Nakatsuka et al. 1999). As shown in Figure 3.1, the garnet structure consists of three different cation-oxygen polyhedra: Y-O dodecahedron, Al-O octahedron, and Al-O tetrahedron (Chenavas et al. 1978; Meagher 1975; Pavese et al. 1995). The Y-O dodecahedron is a

TABLE 3.1

Atomic Coordinates (x, y, z) and Site Occupance Fraction (SOF) for $Y_3Al_5O_{12}$

Atom	Wyckoff	x	y	z	SOF
Y	24c	0.125	0	0.25	1.00
Al1	16a	0	0	0	1.00
Al2	24d	0.375	0	0.25	1.00
O	96h	−0.0318 (3)	0.0511 (3)	0.1498 (3)	1.00

Source: Data from Nakatsuka, A., Yoshiasa, A., and Yamanaka, T., *Acta Cryst. B*, 55, 266–272, 1999.

Note: Space group: Ia3d.
Lattice constant: $a_0 = 12.0062(5)$ Å.
$Z = 8$.
Bond lengths: Y – O: 2.317 Å × 4, Y – O: 2.437 Å × 4, Al1 – O: 1.938 Å × 6, Al2 – O: 1.754 Å × 4.

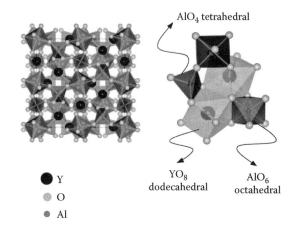

FIGURE 3.1
Crystal structure of YAG viewed from [001], and three different cation-oxygen polyhedra in the garnet structure.

pseudocube (distorted cubic lattice) in between a cube and an antiprism. The octahedra and tetrahedra do not share any edge among themselves, but they share edges with at lease one dodecahedra.

3.3.1.1.2 Photoluminescent Properties

The luminescence of Ce^{3+} in YAG is given in Figure 3.2. As seen, the excitation spectrum is mainly composed of two broad bands that are centered at 340 and 460 nm, respectively, with the spectral tail extending to 550 nm. The excitation spectrum is assigned to the 4f → 5d electronic transitions of Ce^{3+}, and the two excitation bands to the transitions to the two lowest energy levels of the 5d orbital. The position of the two excitation bands does not vary with the composition of the host crystals, but the relative intensity

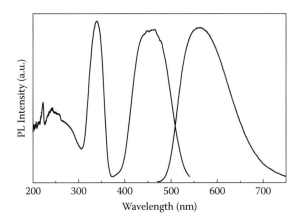

FIGURE 3.2
Excitation and emission spectra of Ce^{3+}-doped $Y_3Al_5O_{12}$.

of these bands does. The strong absorption of blue light by YAG:Ce^{3+} suggests that it matches very well with the blue LED chip. Actually, YAG:Ce^{3+} is one of the most important phosphors and is widely used for white LEDs. The emission spectrum of YAG:Ce^{3+} shows a very broad yellow band with a full width at half maximum (FWHM) of ~120 nm, which can be assigned to the 5d → 4f electronic transitions. The emission peak varies in a broad range of 520–580 nm, depending on the composition of the host lattice.

As discussed in Chapter 2, the Ce^{3+} luminescence strongly depends on the host lattice. Usually, the absorption band of Ce^{3+} ions does not cover the visible spectral region, and their emission colors vary from ultraviolet to blue. The significantly red-shifted excitation and emission spectra of Ce^{3+} in YAG, however, suggest that Ce^{3+} ions are involved in an environment greatly different from that of other host lattices. Figure 3.3 illustrates the energy level diagram of Ce^{3+} in a real inorganic solid. The position of the lowest 5d levels for Ce^{3+} is dependent on two separate factors: the nephelauxetic effect (i.e., the shift in the 5d centroid from the free ion levels) and crystal field splitting of the 5d levels. In Ce^{3+}-activated yttrium-aluminum garnet, the 5d centriod shift (nephelauxetic effect) is estimated as 14,250 cm^{-1}, which is much larger than that of fluoride, phosphate, and borate phosphors. The crystal field splitting consists of two major components if considering the distortion of cubic coordination into a dodecahedron for Ce^{3+}: the splitting between the T_{2g} and E_g levels (i.e., 10 Dq) and the splitting of E_g (i.e, Δd). Setlur et al. (2006) reported that the values of 10 Dq and Δd for YAG:Ce^{3+} were 18,450 and 7,600 cm^{-1}, respectively, leading to the crystal field splitting of ~ 26,050 cm^{-1}.

The Ce^{3+} ion has a 4f^1 configuration, and the ground state consists of a doublet ($^2F_{5/2}$ and $^2F_{7/2}$) with a separation of about 2,000 cm^{-1} due to the spin-orbit interaction (see Figure 3.3). Consequently, the emission of Ce^{3+}

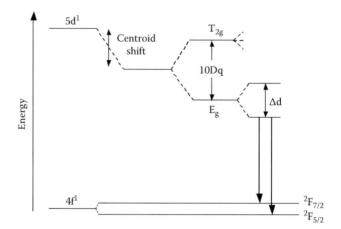

FIGURE 3.3
Energy-level diagram for Ce^{3+}-activated YAG, showing the centroid shift and crystal field splitting contributing to the red-shifted luminescence of Ce^{3+}.

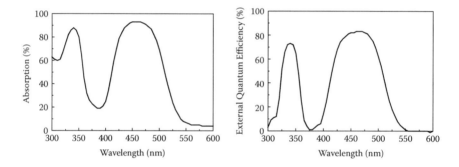

FIGURE 3.4
Absorption (left) and external quantum efficiency (right) of a commercially available YAG:Ce^{3+} measured at varying excitation wavelengths.

generally has a doublet character. This character is the major reason for the large value of FWHM usually observed for Ce^{3+}. On the other hand, the doublet character of the emission band depends on temperature and Ce^{3+} concentration and is not always observed (Bachmann et al. 2009). In Figure 3.2, only one emission band is verified in the visible range.

Commercially available YAG:Ce^{3+} phosphors have high absorption and quantum efficiency, as shown in Figure 3.4. This phosphor has an absorption of ~95% and an external quantum efficiency of ~83% under the excitation by 440–470 nm.

3.3.1.1.3 Thermal Quenching

Thermal quenching of YAG:Ce^{3+} is evaluated by measuring the emission intensity at different temperatures, as shown in Figure 3.5. The emission

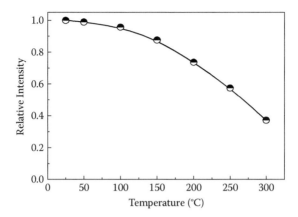

FIGURE 3.5
Temperature-dependent luminescence intensity of YAG:Ce^{3+}.

intensity reduces by about 13% at 150°C, when compared to the initial intensity at room temperature. The quenching temperature (the temperature at which the emission intensity decreases by 50%) of this phosphor is estimated at ~280°C (555 K), which is in good agreement with the intrinsic quenching temperature of 600 K.

Thermal quenching of YAG:Ce^{3+} is greatly affected by two major factors: *the Ce^{3+} concentration* and *the chemical composition of the host lattice*. Bachmann et al. (2009) conducted a detailed investigation of temperature quenching of YAG:Ce^{3+} with varying Ce^{3+} concentrations. They observed that the quenching temperature decreased as the Ce^{3+} concentration increased. The quenching temperature is up to 600 K for a dilute sample (0.033% Ce), and it drops to 440 K for a heavily doped sample (3.33% Ce). The lifetime measurement indicated that the decay time did not change with temperature (up to 600 K) for the dilute sample, but it shortened dramatically for the sample with 3.33% Ce when temperature was over 440 K. The decrease of quenching temperature was interpreted by thermally activated concentration quenching.

Changes in chemical composition of YAG often result in a decrease of quenching temperature. Bachmann et al. (2009) addressed that, upon substitution of Y by Gd, the quenching temperature of $(Y_{1-x}Gd_x)_3Al_5O_{12}$:Ce^{3+} (0.3%) dropped from 600 K (x = 0) to 475 K (x = 0.25), 420 K (x = 0.5), and <400 K (x = 0.75). This decrease in quenching temperature can be explained by the configurational coordinate model (see Figure 3.2.) because of the enhanced Stokes shift with Gd substitution, and the thermal quenching mechanism is a thermally activated crossover from higher vibrational levels of the excited state to the ground state. In the case of substitution of Al by Ga, Hansel et al. (2010) reported that the quenching temperature decreased from 550 K for $Y_3Al_5O_{12}$:Ce^{3+} (1.0%) to 360 K for $Y_3Al_{2.5}Ga_{2.5}O_{12}$:Ce^{3+} (1.0%). With the

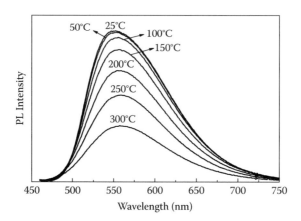

FIGURE 3.6
Temperature-dependent emission spectra of YAG:Ce^{3+}.

Ga substitution, a blue-shift of the emission band is observed because the oxygen atoms surrounding Ce^{3+} are forced to create a more cubic structure. Therefore, the thermal quenching mechanism of Ga-substituted YAG:Ce is attributed to the decrease of the splitting of Eg (Δd) that leads to a thermally activated crossover from the lowest vibrational levels of the E_g state to the high-energy vibrational levels of the ground state.

The change of emission spectra with temperature is given in Figure 3.6. It is seen that the emission band slightly shifts to the long-wavelength side with increasing temperature. In addition, the emission band broadens as temperature rises. These changes in emission spectra with temperature may finally have an influence on the chromaticity coordinates of white LEDs, so that variations in chromaticity of YAG:Ce^{3+} should be minimized.

3.3.1.1.4 Tunable Emission Color

For bichromatic white LEDs using only one yellow phosphor, the color temperature of white LEDs can be achieved only by using YAG:Ce^{3+} with different emission colors. It is especially difficult to generate warm white (correlated color temperature (CCT) ~ 3,000 K) by using general yellow YAG:Ce^{3+} phosphors, because their emission spectra lack enough red components. Therefore, it is necessary and important to tailor the emission color of YAG:Ce^{3+} for practical applications. Two major approaches have been utilized for the tuning of emission color: elemental substitution and codoping with other lanthanide ions.

As stated previously, Al^{3+} ions occupy two different sites (octahedral and tetrahedral sites), and Y^{3+} ions reside in a distorted dodecahedral site. Holloway and Kestigian (1969) reported that substitution of a larger Ga^{3+} ion for Al^{3+} blue-shifted the emission band. The emission band was also shifted toward blue when Y^{3+} ions were substituted by smaller Lu^{3+} ions, but

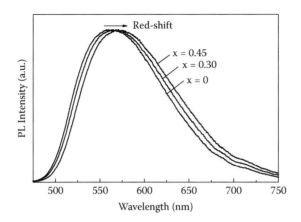

FIGURE 3.7
Emission spectra of $(Y_{2.85-x}Gd_xCe_{0.15})Al_5O_{12}$.

toward red when Y^{3+} ions were replaced by larger ions such as La^{3+}, Gd^{3+}, and Tb^{3+} (Holloway and Kestigian 1969; Jang et al. 2007; Hansel et al. 2010; Bachmann et al. 2008). An example is presented in Figure 3.7, clearly showing the red-shift of the YAG:Ce^{3+} emission band with the substitution of Gd^{3+} for Y^{3+}. The blue-shift occurs with the substitutions by larger ions for Al^{3+} and by smaller ions for Y^{3+}, because the oxygen atoms directly coordinated with Ce^{3+} become more cubic in structure, which reduces the splitting of E_g levels (Setlur et al. 2006; Hansel et al. 2010). Setlur et al. (2006) reported that the splitting of E_g levels was 6,600 and 7,600 cm^{-1} for $Lu_3Al_5O_{12}$:Ce and $Y_3Al_5O_{12}$:Ce, respectively. On the other hand, the substitution of larger ions for Y^{3+} enhances the deviation of the cubic component of the crystal field, and thus increases the splitting of the E_g levels, finally downshifting the lowest-energy levels of the 5d electrons.

The red spectral emission of YAG:Ce^{3+} can be enhanced by codoping with Eu^{3+} or Pr^{3+} (Pan et al. 2004; Jang et al. 2007). Characteristic red emissions of Pr^{3+} (λ_{em} = 610 and 636 nm) are clearly shown in Figure 3.8, suggesting that the energy transfer from Ce^{3+} to Pr^{3+} occurs efficiently in YAG.

3.3.1.2 $Ca_3Sc_2Si_3O_{12}$:Ce^{3+}

Shimomura et al. (2007) reported the crystal structure and photoluminescence of the Ce^{3+}-activated $Ca_3Sc_2Si_3O_{12}$ silicate garnet. This phosphor has a cubic garnet crystal structure with a lattice constant of a = 12.25 Å. Ca^{2+}, Sc^{2+}, and Si^{4+} ions occupy the dodecahedral, octahedral, and tetrahedral sites and are coordinated to eight, six, and four oxygen atoms, respectively. The structural data are presented in Table 3.2. The extended x-ray absorption fine-structure (EXAFS) analysis indicated that the Ce^{3+} ions occupied the Ca^{2+} sites rather than the Sc^{3+} sites. It can also be understood that the ionic radii of Ce^{3+} and

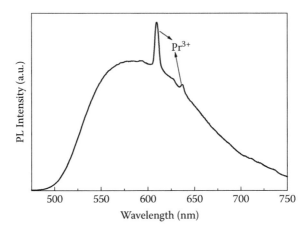

FIGURE 3.8
Emission spectrum of Pr^{3+} codoped YAG:Ce^{3+}, showing characteristic emissions of Ce^{3+} and Pr^{3+}.

TABLE 3.2

Structural Parameters for $Ca_3Sc_2Si_3O_{12}$

Atom	Wyckoff	x	y	z	SOF
Ca	24c	0.125	0	0.25	1.00
Sc	16a	0	0	0	1.00
Si	24d	0.375	0	0.25	1.00
O	96h	−0.0404	0.0501	0.1589	1.00

Source: Data from Shimomura, Y., Honma, T., Shigeiwa, M., Akai, T., Okamoto, K., and Kijima, N., *J. Electrochem. Soc.*, 154, J35–J38, 2007.

Note: Space group: Ia3d (No. 230).
Lattice constant: a = 12.250(2) Å.
Bond lengths: Ca – O: 2.390 Å × 4, Ca – O: 2.532 Å × 4, Sr – O: 2.099 Å × 6, Si – O: 1.645 Å × 4.

Ca^{2+} with a coordination number of 8 are 1.28 and 1.26 Å, whereas they are 1.15 and 0.89 Å for Ce^{3+} and Ca^{2+} with a coordination number of 6.

The $Ca_3Sc_2Si_3O_{12}$:Ce^{3+} phosphor shows one broad excitation band covering the range of 380–500 nm (see Figure 3.10), with the peak wavelength of 455 nm, which is very suitable for blue LED chips. The emission spectrum displays a broad band with a doublet structure, showing a major peak at 505 nm and a shoulder at 560 nm. In comparison with YAG:Ce^{3+}, the observed shorter-emission wavelength is due to weak crystal field splitting or centroid shift, or both. The weak crystal field splitting is expected because $Ca_3Sc_2Si_3O_{12}$ has a larger lattice volume than $Y_3Al_5O_{12}$. In addition, this phosphor exhibits higher quenching temperature (smaller thermal quenching) than YAG:Ce^{3+}, indicating that $Ca_3Sc_2Si_3O_{12}$:Ce^{3+} is a promising green-emitting phosphor for white LEDs.

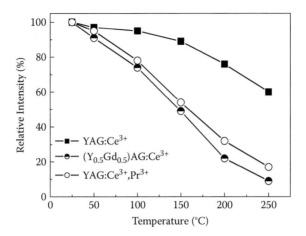

FIGURE 3.9
Temperature-dependent luminescence intensity of red emission color-enhanced YAG:Ce^{3+}.

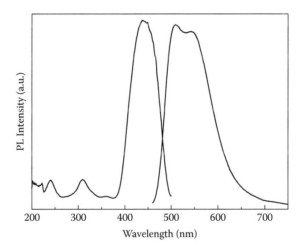

FIGURE 3.10
Excitation and emission spectra of Ca$_3$Sc$_2$Si$_3$O$_{12}$:Ce^{3+}.

Codoping with Mg^{2+} on the Sc^{3+} sites enhances the luminescence of Ca$_3$Sc$_2$Si$_3$O$_{12}$:Ce^{3+}. This is possibly due to the charge compensation for Ce^{3+} substituting for Ca^{2+}.

3.3.1.3 Lu$_2$CaMg$_2$(Si,Ge)$_3$O$_{12}$:Ce^{3+}

Setlur et al. (2006) investigated the crystal structure and photoluminescence of an orange-yellow silicate garnet phosphor based on Lu$_2$CaMg$_2$Si$_{3-x}$Ge$_x$O$_{12}$:Ce^{3+}. The structural data of the nondoped sample are shown in Table 3.3. In comparison with Y$_3$Al$_5$O$_{12}$ and Ca$_3$Sc$_2$Si$_3$O$_{12}$, Lu$_2$CaMg$_2$Si$_3$O$_{12}$ has relatively

TABLE 3.3

Selected Structural Parameters for $Lu_{1.999}CaMg_2Si_3O_{12}$

Atom	Wyckoff	x	y	z	SOF
Mg	16a	0	0	0	1.00
Ca	24c	0.125	0	0.25	0.40
Lu	24c	0.125	0	0.25	0.60
Si	24d	0.375	0	0.25	1.00
O	96h	0.0351(6)	0.0538(6)	0.6578(5)	1.00

Source: Data from Setlur, A. A., Heward, W. J., Gao, Y., Srivastava, A. M., Chandran, R. G., and Shankar, M. V., *Chem. Mater.*, 18, 3314–3322, 2006.

Note: Space group: Ia3d (No. 230).
Lattice constant: a = 11.9758(3) Å.
Bond lengths: {Lu,Ca} – O: 2.315 Å × 4, {Lu,Ca} – O: 2.418 Å × 4, Mg – O: 2.040 Å × 6.

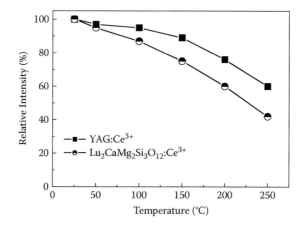

FIGURE 3.11
Temperature-dependent luminescence intensity of $Lu_2CaMg_2Si_3O_{12}$:Ce³⁺ and typical YAG:Ce³⁺.

smaller volumes of the unit cell and the dedocahedra, implying that a larger crystal field splitting will impact the Ce³⁺ ions.

The $Lu_2CaMg_2Si_{3-x}Ge_xO_{12}$:Ce³⁺ phosphor shows a broad excitation band centered at 470 nm, with a red-shift of ~462 cm⁻¹ with respect to YAG:Ce³⁺. The emission spectrum is quite broad (FWHM ~ 150 nm), with a peak wavelength of ~605 nm. The emission is red-shifted about ~1,700 cm⁻¹ compared to YAG:Ce³⁺. These red-shifts are ascribed to the larger crystal field splitting of the Ce³⁺ 5d levels in this silicate garnet. Setlur et al. (2006) reported that the splitting of the lowest E_g levels of $Lu_2CaMg_2Si_3O_{12}$:Ce³⁺ was much larger than that of the YAG:Ce³⁺ levels (i.e., 11,000–12,000 vs. 7,600 cm⁻¹).

As seen in Figure 3.11, $Lu_2CaMg_2Si_3O_{12}$:Ce³⁺ exhibits larger thermal quenching than YAG:Ce³⁺. This is perhaps due to the larger Stokes shift of $Lu_2CaMg_2Si_3O_{12}$:Ce^{3+} than of YAG:Ce^{3+}. On the other hand, the thermal quenching is substantially better than that for the red-spectral-enhanced $Y_{3-x}Gd_xAl_5O_{12}$:Ce^{3+} and $Y_3Al_5O_{12}$:Ce^{3+}, Pr^{3+} phosphors (see Figure 3.9). The promising photoluminescent properties of $Lu_2CaMg_2Si_3O_{12}$:Ce^{3+} enable it to be used as an orange phosphor for making white LEDs with low color temperature and high color rendition.

3.3.2 Silicate Phosphors

Due to their excellent chemical and thermal stability, abundant emission colors, high luminescence, and low cost, alkaline earth silicate phosphors have attracted much attention in recent years, and have great potential for use in white LEDs. In fact, luminescence of Eu^{2+} in silicates was reported much earlier (Blasse et al. 1968a, 1968b; Barry 1968a, 1968b; Yamazaki et al. 1986; Poort et al. 1997). But in early publications, the phosphors were excited only by UV light ($\lambda_{em} \sim 254$ nm), and their photoluminescent properties were not tailored or modified. Recently, silicate phosphors have been extensively investigated for their potential applications in white LEDs, and a number of silicate phosphors thus have been developed with these efforts.

According to their compositions, alkaline earth silicate phosphors can be divided into a binary system, such as MO-SiO_2 (M = Ca, Sr, Ba), and a ternary system, such as MO-$M'O$-SiO_2 (M = Ca, Sr, Ba, M' = Li, Mg, Al, Zn, etc). In the binary system, two important compositions, M_3SiO_5 and M_2SiO_4 (M = Ca, Sr, Ba), have been recognized as the promising host lattice for LED phosphors because they have strong crystal field strengths that lead to significantly red-shifted luminescence. On the other hand, Eu^{2+}-activated ternary silicate phosphors, such as $MMg_2Si_2O_7$, $CaAl_2Si_2O_8$, and $M_3MgSi_2O_8$ (Barry 1968b, 1968c; Poort et al. 1995; Yang et al. 2005; Kim et al. 2004a, 2004b, 2005a, 2005b, 2005c), often have low absorption of blue light, and emit blue-green colors upon UV light excitation. This makes this kind of phosphor uninteresting for use in white LEDs.

Therefore, in this part we concentrate on three silicate phosphors that can be excited by blue light. These phosphors are Eu^{2+}-doped Sr_3SiO_5, Sr_2SiO_4, and Li_2SrSiO_4. Intense green to orange emission colors can be achieved in these phosphors. For trinary silicate phosphors, typically Eu^{2+}-Mn^{2+} codoped phosphors, readers can refer to the literature (Barry 1968b; Poort et al. 1995; Yang et al. 2005; Kim et al. 2004a, 2004b, 2005a).

3.3.2.1 Sr_3SiO_5:Eu^{2+}

3.3.2.1.1 Crystal Structure

Single-crystal x-ray data showed that Sr_3SiO_5 has a tetragonal crystal system (space group P4/ncc, a = 6.934 Å, c = 10.72 Å), as shown in Table 3.4. The crystal structure of Sr_3SiO_5 is related to Ba_3SiO_5 and Ca_3SiO_5, and all are built

TABLE 3.4

Selected Structural Parameters for Sr_3SiO_5

Atom	Wyckoff	x	y	z	SOF
Sr(1)	8f	0.181	0.181	0.25	1.00
Sr(2)	4c	0	0.50	0	1.00
Si	4b	0	0	0	1.00
O(1)	16g	0.169	−0.092	0.088	1.00
O(2)	4c	0	0.50	0.250	1.00

Source: Data from Glasser, L. S. D., and Glasser, F. P.,
Acta Cryst., 18, 453–455, 1965.

Note: Crystal system: Tetragonal.
Space group: P4/ncc (No. 130).
Lattice constant: a = 6.934 Å, c = 10.72 Å.
Bond lengths: Sr1 – O1:2.57 Å, Sr1 – O2:2.54 Å,
Sr2 – O1:2.56 Å, Sr2 – O2:2.68 Å.

of M^{2+} ions (M = Ca, Sr, Ba), an O^{2-} ion, and an isolated $[SiO_4]^{4-}$ tetrahedral (Glasser and Glasser 1965). Differing from Ba_3SiO_5, the SiO_4 tetrahedra in Sr_3SiO_5 are rotated with respect to each other through an angle of 33°, leading to the lower symmetry of space group P4/ncc, instead of I4/mcm for Ba_3SiO_5, and a lower coordination number of 6 for Sr(1) and Sr(2). The coordination of Ba atoms in Ba_3SiO_5, however, is clearly tenfold for Ba(1) and eightfold for Ba(2). In the structure of Sr_3SiO_5, all the O^{2-} ions are surrounded by regular octahedrals of Sr^{2-} ions. The packing of a SiO_4 tetrahedral along the c-axis is clearly seen in Figure 3.12. It also shows the alternating packing of Sr(2) and O(2) along the c-axis.

3.3.2.1.2 Photoluminescent Properties

Blasse et al. (1968a, 1968b) investigated the luminescence of Eu^{2+} in M_3SiO_5 (M = Ca, Sr, Ba) 40 years ago. They reported that the emission band was significantly red-shifted by increasing the ionic size of an alkaline earth element, leading to the maximum emission wavelengths of 510, 545, and 590 nm for Ca, Sr, and Ba, respectively. Figure 3.13 presents typical photoluminescence spectra of Sr_3SiO_5:Eu^{2+} (5 mol%). The excitation spectrum of this phosphor shows a very broad band covering the spectral region of 250–550 nm, indicating large crystal field splitting of Eu^{2+} 5d levels. Park et al. (2004) reported that the crystal field splitting of Sr_3SiO_5:Eu^{2+} was about 8,700 cm^{-1}. Moreover, the phosphor can be efficiently excited by 440–480 nm of blue light, making it very suitable for white LEDs. The phosphor emits a yellow color, having a broad emission band centered at ~580 nm. The FWHM value of the phosphor is around 90 nm, which is the typical case for Eu^{2+}. A small Stokes shift of ~3,700 cm^{-1} is thus estimated from the luminescence spectra.

Phosphors with orange or red emission color are necessary for improving the color rendition of white LEDs. For Sr_3SiO_5:Eu^{2+}, its emission band can

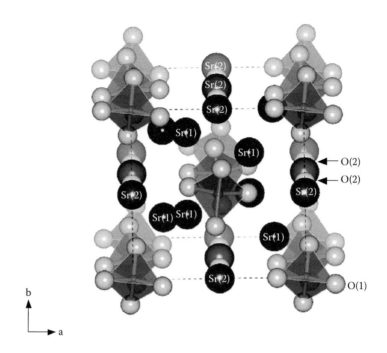

FIGURE 3.12
Crystal structure of Sr_3SiO_5 projected on the ab plane.

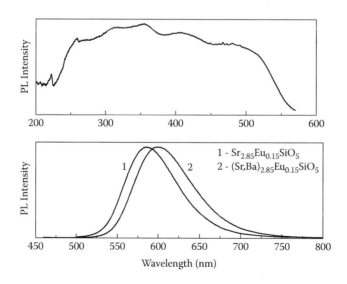

FIGURE 3.13
Excitation and emission spectra of Sr_3SiO_5:Eu^{2+}.

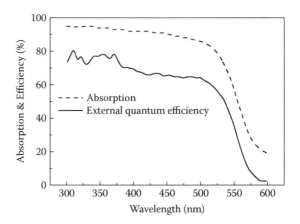

FIGURE 3.14
Quantum efficiency and absorption of a commercial $Sr_3SiO_5:Eu^{2+}$ phosphor.

be further shifted to the long-wavelength side by elemental substitution or compositional tailoring (Park et al. 2004, 2006). As can be seen in Figure 3.13, with partial substitution of Ba for Sr, the emission band shows a substantial red-shift. Park et al. explained this red-shift as a decrease of the preferential orientation of the Eu^{2+} 5d orbital, which was due to the lattice expansion along the c-axis, and octahedral symmetry around Eu^{2+} with Ba substitution. Park et al. (2004) also reported that by increasing the SiO_2 content in the composition (i.e., decreasing the Sr/Si ratio), the emission band was obviously red-shifted. They addressed that this red-shift was ascribed to the increase of covalency that leads to a larger centroid shift.

Park et al. (2004, 2006) reported a quantum efficiency of 68% for Sr_3SiO_5: Eu^{2+}. By optimizing the processing conditions and compositional tailoring, the quantum efficiency can be improved up to 78%. Figure 3.14 presents the quantum efficiency and absorption of a commercial $Sr_3SiO_5:Eu^{2+}$ phosphor.

Doping Sr_3SiO_5 with Ce^{3+} also leads to efficient yellow emissions. Jang et al. (2007) reported that $Sr_3SiO_5:Ce^{3+}$, Li^+ showed a broad emission band covering the spectral region of 465–700 nm, giving a FWHM value of ~122 nm (Jang and Yeon 2007). In comparison to the Eu^{2+}-doped sample, the Ce^{3+}-doped phosphor displays a blue-shifted excitation band, but still has strong absorptions of near-UV or blue light.

3.3.2.1.3 Thermal Quenching

Thermal quenching of a normal $Sr_3SiO_5:Eu^{2+}$ is given in Figure 3.15, where the luminescence intensity is plotted against temperature. The emission intensity is decreased by 32% at 150°C, indicative of large thermal quenching for $Sr_3SiO_5:Eu^{3+}$. Blasse et al. (1968b) reported a quenching temperature of 400 K for $Sr_3SiO_5:Eu^{2+}$, which is in good agreement with the temperature of 460 K measured at our lab.

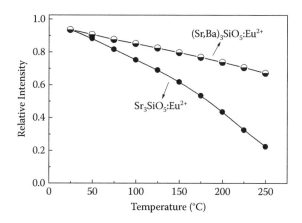

FIGURE 3.15
Temperature-dependent emission intensity of $Sr_3SiO_5:Eu^{2+}$ and $(Sr,Ba)_3SiO_5:Eu^{2+}$.

Thermal quenching of $Sr_3SiO_5:Eu^{2+}$ can be improved by partial substitution of Ba for Sr. As seen in Figure 3.15, the emission intensity is reduced by 10% at 150°C, and the quenching temperature is increased to 550 K for Ba-substituted $Sr_3SiO_5:Eu^{2+}$. Blasse et al. (1968b) also showed that the quenching temperature of $Ba_3SiO_5:Eu^{2+}$ was 55 K higher than that of $Sr_3SiO_5:Eu^{2+}$.

3.3.2.2 $Sr_2SiO_4:Eu^{2+}$

3.3.2.2.1 Crystal Structure

The structure of orthosilicate M_2SiO_4 (M = Ca, Sr, Ba) was discussed in detail more than 30 years ago (Fields et al. 1972; Catti et al. 1983; Barbier and Hyde 1985). Strontium orthosilicate, Sr_2SiO_4, has two crystal forms: α' and β. The low-temperature form, or β, develops at temperatures below 85°C, and the high-temperature form, or α', occurs at temperatures higher than 85°C. β-Sr_2SiO_4 has a monoclinic crystal system, and is isostructural with β-Ca_2SiO_4; α'-Sr_2SiO_4 is orthorhombic and isostructural with Ba_2SiO_4.

There are two sites for Sr atoms: Sr(1) and Sr(2). The Sr(1) atom is coordinated with 10 oxygen atoms, and the Sr(2) atom to 9 oxygen atoms. Catti et al. (1983) reported that the Sr(1) polyhedron has hexagonal pseudosymmetry along the y-axis. It shares the face O(2)-O(3)-O(3,5) and the vertex O(1^i) with two SiO_4 tetrahedra that are vertically above and below it, respectively, and the three edges O($2,2^{ii}$)×××O($3,2^{ii}$), O($2,2^{ii}$)×××O($3,6^{iv}$), and O($3,4^v$)×××O($3,7^v$) with three SiO_4 tetrahedra. The Sr(2) polyhedron is much less symmetrical and shares neither edges nor faces with any SiO_4 tetrehedra. As seen in Table 3.5, the Sr-O distances are quite different for the two polyhedra; the Sr(2) shows a normal value of 2.70 Å on average, but the Sr(1)-O distances are peculiar, with 2.85 Å on average, having one very short bond length with O(1i) (2.386 Å) and all other long bond lengths (2.90 Å on average).

TABLE 3.5

Selected Structural Parameters for α'-$Sr_{1.9}Ba_{0.1}SiO_4$

Atom	x	y	z	SOF
Sr(1)	0.25	0.3403	0.5798	1.00
Sr(2)	0.25	−0.0014	0.3022	1.00
Si	0.25	0.7788	0.5835	1.00
O(1)	0.25	1.004	0.5683	1.00
O(2)	0.25	0.676	0.4339	1.00
O(3)	0.480	0.7070	0.6639	1.00

Source: Data from Catti, M., Gazzoni, G., and Ivaldi, G., *Acta Cryst.*, C39, 29–34, 1983.

Note: Crystal system: Orthorhombic.
Space group: Pnma (No. 53).
Lattice constant: a = 5.674 Å, b = 7.086 Å, c = 9.745 Å.
Bond lengths: Sr(1) – O(1): 2.386 Å, Sr(1) – O(2): 2.771 Å, Sr(1) – O(2,2^{iii}): 2.852 Å × 2, Sr(1) – O(3): 3.021 Å × 2, Sr(1) – O(3,2^{ii}): 2.846 Å × 2, Sr1 – O(3,4^{v}): 2.972 Å × 2, Sr(2) – O(1^{i}): 2.593 Å, Sr(2) – O(2^{i}): 2.622 Å, Sr(2) – O(2,4^{i}): 2.622 Å, Sr(2) – O(1,3^{iii}): 3.105 Å × 2, Sr(2) – O(3,2^{ii}): 2.609 Å × 2, Sr(2) – O(3,3^{vi}): 2.507 Å × 2.

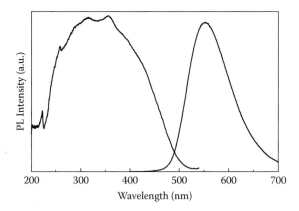

FIGURE 3.16
Excitation and emission spectra of Sr_2SiO_4:Eu^{2+} (5 mol%).

3.3.2.2.2 *Photoluminescent Properties*

The luminescence of Eu^{2+} in orthosilicates was reported by Barry (1968a, 1968b, 1968c) and Blasse et al. (1968a, 1968b) almost 40 years ago; it shows a green-emitting color. Figure 3.16 presents the photoluminescence of Eu^{2+}-doped Sr_2SiO_4. As seen, the excitation spectrum is significantly blue-shifted when compared to Sr_3SiO_5:Eu^{2+}, and the crystal field splitting is calculated as 3,700 cm^{-1}. The emission band is centered at 560 nm. A Stokes shift of ~ 6,000 cm^{-1} is then estimated. Both Blasse et al. (1968b) and Poort et al. (1997) reported that the emission spectrum showed two bands centered at 490 and 560 nm, respectively, when the phosphor was excited by UV light (254 or 320 nm), and they only assigned the 560 nm peak to Eu^{2+}. In fact, both of the emission bands are due to the luminescence of Eu^{2+} in Sr_2SiO_4; the 490 nm emission band is considered the emission of Sr(1), and the 560 nm band the emission of Sr(2), because the average Sr(2)-O distance is much shorter than the Sr(1)-O distance. Park et al. (2003) attempted the fabrication of white LEDs by combining $Sr_2SiO_4:Eu^{2+}$ with a GaN-based LED chip with $\lambda_{em} = 400$ nm, and showed a higher luminous efficiency than for YAG-based white LEDs.

As seen in Figure 3.16, the excitation band of $Sr_2SiO_4:Eu^{2+}$ is not broad enough for achieving high absorption of blue light, making it only suitable for UV or near-UV LED chips. Therefore, engineering of the excitation band is essential. Yoo et al. (2005) and Park et al. (2005) reported that the excitation spectrum of $Sr_2SiO_4:Eu^{2+}$ was dramatically red-shifted with codoping of Ba^{2+} or Ba^{2+}-Mg^{2+}. The Stokes shift of $Sr_2SiO_4:Eu^{2+}$ was reduced from 5,639 cm^{-1} to 3,404 and 3,984 cm^{-1} for Ba^{2+}-Mg^{2+}- and Ba^{2+}-substituted $Sr_2SiO_4:Eu^{2+}$, respectively. A red-shift of the excitation band can also be realized by increasing the SiO_2 content in the composition (Park et al. 2003). The maximum excitation peak shifts from 382 nm for Sr/Si = 2/0.5 to 387 and 394 nm for Sr/Si = 2/1.0 and 2/1.3, respectively. In addition, the emission band is also shifted with the elemental substitution and compositional tailoring. Barry (1968a) observed a red-shift and broadening of the emission band in Ba-substituted $Sr_2SiO_4:Eu^{2+}$. The emission band can be shifted to a long wavelength by increasing the SiO_2 content, the maximum emission peak moving from 523 for Sr/Si = 2/0.5 to 555 nm for Sr/Si = 2/1.3 (Park et al. 2003).

3.3.2.2.3 Thermal Quenching

Blasse et al. (1968b) reported that the quenching temperature of $Sr_2SiO_4:Eu^{2+}$ was 390 K. This is confirmed by Figure 3.17, which shows the temperature-dependent luminescence. The luminescence drops remarkably with rising temperature, which decreases by 62% at 150°C compared to the initial intensity measured at room temperature. At the same time, a red-shift and broadening of the emission band occur with increasing temperature (see Figure 3.17), leading to variations of color points of the phosphors.

Large thermal quenching is a big problem for the practical use of $Sr_2SiO_4:Eu^{2+}$. Blasse et al. (1968b) observed a higher quenching temperature for $Ba_2SiO_4:Eu^{2+}$, which is 425 K. In fact, partial substitution of Ba for Sr, forming $Sr_{2-x}Ba_xSiO_4:Eu^{2+}$, was reported to increase the quenching temperature of $Sr_2SiO_4:Eu^{2+}$ (Kim et al. 2005c). The thermal quenching is related to the lattice phonon. By increasing the Ba concentration, we can slightly increase the bond length due to the larger ionic size of Ba^{2+} compared to Sr^{2+}, so that the force constant decreases and then the vibrational frequency reduces. The reduced phonon energy, as well as the Stokes shift with increasing x, prevents tunneling from an excited state to the ground state, leading to an increase of quenching temperature.

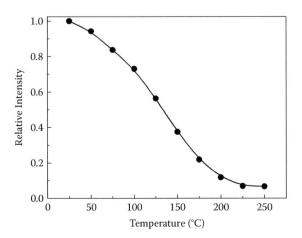

FIGURE 3.17
Thermal quenching of Sr_2SiO_4:Eu^{2+} (5 mol%).

TABLE 3.6

Selected Structural Parameters for Li_2SrSiO_4

Atom	Wyckoff	x	y	z	SOF
Li	6c	0.4186	0.0624	0.0867	1.00
Sr	3b	0	0.4223	0.1667	1.00
Si	3a	0.7282	0.7282	0	1.00
O(1)	6c	0.5494	0.5157	0.1063	1.00
O(2)	6c	0.7615	0.0674	0.0035	1.00

Source: Data from Saradi, M. P., and Varadaraju, U. V., *Chem. Mater.*, 18, 5267–5272, 2006.
Note: Crystal system: Trigonal.
Space group: $P3_121$ (No. 152).
Lattice constant: a = 5.023 Å, c = 12.457 Å.
Bond lengths: Sr – O: 2.720 Å × 2, Sr – O: 2.621 Å × 2,
Sr – O: 2.555 Å × 2, Sr – O: 2.536 Å × 2.

3.3.2.3 Li_2SrSiO_4:Eu^{2+}

3.3.2.3.1 Crystal Structure

Li_2SrSiO_4 is isostructural with Li_2EuSiO_4, having a hexagonal crystal system with space group of $P3_121$. Some selected structural parameters for Li_2SrSiO_4 are summarized in Table 3.6 (Saradhi and Varadaraju 2006). The LiO_4 and SiO_4 tetrahedra share corners and form a space for Sr^{2+} (see Figure 3.18). Sr^{2+} ions are coordinated with the eight nearest oxygen atoms at four distances.

3.3.2.3.2 Photoluminescence Properties

Eu^{2+}-activated Li_2SrSiO_4 shows an orange-yellow emission color (Saradhi and Varadaraju 2006; He et al. 2008). Figure 3.19 illustrates the photoluminescence

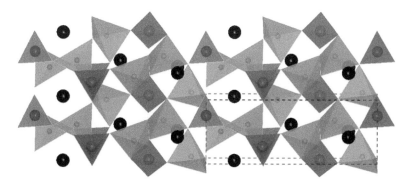

FIGURE 3.18
Crystal structure of Li_2SrSiO_4 projected on the bc plane.

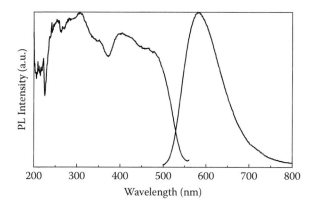

FIGURE 3.19
Excitation and emission spectra of Eu^{2+}-doped Li_2SrSiO_4.

spectra of this phosphor. The excitation spectrum was measured by monitoring at 570 nm, and the emission spectrum was measured upon the 450 nm excitation. As seen, the excitation spectrum covers a broad range of 250–550 nm, indicating a large crystal field splitting of Eu^{2+} 5d energy levels. $Li_2SrSiO_4:Eu^{2+}$ shows high absorptions of 400–470 nm light, making it very attractive for white LEDs using blue LED chips. The phosphor exhibits a broad and intense emission band with a maximum at 570 nm. The Stokes shift of $Li_2SrSiO_4:Eu^{2+}$ is about 4,200 cm^{-1}. The optimal Eu^{2+} concentration was reported to be 0.5 or 0.75 mol%.

The luminescence of $Li_2SrSiO_4:Eu^{2+}$ can be enhanced by codoping of Ce^{3+} (Zhang et al. 2008; He et al. 2010; Kim et al. 2010). Zhang et al. (2008) and He et al. (2010) addressed that this enhancement was due to the energy transfer from Ce^{3+} to Eu^{2+}. On the other hand, Kim et al. (2010) demonstrated that the increase of luminescence by codoping of Ce^{3+} was ascribed to the fact that the Ce^{3+} doping prevents the intrinsic Li vacancy from oxidizing Eu^{2+} to Eu^{3+}, leading to the increase of the Eu^{2+} concentration.

3.3.3 Aluminate Phosphors

Aluminate phosphors, having excellent chemical and thermal stability and high emission efficiency, are important luminescent materials for fluorescent lamps and plasma display panels (PDPs). Aluminate phosphors are usually activated by rare earth ions, such as Eu^{2+}, Ce^{3+}, and Tb^{3+}, and emit blue to green colors upon UV or near ultraviolet (NUV) light excitation. Because aluminate phosphors in general cannot be excited by blue light, they are often used as blue or green candidate phosphors for UV LEDs (Ravichandran et al. 1999; Won et al. 2006; Chen et al. 2006). The notable examples are Eu^{2+}-activated $BaMgAl_{10}O_{17}$ (BAM) and $SrAl_2O_4$; the former is also a key phosphor for PDPs, and the latter is an important long-lasting phosphor. Recently, a yellow-emitting aluminate phosphor that can be excited by blue light was reported (Im et al. 2009b, 2009c), and it is able to generate white light by combining with a blue LED chip.

In this part, we will introduce the newly reported $LaSr_2AlO_5:Ce^{3+}$ yellow phosphor. For other aluminate LED phosphors, interested readers can refer to the literature on the subject.

3.3.3.1 Crystal Structure of LaSr₂AlO₅

$LaSr_2AlO_5$, structurally similar to $EuSr_2AlO_5$, crystallizes in a tetragonal structure with a space group of I4/mcm and the lattice constants of a = b = 6.8856 Å and c = 121.0613 Å (Im et al. 2009b), as given in Table 3.7. The La (Ce) and Sr(1) atoms share the 8h sites, which are coordinated with six O(2) atoms with four equal La-O(2) distances of 2.738 Å and two shorter distances of 2.440 Å, and two O(1) atoms with a La-O(1) distance of 2.527 Å.

TABLE 3.7
Selected Structural Parameters for $La_{0.975}Ce_{0.025}Sr_2AlO_5$

Atom	Wyckoff	x	y	z	SOF
La	8h	0.3198	0.1802	0	0.4875
Ce	8h	0.3198	0.1802	0	0.0125
Sr(1)	8h	0.3198	0.1802	0	0.5
Sr(2)	4a	0	0	0.25	1.00
Al	4b	0.5	0	0.25	1.00
O(1)	4c	0	0	0	1.00
O(2)	16l	0.1346	0.3654	0.1485	1.00

Source: Data from Im, W. B., Fellows, N. N., Denbaars, S. P., Seshadri, R., and Kim, Y. I., *Chem. Mater.*, 21, 2957–2966, 2009b.
Note: Crystal system: Tetragonal.
Space group: I4/mcm (No. 140).
Lattice constant: a = b = 6.8856 Å, c = 11.0613 Å.
Bond lengths: La – O(1): 2.527 Å × 2, La – O(2): 2.440 Å × 2, La – O(2): 2.738 Å × 4.

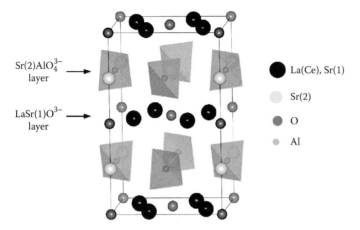

FIGURE 3.20
Crystal structure of $LaSr_2AlO_5$:Ce^{3+} viewed from [011].

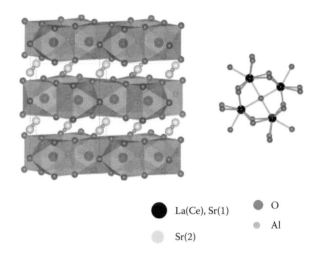

FIGURE 3.21
LaO_8 polyhedra layers along the c-axis, and a ring formed by four La atoms.

As seen in Figure 3.20, the structure of $LaSr_2AlO_5$ contains isolated AlO_4 tetrahedra arranged about the (001) plane, and layers of $LaSr(1)O^{3+}$ and $Sr(2)$ AlO_4^{3-} are alternating along the c-axis. The LaO8 polyhedra are corner shared, forming an alternating layer along the c-axis. A ring formed by four LaO8 polyhedra is clearly seen in Figure 3.21; the distance of La-La is about 3.5746 Å.

3.3.3.2 Photoluminescent Properties of $LaSr_2AlO_5$:Ce^{3+}

The luminescence of Ce^{3+} in $LaSr_2AlO_5$ was reported by Im et al. (2009a, 2009b). The phosphor shows an excitation band with a maximum at 450 nm, enabling it to be combined with blue LED chips. The emission spectrum is a broad band centered at 556 nm, with a FWHM value of 116 nm. The crystal field splitting of Ce^{3+} 5d energy levels in $LaSr_2AlO_5$ is ~10,020 cm^{-1}, and the Stokes shift is estimated to be 1,750 cm^{-1}. This phosphor is reported to have a quantum efficiency of 42% upon the 452 nm excitation (Im et al. 2009b). The promising luminescent properties of $LaSr_2AlO_5$:Ce^{3+} indicate that it can be used for creating high-color-rendering white LEDs. The color rendering index Ra of 80 for white LEDs using this yellow-emitting phosphor is reported (Im et al. 2009b).

The emission color of $LaSr_2AlO_5$:Ce^{3+} can be tuned by elemental substitution. The maximum of the emission band is red-shifted by more than 20 nm when La is partially substituted by Gd. This shift is attributable to the increased crystal field splitting, as the substitution results in the decrease of CeO_8 polyhedra volume. The Ba substitution for Sr also leads to a red-shift of the emission band by about 20 nm, which is ascribed to the increased lattice covalency. On the other hand, the substitution of Ga for Al blue-shifts the emission band, and also decreases the luminescence intensity.

A solid solution between $LaSr_2AlO_5$ and Sr_3SiO_5 ($La_{1-x-0.025}Ce_{0.025}Sr_{2+x}Al_{1-x}Si_xO_5$) is reported by Im et al. (2009c). The peak of the emission band varies in the range of 527–556 nm, depending on x in the solid solution. The quantum efficiency of the solid solution with x = 0.8 is reported to be 55% upon the 457 nm excitation. The thermal quenching of this yellow phosphor is not as good as that of YAG:Ce^{3+}. The luminescence intensity is reported to be decreased by 52% of the initial intensity when the temperature rises from room temperature to 170°C. The quenching temperature is estimated to be 410 K for this solid-solution phosphor. Perhaps the compositional disordering of La^{3+}/Sr^{2+} and Al^{3+}/Si^{4+} has a negative impact on the thermal quenching, as the disorder provides an alternative quenching path in the lattice (Im et al. 2009c).

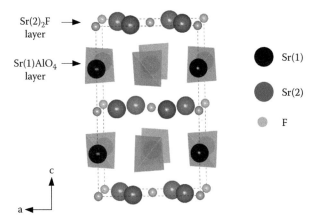

FIGURE 3.22
Crystal structure of Sr_3AlO_4F, which is made up of alternating $[Sr(2)_2F]^{3+}$ and $[Sr(1)AlO_4]^{3-}$ layers.

3.3.4 Oxyfluoride Phosphors

Fluorine-containing compounds are usually not considered as hosts for LED phosphors because the ionic nature of metal-F^{-1} bonds may reduce the covalency of the lattice. Moreover, it is reported that phosphors are more moisture sensitive with the introduction of halide ions. Very recently, air-stable and highly efficient blue-green $Sr_3AlO_4F:Ce^{3+}$-based phosphors have been reported, and show promising luminescent properties that are suitable for white LEDs (Im et al. 2010).

In this section, we will summarize the structure and luminescence of $Sr_3AlO_4F:Ce^{3+}$-based phosphors based on the literature data.

3.3.4.1 Crystal Structure of Sr_3AlO_4F

Vogt et al. (1999) and Prodjosantoso et al. (2003) reported that Sr_3AlO_4F is an ordered oxyfluoride, which crystallizes in a tetragonal system with the I4/mcm space group and lattice constants of a = b = 6.7822(9) and c = 11.1437(2) Å. Some structural parameters are summarized in Table 3.8. The structure of Sr_3AlO_4F is layered, in which isolated AlO_4 tetrahedra whose oxygens coordinate the Sr(1) atoms are separated by $Sr(2)_2F$ layers, as seen in Figure 3.23. An alternative way to describe the structure is to consider it as being built up of alternating $[Sr(2)_2F]^{3+}$ and $[Sr(1)AlO_4]^{3-}$ layers. The large formal charge associated with each layer enables the holding of the layers together tightly. The Sr atoms occupy two different crystallographic sites. The Sr(1) atoms are coordinated with eight oxygen atoms located on adjacent tetrahedral and to two apical fluorine atoms, forming the 10-coordination polyhedra of Sr(1) (see Figure 3.22, right). The Sr(2) atoms are coordinated by

TABLE 3.8

Selected Structural Parameters for Sr_3AlO_4F

Atom	Wyckoff	x	y	z	SOF
Sr(1)	4a	0	0	0.25	1.00
Sr(2)	8h	0.1696	0.6696	0	1.00
Al	4b	0	0.5	0.25	1.00
O	16l	0.1418	0.6418	0.6496	1.00
F	4c	0	0	0	1.00

Source: Data from Vogt T., Woodward P. M., Hunter B. A., Prodjosantoso A. K., and Kennedy B. J., *J. Solid State Chem.*, 144, 228–231, 1999.

Note: Crystal system: Tetragonal.
Space group: I4/mcm (No. 140).
Lattice constant: a = b = 6.7822(9) Å, c = 11.1437(2) Å.
Bond lengths: Sr(1) – O: 2.842(4) Å × 8, Sr(1) – F: 2.768(1) Å × 2, Sr(2) – O: 2.797(4) Å × 4, Sr(2) – O: 2.460(4) Å × 2, Sr(2) – F: 2.5188(8) Å × 2.

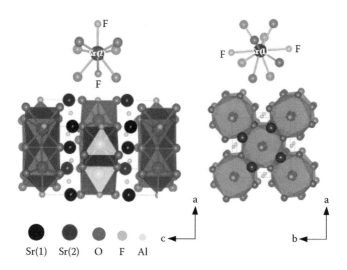

FIGURE 3.23
Sr atoms occupying two crystallographic sites, with Sr(1) coordinated by eight oxygen and two fluorine atoms (right) and Sr(2) coordinated by six oxygen and two fluorine atoms (left).

two fluorine atoms in the ab plane, and by three oxygen atoms above and another three oxygen atoms below the Sr(2)$_2$F plane (see Figure 3.22, left).

3.3.4.2 Photoluminescent Properties of Sr$_3$AlO$_4$F:Ce^{3+}

The luminescence of Ce^{3+}-doped Sr$_3$AlO$_4$F was reported by Im et al. (2010) and Chen et al. (2010). The excitation spectrum of Sr$_3$AlO$_4$F:Ce^{3+} displays a broad band with a maximum at 400 nm, indicating that it is an interesting phosphor that can be pumped by near-UV LED chips. The emission spectrum shows a broad band covering from 400 to 650 nm, with two maxima at 460 and 502 nm, respectively. These two emission bands are ascribed to two different Ce^{3+} sites in the lattice, with the blue band for Ce(1) occupying the Sr(1) sites and the green band for Ce(2) at the Sr(2) sites. The green band enhances as the Ce^{3+} concentration increases, possibly due to (1) the increased number of Ce(2) and (2) energy transfer from Ce(1) to Ce(2). The 502 nm green emission band dominates when the Ce^{3+} concentration is higher than 3 mol%. The quantum efficiency of optimal Sr$_3$AlO$_4$F:Ce^{3+} was reported to be 83% (Im et al. 2010), which is higher than that of YAG:Ce^{3+} (75%).

Substitution of Ba for Sr in Sr$_3$AlO$_4$F:Ce^{3+} does not change the shape of the luminescence spectra, but significantly improves the luminescence efficiency. Im et al. (2010) reported that the quantum efficiency of Sr$_{1.975}$BaCe$_{0.025}$AlO$_4$F was 95%, which is double that of LaSr$_2$AlO$_5$:Ce^{3+}. Although they have the same crystal system and space group, Sr$_{1.975}$BaCe$_{0.025}$AlO$_4$F shows a blue-shifted luminescence spectra in comparison to yellow-emitting LaSr$_2$AlO$_5$:Ce^{3+}. The energy difference in luminescence spectra was explained as the smaller centroid shift of $Sr_{1.975}BaCe_{0.025}AlO_4F$ compared to $LaSr_2AlO_5:Ce^{3+}$ (5,000 cm^{-1} vs. 14,000 cm^{-1}).

Im et al. (2009d) reported the luminescence of a solid solution between $GdSr_2AlO_5$ and Sr_3AlO_4F: $Gd_{1-x}Sr_{2+x}AlO_{5-x}F_x$. It is addressed that the emission band is gradually blue-shifted from 574 nm (x = 0) to 474 nm (x = 1), by increasing the solid-solution amount x. This shift is attributable to (1) an increased distance of Ce-(O,F) bonds that results in a weak crystal field splitting of Ce^{3+} 5d levels, and (2) a decreased covalency of the lattice that reduces the centriod shift when the Gd^{3+}-O^{2-} pair is substituted by the Sr^{2+}-F^- pair. The quantum efficiencies of the solid-solution phosphors with x = 0, 0.3, and 1.0 were reported to be about 39, 50, and 83%, respectively (Im et al. 2009d).

Substitution of Ca^{2+} and Si^{4+}-O^{2-} for Sr^{2+} and Al^{3+}-F^- in $Sr_3AlO_4F:Ce^{3+}$ results in a significant red-shift of the excitation and emission bands. The maximum emission band shifts from 502 nm for $Sr_{2.975}Ce_{0.025}AlO_4F$ to 550 nm for $(Sr_{0.595}Ca_{0.4}Ce_{0.005})_3Al_{0.6}Si_{0.4}O_{4.415}F_{0.585}$. This red-shift is due to the fact that (1) the Ce-(O,F) bond lengths are reduced with the substitution of smaller Ca^{2+}/Sr^{4+} ions for larger Sr^{2+}/Al^{3+}, leading to large crystal field splitting of Ce^{3+} 5d energy levels, and (2) the covalency of Ce-(O,F) bonds and anion polarization is enhanced by substituting F^- for O^{2-}, which further decreases the energy of the lowest 5d levels. The quantum efficiency of these substituted phosphors is reported to be nearly equal to that of YAG:Ce^{3+} phosphors (~80%).

3.3.4.3 Thermal Quenching of $Sr_3AlO_4F:Ce^{3+}$

The thermal quenching of $Sr_3AlO_4F:Ce^{3+}$ is better than that of $LaSr_2AlO_5:Ce^{3+}$, and is equivalent to that of YAG:Ce^{3+}. The luminescence intensity at 150°C decreases by 15% of the initial intensity for $Sr_3AlO_4F:Ce^{3+}$, whereas it decreases by >50% for $LaSr_2AlO_5:Ce^{3+}$. The great difference in thermal quenching of isotypic host compounds Sr_3AlO_4F and $LaSr_2AlO_5$ is due to the fact that $LaSr_2AlO_5:Ce^{3+}$ has "free" O^{2-} anions in its lattice, which leads to ionization-based quenching, whereas a larger barrier for ionization is generated by substituting F for "free" O^{2-} anions in $Sr_3AlO_4F:Ce^{3+}$. As addressed by Setlur et al. (2006), strong thermal quenching is observed when only Al^{3+}-F^- is replaced by Si^{4+}-O^{2-}, with the luminescence intensity at 150°C decreasing by 40% of the initial intensity. This indicates that fluorine plays a dominant role in enhancing the thermal quenching.

3.3.5 Sulfide and Oxysulfide Phosphors

Sulfide phosphors are widely used for cathodoluminescence and electroluminescence applications due to their high luminous efficiency and abundant emission colors. Before highly efficient and blue light excitable red and green LED phosphors were available, alkaline earth sulfides, such as red MS:Eu^{2+} and green MGa_2S_4:Eu^{2+} (M = Ca, Sr, Ba) were considered valuable phosphors for LEDs, because they show broad emission bands covering the red and green part of the visible range very well, as well as broad excitation bands matching very well with the emission wavelength of blue LED chips. On the other hand, sulfide phosphors have fatal weaknesses, such as chemically unstable and large thermal quenching, that prevent their wide use in LEDs. Although coating sulfide phosphors with a protective layer is an effective way to enhance their chemical stability, the excellent luminescent properties and high reliability of red and green nitride phosphors make sulfide phosphors lose their competitiveness.

3.3.5.1 MS:Eu²⁺ (M = Ca, Sr)

3.3.5.1.1 Crystal Structure

Alkaline earth sulfide MS (M = Ca, Sr) has a cubic rock-salt structure with a space group of Fm-3m (No. 225) and a lattice constant of 5.6903 Å. As shown in Figure 3.24, the Ca atoms are coordinated to eight S atoms at an equal distance of 2.8452 Å.

3.3.5.1.2 Photoluminescence Properties of MS:Eu²⁺ (M = Ca, Sr)

CaS:Eu²⁺ is an efficient deep red phosphor that has a broad emission band with a maximum at ~650 nm (Hu et al. 2005a; Poelman et al. 2009). Its excitation spectrum displays a broad band covering the range of 400–550 nm, which matches well with the blue LEDs. The significantly red-shifted excitation and emission spectra of Eu²⁺ in CaS are caused by large crystal field splitting, where Eu²⁺ occupies the Ca²⁺ sites in a cubic structure and has *Oh* symmetry.

The emission color of CaS:Eu²⁺ is so red that the eye sensitivity is quite low. A blue-shift of the emission band of CaS:Eu²⁺ can be achieved by substituting Sr for Ca (Hu et al. 2005a; Poelman et al. 2009). A complete solid solution can be formed between CaS and SrS, as they have the same crystal structure. The emission color of the solid-solution phosphor $Ca_{1-x}Sr_xS:Eu^{2+}$ shifts from 650 nm (x = 0) to 610 nm (x = 1).

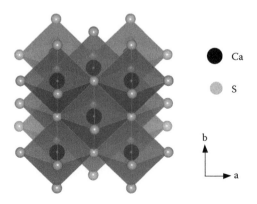

FIGURE 3.24
Crystal structure of cubic CaS viewed from [001].

It is reported that $Ca_{1-x}Sr_xS:Eu^{2+}$ has large thermal quenching, and the luminescence of $SrS:Eu^{2+}$ at 150°C decreases by more than 40% of the initial intensity (Kuo et al. 2010).

3.3.5.2 $MGa_2S_4:Eu^{2+}$ (M = Ca, Sr)

The Sr and Ca thiogallate compounds MGa_2S_4 (M = Ca, Sr) have an ortho-rhombic crystal system with a space group of D^{24} 2h (Fddd). The Sr atoms occupy square antiprismatic sites formed by eight sulfur atoms (symmetry group D4d), forming Sr(S)8 units with C2 and D2 symmetry (Eisenmann et al. 1983). The Sr atoms are situated in three slightly different sites: 8a, 8b, and 16c. Gallium atoms are tetrahedrally coordinated to four sulfur atoms, forming Ga(S)4 units (symmetry group Td). The sulfur atoms are at the center of deformed Sr2Ga2 tetrahedrons, forming (S)Sr2Ga2 units. The assembly of the Sr(S)8 antiprismatic units with common edges forms chains parallel to the a-axis of the unit cell. Each chain is linked to four chains by corner shar-ing. Gallium atoms are located between two consecutive chains. The unit cell consists of four layers along the c-axis. $BaGa_2S_4$ has a different crystal structure from $SrGa_2S_4$ and $CaGa_2S_4$, which crystallizes in the cubic struc-ture with a space group of Th6-Pa3.

The photoluminescence of Eu^{2+}-doped thiogallate compounds MGa_2S_4 (M = Ca, Sr, Ba) was reported by Peter and Baglio (1972). The excitation spectra consist of an extremely broad band that spreads into the visible region. The significant red-shift in $MGa_2S_4:Eu^{2+}$ is interpreted as the strong nephelauxetic effect (large centroid shift). The emission spectra of thiogallate compounds consist of a broad band, the position of which is dependent on the choice of the host cation. The $BaGa_2S_4:Eu^{2+}$ is a blue-green phosphor that emits at ~500 nm, and $SrGa_2S_4:Eu^{2+}$ and $CaGa_2S_4:Eu^{2+}$ are green- and yellow-emitting phosphors, respectively. In addition, the isostructural Ca and Sr thiogallates form a complete series of solid-solution $Sr_{1-x}Ca_xGa_2S_4:Eu^{2+}$. The emission spectra of the solid solution vary uniformly from the green of $SrGa_2S_4:Eu^{2+}$ (535 nm) to the yellow of $CaGa_2S_4:Eu^{2+}$ (555 nm).

Because the excitation spectra of Eu^{2+}-doped thiogallate compounds match well with the emission spectra of blue LED chips, these phosphors are very promising green-emitting materials for high-color-rendition white LEDs (Mueller-March 2002; Do et al. 2006; Yoo et al. 2008). One of the serious short-comings of these phosphors is that they will undergo hydrolysis, indicative of their poor chemical stability.

3.3.5.3 $Sr_8Al_{12}O_{24}S_2:Eu^{2+}$

$Sr_8Al_{12}O_{24}S_2$ crystallizes in a cubic structure with a space group of I-43m (No. 217) and a lattice constant of 9.2572 Å (Brenchley and Weller 1992). Its crystal structure is illustrated in Figure 3.25.

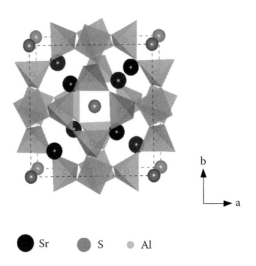

FIGURE 3.25
Crystal structure of $Sr_8Al_{12}O_{24}S_2$ viewed from [001].

The luminescence of Eu^{2+}-doped oxyfulide $Sr_8Al_{12}O_{24}S_2$ was reported by Kuo et al. (2010). $Sr_8Al_{12}O_{24}S_2$:Eu^{2+} shows a broad emission band centered at 605 nm, indicative of an orange-yellow phosphor. Its excitation spectrum covers a wide range of 400–500 nm, matching well with the emission spectral range of blue LED chips.

The thermal quenching of $Sr_8Al_{12}O_{24}S_2$:Eu^{2+} is better than that of SrS:Eu^{2+}. The luminescence at 150°C reduces by ~35% of the initial intensity.

3.3.6 Scheelite Phosphors

Eu^{3+}-doped inorganic compounds with the scheelite structure, such as molybdates, tungstates, and vanadates, have gained much attention because they show very intense and characteristic sharp-line red emission spectra of Eu^{3+}, and can be excited efficiently by near UV (394 nm) or blue light (464 nm). The Eu^{3+} luminescence is interesting because (1) the color purity of Eu^{3+} emission is high due to its sharp-line spectra; (2) the magnetic dipole $5D^0$-$7F^2$ transition is allowed, resulting in very intense ~613 nm red emission, when Eu^{3+} is presenting a noncentrosymmetric site; and (3) the emitting level $5D^0$ and the ground state $7F^0$ are nondegenerate.

Eu^{3+}-doped scheelite phosphors containing WO_4, MO_4, and VO_4 complex ions are interesting red-emitting phosphors for near-UV or blue light LEDs. They basically show similar emission spectra, which are usually composed of several groups of sharp lines. Their excitation spectra consist of a broad charge-transfer band (O → W, Mo, or V) along with sharp lines at 394 and

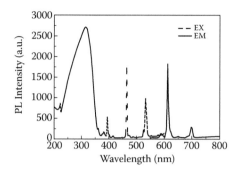

FIGURE 3.26
Excitation and emission spectra of a Eu^{3+}-doped $Ca_9La(VO_4)_7$ red-emitting phosphor.

463 nm due to f-f transitions of Eu^{3+}, which have luminescence intensity comparable to that of the broad change-transfer (CT) band.

For better understanding of the luminescence of scheelite phosphors, we present the photoluminescence spectra of Eu^{3+}-activated $Ca_9La(VO_4)_7$ in Figure 3.26.

The excitation spectrum consists of a very strong broad band (200–370 nm) and some narrow lines in the spectral region of 370–600 nm. The broad excitation band is due to the charge transfer of V^{5+}-O^{2-} in VO_4^{3-} but is not assigned to the charge transfer of Eu^{3+}-O^{2-}. The emission spectrum, measured under 465 nm excitation, is composed of four groups of sharp lines, which is due to the $^5D_0 \rightarrow {}^7F_J$ (J = 1, 2, 3, 4) electronic transitions of Eu^{3+}. Generally, the $^5D_0 \rightarrow {}^7F_{2,4}$ are electric-dipole transitions, whereas the $^5D_0 \rightarrow {}^7F_{1,3}$ are magnetic-dipole transitions. The major emission peak at 613 nm due to the $^5D_0 \rightarrow {}^7F_2$ transition of Eu^{3+} reveals that the electric-dipole transition is dominant. Since this electric-dipole transition is a parity-forbidden f-f configuration transition, it can happen only when the Eu^{3+} ions locate in the lattice sites without inversion symmetry. As seen from the photoluminescence spectra, $Ca_9La(VO_4)_7$:Eu^{3+} is an interesting red phosphor suitable for UV or blue LED chips.

Table 3.9 summarizes typical red-emitting scheelite phosphors that have been reported in the literature.

3.3.7 Phosphate Phosphors

Several kinds of Eu^{2+}-doped phosphate phosphors were reported to have potential applications in white LEDs, such as $KMPO_4$ (M = Ca, Sr, Ba) (Tang et al. 2007; Lin et al. 2010). In general, the phosphate phosphors emit blue colors, and can only be excited by UV light (350–380 nm). Moreover, they show large thermal quenching. Therefore, in comparison with the blue-emitting BAM:Eu^{2+}, which has a quantum efficiency of >90% and smaller thermal quenching, phosphate phosphors do not have a real value for white LEDs.

TABLE 3.9

Typical Eu^{3+}-Activated Red-Emitting Scheelite Phosphors

Materials	Phosphors	References
Molybdate	$CaMoO_4:Eu^{3+}$	Hu et al. 2005b; Liu et al. 2007
	$M_5Eu(MoO_4)_4$ (M = Na, K)	Pan et al. 1988
	$Ca_{1-2x}Eu_xLi_xMoO_4$	Wang et al. 2005
Tungstate	$Gd_{2-y}Eu_y(WO_4)_{3-x}(MoO_4)_x:Eu^{3+}$	Wang et al. 2007
	$NaM(WO_4)_{2-x}(MoO_4)_x:Eu^{3+}$ (M = Gd, Y, Bi)	Neeraj et al. 2004
	$AgGd(WO_4)_{2-x}(MoO_4)_x:Eu^{3+}$	Sivakumar and Varadaraju 2005
	$K_5Eu_x(WO_4)_{2.5+1.5x}$	Do and Huh 2000
Vanadate	$YVO_4: Eu^{3+}, Bi^{3+}$	Nguyen et al. 2009; Xia et al. 2010
	$Ca_3Sr_3(VO_4)_4:Eu^{3+}$	Choi et al. 2009
	$Ca_3La(VO_4)_3:Eu^{3+}$	Rao et al. 2010
	$Bi_xLn_yNa_2Mg_2(VO_4)_3:Eu^{3+}$ (Ln = Y, La, Gd)	Shimomura et al. 2008

References

Bachmann, V., Ronda, C., and Meijerink, A. 2009. Temperature quenching of yellow Ce^{3+} luminescence in YAG:Ce. *Chem. Mater.* 21:2077–2084.

Barbier, J., and Hyde, B. G. 1985. The structures of the polymorphs of dicalcium silicate, Ca_2SiO_4. *Acta Cryst. B* 41:383–390.

Barry, T. L. 1968a. Fluorescence of Eu^{2+} activated phase in binary alkaline earth ortho-silicate system. *J. Electrochem. Soc.* 115:1181–1183.

Barry, T. L. 1968b. Luminescent properties of Eu^{2+} and $Eu^{2+}+Mn^{2+}$ activated $BaMg_2Si_2O_7$. *J. Electrochem. Soc.* 117:381–385.

Barry, T. L. 1968c. Equilibria and Eu^{2+} luminescence of subsolidus phase bounded by $Ba_3MgSi_2O_8$, $Sr_3MgSi_2O_8$, $Ca_3MgSi_2O_8$. *J. Electrochem. Soc.* 115:733–738.

Blasse, G. Wanmaker, W. L., and Vrugt, J. W. 1968a. Some new classes of efficient Eu2+-activated phosphors. *J. Electrochem. Soc.* 115:673.

Blasse, G., Wanmaker, W. L., Vrugt, J. W., and Bril, A. 1968b. Fluorescence of Eu2+-activated silicates. *Philips Res. Rep.* 23:189–200.

Brenchley, M. E., and Weller, M. T. 1992. Synthesis and structure of sulfide aluminate sodalites. *J. Mater. Chem.* 2:1003–1005.

Catti, M., Gazzoni, G., and Ivaldi, G. 1983. Structures of twinned β-Sr_2SiO_4 and of α'-$Sr_{1.9}Ba_{0.1}SiO_4$. *Acta Cryst. C* 39:29–34.

Chen, L. T., Hwang, C. S., Sun, I. L., and Chen, I. G. 2006. Luminescence and chromaticity of alkaline earth aluminate $M_xSr_{1-x}Al_2O_4:Eu^{2+}$ (M: Ca,Ba). *J. Lumin.* 118:12–20.

Chen, W., Liang, H., Ni, H., He, P., and Su, Q. 2010. Chromaticity-tunable emission of $Sr_3AlO_4F:Ce^{3+}$ phosphors: Correlation with matrix structure and application in LEDs. *J. Electrochem. Soc.* 157:J159–J163.

Chenavas, J., Joubert, J. C., and Marezio, M. 1978. On the crystal symmetry of the garnet structure. *J. Less-Common Metals* 62:373–380.

Chiang, C. C., Tsai, M. S., and Hon, M. H. 2008. Luminescent properties of cerium-activated garnet series phosphors: Structure and temperature effects. *J. Electrochem. Soc.* 155:B517–B520.

Choi, S., Moon, Y. M., Kim, K., Jung, H. K., and Nahm, S. 2009. Luminescent properties of a novel red-emitting phosphor: Eu^{3+}-activated $Ca_3Sr_3(VO_4)_4$. *J. Lumin.*
129:988–990.

Do, Y. R., and Huh, Y. D. 2000. Optical properties of potassium europium tungstate
phosphors. *J. Electrochem. Soc.* 147:4385–4388.

Do, Y. R., Ko, K. Y., Na, S. H., and Huh, Y. D. 2006. Luminescence properties of potential $Sr_{1-x}Ca_xGa_2S_4$:Eu green and greenish-yellow-emitting phosphors for white
LED. *J. Electrochem. Soc.* 153:H142–H146.

Eisenmann, B., Jakowski, M., Klee, W., and Schaefer, H. 1983. Luminescence and
structural properties of $BaGa_2S_4$ and $BaAl_2S_4$. *Rev. Chim. Miner.* 20:255–261.

Fields, J. M., Dear, P. S., and Brown, J. J. 1972. Phase equilibria in the system $BaO-SrO-SiO_2$.
J. Am. Ceram. Soc. 55:585–588.

Glasser, L. S. D., and Glasser, F. P. 1965. Silicates M_3SiO_5. I. Sr_3SiO_5. *Acta Cryst.* 18:453–455.

Hansel, R., Allison, S., and Walker, G. 2010. Temperature-dependent luminescence of
gallium-substituted YAG:Ce. *J. Mater. Sci.* 45:146–150.

He, H., Fu, R., Cao, Y., Song, X., Pan, Z., Zhao, X., Xiao, Q., and Li, R. 2010. $Ce^{3+} \rightarrow Eu^{2+}$
energy transfer mechanism in the Li_2SrSiO_4:Eu^{2+},Ce^{3+} phosphor. *Opt. Mater.*
32:632–636.

He, H., Fu, R., Wang, H., Song, X., Pan, Z., Zhao, X., Zhang, X., and Cao, Y. 2008.
Li_2SrSiO_4:Eu^{2+} phosphor prepared by the Pechini method and its application in
white light emitting diode. *J. Mater. Res.* 23:3288–3294.

Holloway, W. W., and Kestigian, M. 1969. Optical properties of cerium-activated
garnet crystals. *J. Opt. Soc. Am.* 59:60–63.

Hu, Y., Zhuang, W., Ye, H., Zhang, S., Fang, Y., and Huang, X. 2005a. Preparation
and luminescent properties of $(Ca_{1-x}Sr_x)S$:Eu^{2+} red-emitting phosphor for white
LED. *J. Lumin.* 111:139–145.

Hu, Y. S., Zhuang, W. D., Ye, H. Q., Wang, D. H., Zhang, S. S., and Huang, X. W. 2005b.
A novel red phosphor for white light emitting diodes. *J. Alloys Compd.* 390:226–229.

Im, W. B., Brinkley, S., Hu, J., Mikhailovsky, A., Denbaars, S. P., and Seshadri, R. 2010.
$Sr_{2.975-x}Ba_xCe_{0.025}AlO_4F$: A highly efficient green-emitting oxyfluoride phosphor
for solid state white lighting. *Chem. Mater.* 22:2842–2849.

Im, W. B., Fellows, N. N., Denbaars, S. P., and Seshadri, R. 2009c. $La_{1-x-0.025}Ce_{0.025}Sr_{2+x}$
$Al_{1-x}Si_xO_5$ solid solutions as tunable yellow phosphors for solid state white
lighting. *J. Mater. Chem.* 19:1325–1330.

Im, W. B., Fellows, N. N., Denbaars, S. P., Seshadri, R., and Kim, Y. I. 2009b. $LaSr_2AlO_5$,
a versatile host compound for Ce^{3+}-based yellow phosphors: Structural tuning
of optical properties and use in solids state white lighting. *Chem. Mater.*
21:2957–2966.

Im, W. B., Fourre, Y., Brinkley, S., Sonoda, J., Nakamura, S., Denbaars, S. P., and
Seshadri, R. 2009d. Substitution of oxygen by fluorine in the $GdSr_2AlO_5$:Ce^{3+}
phosphors: $Gd_{1-ph}Sr_{2+x}AlO_{5-x}F_x$ solid solutions for solid state white lighting. *Opt.
Express* 17:22673–22679.

Im, W. B., Page, K., Denbaars, S. P., and Seshadri, R. 2009a. Probing local structure
in the yellow phosphor $LaSr_2AlO_5$:Ce^{3+}, by the maximum entropy method and
pair distribution function analysis. *J. Mater. Chem.* 19:8761–8766.

Jang, H. S., Im, W. B., Lee, D. C., Jeon, D. Y., and Kim, S. S. 2007. Enhancement of red
spectral emission intensity of $Y_3Al_5O_{12}$:Ce^{3+} phosphor via Pr co-doping and Tb
substitution for the application to white LEDs. *J. Lumin.* 126:371–377.

Jang, H. S., and Yeon, D. Y. 2007. Yellow-emitting $Sr_3SiO_5:Ce^{3+}$, Li^+ phosphor for white-light-emitting diodes and yellow-light-emitting diodes. *Appl. Phys. Lett.* 90:041906.

Jiang, Z. Q., Wang, Y. H., and Wang, L. S. 2010. Enhanced yellow-to-orange emission of Si-doped $Mg_3Y_2Ge_3O_{12}:Ce^{3+}$ garnet phosphors for warm white light-emitting diodes. *J. Electrochem. Soc.* 157:J155–J158.

Kim, J. S., Jeon, P., and Choi, J. C. 2004a. Warm-white-light emitting diode utilizing a single-phase full-color $Ba_3MgSi_2O_8:Eu^{2+},Mn^{2+}$ phosphor. *Appl. Phys. Lett.* 84:2931–2933.

Kim, J. S., Jeon, P. E., and Park, Y. H. 2004b. White-light generation through ultraviolet-emitting diode and white-emitting phosphor. *Appl. Phys. Lett.* 85:3696–3698.

Kim, J. S., Park, Y. H., and Choi, J. C. 2005a. Temperature-dependent emission spectrum of $Ba_3MgSi_2O_8:Eu^{2+},Mn^{2+}$ phosphor for white-light-emitting diode. *Electrochem. Solid State Lett.* 8:H65–H67.

Kim, J. S., Park, Y. H., Choi, J. C., and Park, H. L. 2005b. Optical and structural properties of Eu^{2+}-doped $(Sr_{1-x}Ba_x)_2SiO_4$ phosphors. *J. Electrochem. Soc.* 152:H135.

Kim, J. S., Park, Y. H., and Kim, S. M. 2005c. Temperature-dependent emission spectra of $M_2SiO_4:Eu^{2+}$ (M = Ca, Sr, Ba) phosphors for light-emitting diode. *Solid State Comm.* 133:187–190.

Kim, T. G., Lee, H. S., Lin, C. C., Kim, T., Liu, R. S., Chan, T. S., and Im, S. J. 2010. Effects of additional Ce^{3+} doping on the luminescence of $Li_2SrSiO_4:Eu^{2+}$ yellow phosphor. *Appl. Phys. Lett.* 96:061904-1–061904-3.

Kuo, T. W., Huang, C. H., and Chen, T. M. 2010. Novel yellowish-orange $Sr_8Al_{12}O_{24}S_2:Eu^{2+}$ phosphor for application in blue light-emitting diode based white LED. *Opt. Express* 18:A231–A236.

Lin, C. C., Xiao, Z. R., Guo, G. Y., Chan, T. S., and Liu R. S. 2010. Versatile phosphate phosphors ABPO(4) in white light-emitting diodes: Collocated characteristic analysis and theoretical calculations. *J. Am. Chem. Soc.* 132:3020–3028.

Liu, J., Lian, H. Z., and Shi, C. S. 2007. Improved optical photoluminescence by charge compensation in the phosphor system $CaMoO_4:Eu^{3+}$. *Opt. Mater.* 29:1591–1594.

Meagher, E. P. 1975. The crystal structures of pyrope and grossularite at elevated temperatures. *Am. Miner.* 60:218–228.

Mueller-March, R., Mueller, G. O., Krames, M. R., and Trottier, T. 2002. High-power phosphor-converted light-emitting diodes based on III-nitrides. *IEEE J. Selected Topics Quantum Electron* 8:339–345.

Nakatsuka, A., Yoshiasa, A., and Yamanaka, T. 1999. Cation distribution and crystal chemistry of $Y_3Al_{5-x}Ga_xO_{12}$ (0 ≤ x ≤ 5) garnet solid solutions. *Acta Cryst. B* 55:266–272.

Neeraj, S., Kijima, N., and Cheetham, A. K. 2004. Novel red phosphors for solid-state lighting: The system $NaM(WO_4)_{2-x}(MoO_4)_x:Eu^{3+}$ (M = Gd, Y, Bi). *Chem. Phys. Lett.* 387:2–6.

Nguyen, H. D., Mho, S., and Yeo, I. H. 2009. Preparation and characterization of nano-sized (Y, Bi)$VO_4:Eu^{3+}$ and $Y(V,P)O_4:Eu^{3+}$ red phosphors. *J. Lumin.* 129:1754–1758.

Pan, J., Liu, L. Z., Chen, L. G., Zhao, G. W., Zhou, G. E., and Guo, C. X. 1988. Studies on spectra properties of $Na_5Eu(WO_4)_4$ luminescent crystal. *J. Lumin.* 40–41:856–857.

Pan, Y., Wu, M., and Su, Q. 2004. Tailored photoluminescence of YAG:Ce phosphor through various methods. *J. Phys. Chem. Solids* 65:845–850.

Park, J. K., Choi, K. J., and Park S. H. 2005. Application of Ba^{2+}-Mg^{2+} co-doped $Sr_2SiO_4:Eu$ yellow phosphor for white-lighting-emitting diodes. *J. Electrochem. Soc.* 152:H121–H123.

Park, J. K., Choi, K. J., Yeon, J. H., Lee, S. J., and Kim, C. H. 2006. Embodiment of the warm white-light-emitting diodes by using a Ba^{2+} codoped Sr_3SiO_5:Eu phosphor. *Appl. Phys. Lett.* 88:043511.

Park, J. K., Kim, C. H., Park, S. H., Park, H. D., and Chois, S. Y. 2004. Application of strontium silicate yellow phosphor for white light-emitting diodes. *Appl. Phys. Lett.* 84:1647–1649.

Park, J. K., Lim, M. A., Kim, C. H., Park, H. D., Park, J. T., and Choi, S. Y. 2003. White light-emitting diodes of Gan-based Sr_2SiO_4:Eu and the luminescent properties. *Appl. Phys. Lett.* 82:683–685.

Pavese, A., Artioli, G., and Prencipe, M. 1995. X-ray single-crystal diffraction study of pyrope in the temperature 30–973 K. *Am. Miner.* 80:457–464.

Peter, T. E., and Baglio, J. A. 1972. Luminescence and structural properties of thio-gallate phosphors Ce^{3+} and Eu^{2+} activated phosphors. Part I. *J. Electrochem. Soc.* 119:230–236.

Poelman, D., Haecke, J. E., and Smet, P. F. 2009. Advances in sulfide phosphors for displays and lighting. *J. Mater. Sci. Mater Electron.* 20:S134–S138.

Poort, S. H. M., Blokpoel, P. W., and Blasse, G. 1995. Luminescence of Eu^{2+} in barium and strontium aluminate and gallate. *Chem. Mater.* 7:1547–1551.

Poort, S. H. M, Hanssen, W., and Blasse, G. 1997. Optical properties of Eu^{2+}-activated orthosilicates and orthophosphates. *J. Alloy Compds.* 260:93–97.

Prodjosantoso, A. K., Kennedy, B. J., Vogt, T., and Woodward, P. M. 2003. Cation and anion ordering in the layered oxyfluorides $Sr_{3-x}AxAlO_4F$ (A= Ba, Ca). *J. Solid State Chem.* 172:89–94.

Rao, B. V., Jang, K., Lee, H. S., Yi, S. S., and Jeong, J. H. 2010. Synthesis and photo-luminescence characterization of RE^{3+} (=Eu^{3+}, Dy^{3+})-activated $Ca_3La(VO_4)_3$ phosphors for white light-emitting diodes. *J. Alloys Compd.* 496:251–255.

Ravichandran, D., Johnson, S. T., Erdei, S., Roy, R., and White, W. B. 1999. Crystal chemistry and luminescence of the Eu^{2+}-activated alkaline earth aluminate phosphors. *Display* 19:197–203.

Saradi, M. P., and Varadaraju, U. V. 2006. Photoluminescence studies on Eu^{2+}-activated Li_2SrSiO_4—A potential orange-yellow phosphor for solid-state lighting. *Chem. Mater.* 18:5267–5272.

Setlur, A. A., Heward, W. J., Gao, Y., Srivastava, A. M., Chandran, R. G., and Shankar, M. V. 2006. Crystal chemistry and luminescence of Ce^{3+}-doped $Lu_2CaMg_2(Si,Ge)_3O_{12}$ and its use in LED based lighting. *Chem. Mater.* 18:3314–3322.

Shimomura, Y., Honma, T., Shigeiwa, M., Akai, T., Okamoto, K., and Kijima, N. 2007. Photoluminescence and crystal structure of green-emitting $Ca_3Sc_2Si_3O_{12}$:Ce^{3+} phosphors for white light emitting diodes. *J. Electrochem. Soc.* 154:J35–J38.

Shimomura, Y., Kijima, N., and Cheetman, A. K. 2008. Novel red phosphors based on vanadate garnets for solid state lighting applications. *Chem. Phys. Lett.* 455:279–283.

Sivakumar, V., and Varadaraju, U. V. 2005. Intense red-emitting phosphors for white light emitting diodes. *J. Electrochem. Soc.* 152:H168–H171.

Tang, Y. S., Hu, S. F., Lin, C. C., Bagkar, N. C., and Liu, R. S. 2007. Thermally stable luminescence of $KSrPO_4$:Eu^{2+} phosphor for white light UV light-emitting diodes. *Appl. Phys. Lett.* 90:151108.

Vogt T., Woodward P. M., Hunter B. A., Prodjosantoso A. K., and Kennedy B. J. 1999. Sr_3MO_4F (M= Al, Ga)—A new family of ordered oxyfluorides. *J. Solid State Chem.* 144:228–231.

Wang, J., Jing, X., Yan, C., and Lin, J. 2005. $Ca_{1Cag}Eu_xLi_xMoO_4$: A novel red phosphor for solid state lighting based on a GaN LED. *J. Electrochem. Soc.* 152:G186–G188.

Wang, X. X., Xian, Y. L., Shi, J. X., Su, Q., and Gong, M. L. 2007. The potential red emitting $Gd_{2-y}Eu_y(WO_4)_{3-x}(MoO_4)_x$ phosphors for UV InGaN-based light-emitting diodes. *Mater. Sci. Eng. B* 140:69–72.

Won, Y. H., Jang, H. S., Im, W. B., Jeon, D. Y., and Lee, J. S. 2006. Tunable full-color-emitting $La_{0.827}Al_{11.9}O_{19.09}$:$Eu^{2+}$,$Mn^{2+}$ phosphor for application to warm white-light-emitting diodes. *Appl. Phys. Lett.* 89:231909-1–231909-3.

Xia, Z. G., Chen, D. M., Yang, M., and Ying, T. 2010. Synthesis and luminescence properties of YVO_4:Eu^{3+}, Bi^{3+} phosphor with enhanced photoluminescence by Bi^{3+} doping. *J. Phys. Chem. Solids* 71:175–180.

Yamazaki, K, Nakabayashi, H., Kotera, Y., and Ueno, A. 1986. Fluorescence of Eu^{2+}-activated binary alkaline earth silicate. *J. Electrochem. Soc.* 133:657–660.

Yang, W. J., Luo, L. Y., and Chen, T. M. 2005. Luminescence and energy transfer of Eu- and Mn-coactivated $CaAl_2Si_2O_8$ as a potential phosphor for white-light UVLED. *Chem. Mater.* 17:3883–3888.

Yoo, H. S., Im, W. B., Vaidyanathan, S., Park, B. J., and Jeon, D. Y. 2008. Effects of Eu^{2+} concentration variation and Ce^{3+} codoping on photoluminescence properties of $BaGa_2S_4$:Eu^{2+} phosphor. *J. Electrochem. Soc.* 155:J66–J70.

Yoo, J. S., Kin, S. H., and Yoo, W. T. 2005. Control of spectral properties of strontium-alkaline earth-silicate-europium phosphors for LED applications. *J. Electrochem. Soc.* 152:G382–G385.

Zhang, X., He, H., Li, Z., Yu, T., and Zou, Z. 2008. Photoluminescence studies on Eu^{2+} and Ce^{3+}-doped Li_2SrSiO_4. *J. Lumin.* 128:1876–1879.

4

Nitride Phosphors in White LEDs

In Chapter 3 we presented and discussed photoluminescent properties of traditional phosphors for white LEDs, such as garnets, silicates, aluminates, sulfides, and scheelite phosphors. In this chapter, we will turn our attention to a new type of luminescent material—nitride phosphors, which are very attractive for white LEDs.

Nitride compounds, typically nitridosilicates (M-Si-N, M = alkaline earth metals and lanthanide metals), nitridoaluminosilicates (M-Si-Al-N), oxynitridosilicates (M-Si-O-N), and oxynitridoaluminosilicates (M-Si-Al-O-N), are structurally built upon three-dimensional SiN_4, $Si(O,N)_4$, AlN_4, or $Al(O,N)_4$ tetrahedral networks, with M metals residing in the channels or voids formed by the network. When activators ions (e.g., Eu^{2+} and Ce^{3+}) are introduced into the nitride structures, they are coordinated to nitrogen atoms, forming covalent chemical bonds with relatively short distances. This then lowers the 5d energy levels of activator ions, and significantly red-shifts the excitation and emission spectra of rare-earth-activated nitride phosphors. Therefore, the excitation spectra of nitride phosphors usually display a very broad band and extend to the visible light spectral region, enabling them to absorb near-UV or blue light efficiently. In addition, nitride compounds have a great diversity of crystal structures and varying local structures surrounding the activator ions, which can make changes in the centroid shift and crystal field splitting of 5d energy levels of rare earth ions, and lead to abundant emission colors of nitride phosphors. Finally, nitride host lattices are structurally stable against chemical or thermal attacks, which generally results in small thermal quenching or degradation of nitride phosphors. With these promising photoluminescent properties, nitride phosphors are very suitable for use as downconversion luminescent materials in white LEDs, forming a fairly new and important family of phosphor materials for lighting. A number of nitride phosphors, for example, Eu^{2+}-activated α-sialon, β-sialon, and $CaAlSiN_3$, have been commercialized, and play key roles in producing highly efficient and reliable white LED products.

Extensive investigations have been carried out on the electroluminescence of binary or ternary semiconductor nitride thin films (e.g., GaN, InGaN, AlN), so the reader should be quite familiar with optoelectronic properties of well-known III-V group nitride materials. On the other hand, ternary or multinary (oxy)nitridosilicates and (oxy)nitridoaluminosilicates are usually considered as structural ceramic materials; they were not known to be a host lattice for luminescent materials until about 10 years ago, when they were reported to show interesting photoluminescence, and they have recently found their superior suitability in white LEDs (van Krevel et al. 1998; Hoppe et al. 2000; Xie et al. 2004c, Xie and Hirosaki 2007a). In this chapter we first will give an overview of the crystal chemistry of nitride materials, and then investigate the crystal structure and photoluminescent properties of rare-earth-activated nitride phosphors. Finally, we discuss the synthetic approaches of nitride phosphors.

4.1 Classification and Crystal Chemistry of Nitride Compounds

4.1.1 Types of Nitride Materials

Nitrogen has a trivalent state with a −3 more or less formal charge, and its Pauling's electronegativity is 3.04. Nitride materials are compounds that are usually formed by the combination of nitrogen with elements that are less electronegative than nitrogen itself. Those elements consist of nonmetal elements (Si, B, P), alkaline earth and alkali metals, transition metals, and lanthanide elements. Although nitrides are less numerous than oxide due to the thermodynamic factors (e.g., larger triple bond energy of 941 kJ/mol dinitrogen vs. 499 kJ/mol oxygen), they have a fairly large number of compounds, which therefore leads to an extremely rich nitride chemistry.

The chemistry of nitrides has advanced dramatically over the last decades, resulting in lots of novel materials embracing transition metal nitrides and nitridosilicates (Gregory 1999; Schnick et al. 1999; Schnick 2001; Tessier and Marchand 2003; Metselaar 1994). This makes it difficult to classify such a varied group of nitride compounds. Marchand and coworkers (1991, 1998) grouped the nitride and oxynitride materials in terms of their crystal structure. Niewa and DiSalvo (1998) overviewed the nitride chemistry with emphasis on the coordination number of metals and on the structural features. It is well known that nitrogen is located in the periodic table between carbon and oxygen, giving rise to two major classes of binary nitrides having similarities either with carbides (metallic bonding) or with oxides (ionic covalent bonding) (Marchand 1998). For simplicity, the nitride compounds are classified into three main types—ionic, metallic, and covalent nitrides in this work, depending on the nature of the chemical bonds between nitrogen and coordinating elements.

Ionic nitrides are compounds by the combination of nitrogen with alkaline earth, alkali, and rare earth metals. The electronegativity values of alkaline earth, alkali, and rare earth metals are 0.89–1.00, 0.82–1.23, and 1.1–1.5, respectively. The large difference in electronegativity between nitrogen and metals

TABLE 4.1

Binary and Ternary Ionic Nitrides

	Binary Nitrides	Ternary Nitrides
Alkaline earth or alkali metals	Li_3N, Be_3N_2, Mg_3N_2, Ca_3N_2, Sr_2N, SrN	$LiMgN$, $LiZnN$, Li_3AlN_2, Li_3GaN_2, Li_5SiN_3, Li_5TiN_3, Li_7VN_4, Li_7MnN_4
Rare earth metals	LaN, CeN, EuN, etc.	Li_2ZrN_2, Li_2CeN_2

explains the great affinity of alkaline earth, alkali, and rare earth metals for nitrogen, and thus the predominant ionic character of the chemical bonding. Some typical binary and ternary ionic nitrides are shown in Table 4.1. Most binary ionic nitrides are very sensitive to hydrolysis, releasing ammonia associated with the loss of nitrogen when they are not protected carefully. Ionic nitride, especially lithium-containing ionic nitrides, exhibits promising ionic conductivity due to the mobility of Li^+ ions, so that it has been studied as an anode material for lithium ion batteries, such as Li_3N (Prosini 2003), Li_7MnN_4 (Suzuki and Shodai 1999), $Li_{2.6}(Co,Cu,Ni)_{0.4}N$ (Takeda et al. 2000; Liu et al. 2004), and LiNiN (Stoeva et al. 2007), or a lithium solid electrolyte, such as Li_3AlN_2 (Yamane et al. 1985) and Li_5SiN_3 (Yamane et al. 1987). In addition, Li_3N and Ca_3N_2 have been investigated as hydrogen storage materials (Chen et al. 2002; Xiong et al. 2003; Kojima and Kawai 2004).

Metallic nitrides are formed by the reaction between nitrogen and transition metals; most of them are binary nitrides, including TiN, TaN, ZrN, VN, NbN, and CrN. These transition metal nitrides are closely related to transition metal carbides. They are considered insertion compounds of nitrogen in the metal network, so that they are essentially metallic in character. The binary transition metal nitrides are usually chemically stable and show high melting points, metallic conductance, and high hardness, enabling them to become important structural materials and to be used as ceramic coatings (Su and Yao 1997; Starosvetsky and Gotman 2001; Bemporad et al. 2004; Feng et al. 2005; Caicedo et al. 2010).

Covalent nitrides or oxynitrides are formed by the combination of nitrogen with nonmetal elements (such as B, Si, C, P) or with transition metals, lanthanide metals, and alkaline earth metals, the structure of which is generally characterized by the network of edge- or corner-sharing [MN4] tetraheda or edge-sharing [MN6] octahedra. According to Marchand and coworkers (1991, 1998) and Niewa and DiSalvo (1998), we summarize the covalent nitrides or oxynitrides with coordination numbers of 4 and 6 for metal M in Table 4.2.

In fact, the covalent nitrides are rarely all covalently bonded; they usually consist of a mix of ionic and covalent chemical bonds. For example, Si_3N_4 consists of approximately 30% ionic bonds and 70% covalent bonds. The covalent nitrides or oxynitrides generally show interesting physical and chemical properties, such as excellent chemical stability, high thermomechanical

TABLE 4.2

Some Typical Covalent Nitrides and Oxynitrides

Types	Structures and Materials	
Binary	Group III: BN, AlN, GaN, InN	
	Group IV: Si_3N_4, Ge_3N_4	
Tetrahedral unit (CN = 4)	Wurtzite type (with Si, Ge, P)	245_2 family: $BeSiN_2$, $MgSiN_2$, $MnSiN_2$, $MgGeN_2$, $ZnGeN_2$
		14_25_3 family: $LiSi_2N_3$, $LiGe_2N_3$
		2_255_3 family: Mg_2PN_3, Mn_2PN_3
		1465 family: LiSiON, NaSiON, KGeON
	Cristobalite type (with Ge, P)	$CaGeN_2$, $LiPN_2$, $NaPN_2$
	Scheelite type (with Os, W)	$KOsO_3N$
		$LnWO_3N$ (Ln = Nd, Sm, Gd, Dy)
	Silicate-derived structure (with Si)	$Ln_{10}Si_6O_{24}N_2$ (Ln = La, Ce, Nd, Sm, Gd, Y)
		$Ln_2Si_3O_3N_4$ (Ln = La-Yb, Y)
		$Ln_4Si_2O_7N_2$ (Ln = Nd-Yb, Y)
		$LnSiO_2N$ (Ln = La, Ce, Y)
		$Ln_3Si_6N_{11}$ (Ln = Y, La-Nd, Sm)
		$LnSi_3N_5$ (Ln = Y, La-Nd)
		$AEYbSi_4N_7$ (AE = Sr, Ba, Eu)
		$BaSi_7N_{10}$
		$AE_2Si_5N_8$ (AE = Ca, Sr, Ba)
Octahedral unit (CN = 6)	Perovskite type (with Ta, Nb, W, Mo, V)	$MTaO_2N$ (M = Ca, Sr, Ba)
		$MNbO_2N$ (M = Sr, Ba)
		$LnTiO_2N$ (Ln = La, Dy)
		$LnTaON_2$ (Ln = La-Dy)
	K_2NiF_4 type (with Al, Ta, Nb)	Ln_2AlO_3N (Ln = La, Nd, Sm)
		$A_2^{II}B^VO_3N$ (A^{II} = Ca, Sr, Ba; B^V = Ta, Nb)
	Spinel type (with Al)	γ-alon, $LnAl_{12}O_{18}$ (Ln = La-Gd)

properties, and a wide band gap. Consequently, covalent nitrides can have a broad range of applications, as shown below:

- *Semiconductor materials.* Binary nitrides and their solid solutions formed by nitrogen and II-V group metals are long known as important semiconductor materials. These nitrides include GaN, AlN, InN, and Ge_3N_4 (Strite and Morkoc 1992; Morkoc et al. 1994; Ponce and Bour 1997).

- *Structural ceramics.* AlN, BN, and Si_3N_4 materials are very important nonoxide structural ceramics due to their superior strength, hardness, and chemical stability. Si_3N_4-based ceramics are widely used as components in ceramic engines and turbines and as cutting tools, bearing balls, and abrasive particles (Cao and Metselaar 1991; Ekstrom and Nygen 1992; Mandal 1999). Furthermore, AlN and Si_3N_4

exhibit high thermal conductivity and are used as ceramic substrates for electronic devices. BN has a layered structure that is similar to that of graphite, so that it finds applications as a lubricant material.

- *Luminescent materials.* Rare-earth-doped (oxy)nitridosilicates and (oxy)nitridoaluminosilicates are extensively investigated as inorganic phosphors (Hoppe et al. 2000; van Krevel et al. 2002; Xie et al. 2002; Xie and Hirosaki 2007a). These new phosphors show promising luminescent properties, and thus are suitable wavelength conversion materials in white LEDs.

- *Photocatalysts and pigments.* Perovskite nitrides, such as $AEMO_2N$ (AE = Ca, Sr, Ba; M = Ta, Nb) and $LaTiO_2N$, have been studied as visible-light-driven photocatalysts or nontoxic pigments (Jansen and Letschert 2000; Kim et al. 2004; Liu et al. 2006; Higashi et al. 2009).

4.1.2 Crystal Chemistry of Nitridosilicates

Of the large quantities of nitride compounds, covalent (oxo)nitridosilicates and (oxo)nitridoaluminosilicates are suitable host lattices for luminescent materials because they have a wide band gap and are transparent over the UV-to-visible light spectral region. Over the last several years, the research on (oxo)nitridosilicates has significantly advanced, and a number of novel materials and crystal structures have been found. It is well known that the luminescence of activator ions is strongly related to the structure of the host lattice, typically for those ions involved in the f-d electronic transitions. Therefore, it is necessary to understand the crystal chemistry of nitridosilicates before we can properly explain the luminescent properties of nitride phosphors.

The crystal chemistry of nitridosilicates (α- and β-Si_3N_4), oxonitridosilicates (Ln-Si-O-N), and oxonitridoaluminosilicates (α- and β-sialons) has been overviewed by Mestalaar (1994), Cao and Metselaar (1991), and Ekstrom and Nygen (1992). With developing a unique high-temperature synthetic route by the use of metals and silicon diimide, Schnick and coworkers (1997, 1999, 2001) have carried out extensive investigations of the solid-state chemistry of nitridosilicates and sialons in recent years. Nitridosilicates can be simply considered to be formed by the integration of nitrogen in silicates. Most oxosilicates are constructed from SiO_4 tetrahedra, so that oxygen may be terminally bound to Si ($O^{[1]}$), or it may bridge two Si ($O^{[2]}$). Moreover, the SiO_4 tetrahedra are always connected through common vertices, and the edge sharing of SiO_4 is unlikely in oxosilicates. On the other hand, nitridosilicates are structurally built up on highly dense SiN4 tetrahedra, and nitrogen can be bound to either one ($N^{[1]}$), two ($N^{[2]}$), three ($N^{[3]}$), or even four ($N^{[4]}$) neighboring Si. Furthermore, except for the corner sharing of SiN4 tetrahedra, edging sharing of SiN4 tetrahedra is also observed in nitridosilicates, such as in $MYbSi_4N_7$ (M = Sr, Ba) and $Ba_5Si_2N_6$. Consequently, with highly cross-linked $N^{[3]}$ and $N^{[4]}$ as well as edge sharing of SiN4 tetrahedra, the nitridosilicates represent a significant

TABLE 4.3

Nitridosilicates and Their Structural Characterization

Materials	Bound to Si	Si:N	Structural Nature	Reference
α- or β-Si_3N_4	$N^{[3]}$	0.75	Corner-sharing SiN_4	Grun 1979; Kohatsu and McCauley 1974
$BaSi_7N_{10}$	$N^{[2]}$, $N^{[3]}$	0.7	Edge- and corner-sharing SiN_4	Huppertz and Schnick 1997a
$LiSi_2N_3$	$N^{[3]}$	0.667	Corner-sharing SiN_4	Orth and Schnick 1999
$M_2Si_5N_8$ (M = Ca, Sr, Ba)	$N^{[2]}$, $N^{[3]}$	0.625	Corner-sharing SiN_4	Schlieper and Schnick 1995a, 1995b
$LaSi_3N_5$	$N^{[3]}$	0.6	Corner-sharing SiN_4	Inoue et al. 1980
$SrAlSi_4N_7$	$N^{[2]}$, $N^{[3]}$	0.571	Corner-sharing SiN_4, edge-sharing AlN_4	Hecht et al. 2009
$MYbSi_4N_7$ (M = Sr, Ba)	$N^{[2]}$, $N^{[4]}$	0.571	Edge-sharing SiN_4	Huppertz and Schnick 1997b
$MYSi_4N_7$ (M = Sr, Ba)	$N^{[2]}$, $N^{[4]}$	0.571	Edge-sharing SiN_4	Li et al. 2004
$Ba_2AlSi_5N_9$	$N^{[2]}$, $N^{[3]}$	0.556	Corner-sharing SiN_4, edge-sharing AlN_4	Kechele et al. 2009a
$Ln_3Si_6N_{11}$ (Ln = La, Ce, Pr)	$N^{[2]}$, $N^{[3]}$	0.545	Corner-sharing SiN_4	Woike and Jeitschko 1995
$Ln_5Si_3N_9$ (Ln = La, Ce))	$N^{[1]}$, $N^{[2]}$, $N^{[3]}$	0.333	Corner-sharing SiN_4	Schmolke et al. 2009
$M_5Si_2N_6$ (M = Ca, Ba)	$N^{[1]}$, $N^{[2]}$	0.333	Edge-sharing SiN_4	Yamane and DiSalvo 1996

structural extension of traditional oxosilicates, forming a large family of new compounds with interesting crystal chemistry.

The condensation degree in nitridosilicates, which are constructed from three-dimensional and highly dense SiN4 tetrahedral networks, can be evaluated by the molar ratio Si:N. Oxosilicates have the maximal degree of condensation of 0.5 in SiO_2, whereas nitridosilicates have a degree of condensation varying in the range of 0.25–0.75. The highest degree of condensation (Si:N = 0.75) is seen in Si_3N_4 (see Table 4.3). The enhanced cross-linking at the N atoms, together with the higher condensation degree in nitridosilicates, leads to their excellent thermomechanical properties and chemical stability.

Similar to nitridosilicates, oxonitridosilicates or oxonitridoaluminosilicates (sialons) are derived from oxosilicates or oxoaluminosilicates by formal exchanges of oxygen by nitrogen and of silicon by aluminum. The notable example is α-sialon, which is formed by the partial substitution of oxygen for nitrogen and of aluminum for silicon in α-Si_3N_4. These substitutions therefore lead to a significant structural extension of nitridosilicates. Oxonitridosilicates or oxonitridoaluminosilicates are constructed from a three-dimensional and highly dense $(Si,Al)(O,N)_4$ tetrahedral network. The

TABLE 4.4

Oxonitridosilicates and Oxonitridoaluminosilicates

Materials	Bound to Si	(Si,Al)/ (O,N)	Structural Nature	Reference
α-sialon	N[3]	0.75	Corner-sharing $(Si,Al)(O,N)_4$	Hampshire et al. 1978
β-sialon	N[3]	0.75	Corner-sharing $(Si,Al)(O,N)_4$	Ekstrom and Nygen 1992
$SrSiAl_2O_3N_2$	N[3]	0.6	Corner-sharing $SiON_3$, AlO_3N, AlO_2N_2	Lauterbach and Schnick 1998
$Sr_3Ln_{10}Si_{18}Al_{12}O_{18}N_{36}$	N[2], N[3]	0.556	Corner-sharing $SiON_3$, $AlON_3$, SiN_4	Lauterbach et al. 2000
$La_3Si_{6.5}Al_{1.5}N_{9.5}O_{5.5}$	N[2], N[3]	0.533	Corner-sharing $(Si,Al)(O,N)_4$	Grins et al. 2001
$MSi_2O_2N_2$ (M = Ca, Sr, Ba)	N[3]	0.5	Corner-sharing $SiON_3$	Kechele et al. 2009b Oeckler et al. 2007
$Ba_3Si_6O_{12}N_2$	N[3]	0.429	Corner-sharing SiO_3N	Uheda et al. 2008
$La_4Si_2O_7N_2$	N[2]	0.222	Corner-sharing SiO_3N and SiO_2N_2	Takahashi et al. 2003
$Y_3Si_5N_9O$		0.2	Corner-sharing SiN_3O and SiN_4	Liddell and Thompson 2001

nitridosili can be SiN_4, $SiON_3$, SiO_2N_2, SiO_3N, AlO_3N, AlO_2N_2, $AlON_3$, or AlN_4 (see Table 4.4). Moreover, these nitridosili are generally connected by common vertices (corner shared).

4.1.3 Solid Solution Nitridosilicate Materials

Phosphors with solid solution compositions are so important that their luminescent properties, such as the emission color, excitation spectra, quantum efficiency, and thermal quenching, can be altered or modified just by tailoring their compositions, while their crystal structures remain unchanged. A notable example is $Y_3Al_5O_{12}$-$Y_3Ga_3O_{12}$ or $Y_3Al_5O_{12}$-$Tb_3Al_5O_{12}$ solid solutions that can blue-shift or red-shift the emission color of YAG phosphors (Nakatsuka 1999; Jang et al. 2007; Pan 2004).

Solid solutions may form by either a substitutional or interstitial mechanism. If the guest atoms or ions enter into the host crystal structure at regular crystallographic positions, this forms substitutional solid solutions, and if the guest atoms or ions are too small to occupy any regular crystallographic positions, they enter into the interstices of the host crystal structure to form an interstitial solid solution. Compared to oxidic compounds, this allows nitride materials more flexibility to produce solid solutions. On the one hand, like oxidic materials, nitridosilicates form solid solutions by replacing cation elements with guest atoms or ions that have an ionic size of more or less 15% of the replaced cations. On the other hand, nitridosilicates can form solid solutions by substituting nitrogen atoms with oxygen (i.e., substitution of anion elements). For example, from the viewpoint of the crystal chemistry of nitride compounds discussed above, oxonitridosilicates, nitridoalumino-silicates, or oxonitridoaluminosilicates can be produced by partially substi-tuting Si-N bonds with Al-O or Al-N bonds in nitridosilicates.

The examples of the first case (cation substitution) can be found in some important nitride phosphors, such as Ca-α-sialon, $SrSi_2O_2N_2$, and $CaAlSiN_3$. Complete or limited solid solutions between the end members of Ca-α-sialon and Li-α-sialon (Xie et al. 2006b), $SrSi_2O_2N_2$ and $BaSi_2O_2N_2$ (Bachmann 2009), as well as $CaAlSiN_3$ and $SrAlSiN_3$ (Watanabe et al. 2008b), can be formed, enabling them to tune the emission color in a broad range. For the second case (anion substitution), the notable example is Ca-α-sialon, in which the ratio of O/N can be varied in the range of 0–0.2. Another example is nitrido-YAG, which is formed by substituting Al-O with Si-N in YAG. The emission color of nitrido-YAG is significantly red-shifted compared to YAG (Setlur et al. 2008).

4.2 Nitride Phosphors

Photoluminescence of rare-earth-doped nitride phosphors was reported in 1997 for $CaSiN_2:Eu^{2+}$ (Lee 1997). Later, van Krevel et al. (1998), Uheda (2000), and Hope (2000) presented the photoluminescence of Y-Si-O-N:Ce^{3+}, $LaSi_3N_5:Eu^{2+}$, and $Ba_2Si_5N_8:Eu^{2+}$, respectively. They observed significant red-shifting of excitation and emission spectra in oxynitride and nitride phosphors, and contributed it to the large crystal field splitting and nephelauxetic effect. Due to their very interesting photoluminescence and high chemical stability, nitride phosphors have attracted increasing atten-tion during the last 10 years (Xie and Hirosaki 2007a). Typically, in 1993 Nakamura developed highly efficient blue LEDs, and later produced the first commercially available white LEDs by combining a blue LED chip with a yellow-emitting phosphor. This kind of white LEDs, which is also called *phosphor-converted white LEDs*, generally requires phosphors that can downconvert light from LED chips into visible light efficiently. Since rare-earth-doped nitride phosphors have broad excitation spectra extend-ing from UV to the visible spectral region, they are very suitable for use as downconversion luminescent materials in white LEDs. This leads to exten-sive studies of luminescent properties of nitride materials, and numerous nitride phosphors have been reported in recent years (Xie and Hirosaki 2007a). In this section, we will make an overview of nitride luminescent materials suitable for white LED applications. These phosphors are grouped and discussed according to their emission spectra or emission colors.

TABLE 4.5

Structural Parameters for $LaAl(Si_{6-z}Al_z)(N_{10-z}O_z)$; $M = (Si_{5/6}Al_{1/6})$ and $X = (N_{9/10}O_{1/10})$

Atom	Wyckoff	x	y	z	SOF
La	8d	0.0533	0.0961	0.1824	0.5
Al	4c	0	0.427	0.25	1.00
M(1)	8d	0.434	0.185	0.057	1.00
M(2)	8d	0.270	0.082	0.520	1.00
M(3)	8d	0.293	0.333	0.337	1.00
X(1)	8d	0.344	0.320	0.140	1.00
X(2)	8d	0.383	0.210	0.438	1.00
X(3)	8d	0.340	0.485	0.410	1.00
X(4)	8d	0.110	0.314	0.363	1.00
X(5)	8d	0.119	0.523	0.127	1.00

Source: Data from Grins, J., Shen, Z., Nygren, M., and Ekstrom, T., *J. Mater. Chem.*, 5, 2001–2006, 1995.

Note: Crystal system: Orthorhombic.
Space group: Pbcn.
Lattice constant: a = 9.4303(7) Å, b = 9.7689(8) Å, c = 8.9386(6) Å.
Bond lengths: La – X(1): 2.88 Å, La – X(2): 2.95 Å, La – X(3): 2.45 Å, La – X(3): 2.51 Å, La – X(3): 2.74 Å, La – X(4): 2.67 Å, La – X(4): 2.72 Å.

4.2.1 Blue-Emitting Phosphors

4.2.1.1 *LaAl(Si$_{6-z}$Al$_z$)(N$_{10-z}$O$_z$):Ce^{3+} (z = 1)*

4.2.1.1.1 *Crystal Structure*

The crystal structure of $LaAl(Si_{6-z}Al_z)N_{10-z}O_z$ was reported by Grins et al. (1995). $LaAl(Si_{6-z}Al_z)N_{10-z}O_z$, which is also called the *JEM phase*, crystallizes in the orthorhombic system with the space group of Pbcn (Table 4.5). As seen in Figure 4.1, the La atoms are located in the channels along the [001] direction and are irregularly coordinated by seven (O,N) atoms at an average distance of 2.70 Å. The Al and (Si,Al) atoms are tetrahedrally coordinated by (O,N) atoms, forming an $Al(Si,Al)_6(O,N)_{10}^{3-}$ network.

4.2.1.1.2 *Photoluminescence Spectra*

Hirosaki et al. (2006) reported the luminescence of Ce^{3+}-doped JEM. The sample was prepared by firing the powder mixture of La_2O_3, Si_3N_4, AlN, and CeO_2 at 1,700°C under 0.5 Mpa N_2. As shown in Figure 4.2, JEM:Ce^{3+} shows a broad and asymmetric emission band covering from 400 to 700 nm when excited under 386 nm. The emission peak is located at 475 nm for the Ce concentration of 5% with respect to La, indicative of a blue-emitting phosphor. The excitation spectrum of JEM:Ce^{3+} displays a broad band extending from 200 to 450 nm, with the peak wavelength at 386 nm. The broad excitation and emission band are due to the allowed 4f ↔ 5d electronic transition of Ce^{3+}. It is seen from the excitation spectrum that this blue phosphor can be combined with UV LEDs (360–400 nm).

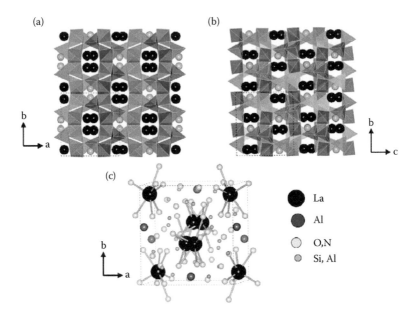

FIGURE 4.1
Crystal structure of LaAl(Si$_{6-z}$Al$_z$)(N$_{10-z}$O$_z$) viewed from (a) the ab plane and (b) the bc plane.
The La-X coordination is shown in (c).

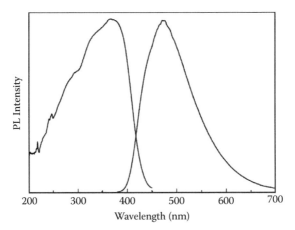

FIGURE 4.2
Excitation and emission spectra of Ce^{3+}-doped JEM. The emission spectrum was measured
under the 368 nm excitation, and the excitation spectrum was monitored at 475 nm.

To make it possible to use JEM:Ce^{3+} with near-UV LEDs, Takahashi et al.
(2007) investigated the effect of Ce concentration on the luminescence of
La$_{1-x}$Ce$_x$Al(Si$_{6-z}$Al$_z$)(N$_{10-z}$O$_z$) (z = 1). A complete solid solution was formed
between the end members of LaAl(Si$_{6-z}$Al$_z$)(N$_{10-z}$O$_z$) and CeAl(Si$_{6-z}$Al$_z$)
(N$_{10-z}$O$_z$), indicating that the substitution of La by Ce does not change the

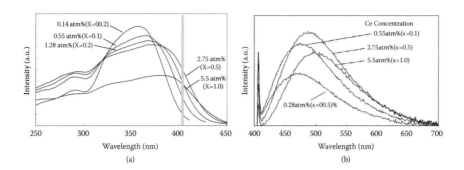

FIGURE 4.3

Excitation (a) and emission (b) spectra of $La_{1-x}Ce_xAl(Si_{6-z}Al_z)(N_{10-z}O_z)$. (Reprinted from Takahashi, K., Hirosaki, N., Xie, R.-J., Harada, M., Yoshimura, K., and Tomomura, Y., *Appl. Phys. Lett.*, 91, 091923, 2007. With permission.)

crystal structure of $LaAl(Si_{6-z}Al_z)(N_{10-z}O_z)$. The significant red-shift of both excitation and emission spectra was observed when the Ce concentration increased. As seen in Figure 4.3, the shift of the excitation spectra toward the long-wavelength side enhances the shoulder at the blue spectral part, enabling JEM:Ce^{3+} to absorb the near-UV light efficiently. The external quantum efficiency of the red-shifted JEM:Ce^{3+} (x = 0.5) is 48% when it is measured upon the excitation at 405 nm. Therefore, the red-shifted JEM:Ce^{3+} is an interesting blue phosphor for use with the combination of a near-UV LED (λ_{em} = 405 nm).

4.2.1.2 AlN:Eu²⁺

4.2.1.2.1 Preparation

The AlN:Eu^{2+} powder samples were prepared by firing the powder mixture of AlN, Si_3N_4, and Eu_2O_3 at 1,700–2,050°C under 1.0 Mpa N_2. The role of Si_3N_4 is to supply the Si source to improve the solubility of Eu in the AlN lattice (Hirosaksaki et al. 2007; Dierre et al. 2009). Inoue et al. (2009) reported that the doping of 2.2 mol% Si led to a phase-pure AlN:Eu^{2+}, whereas an impurity phase was identified without the Si doping even if the Eu concentration was as low as 0.1 mol%. We suggest that the codoping of Si with Eu is to form some local structures that are similar to the layer structure containing the $Sr(O,N)_{12}$ cubo-octahedron and $(Si,Al)(O,N)_4$ tetrahedron in the Sr-containing AlN polytypoid (Grins et al. 1999).

The particle size and morphology of AlN:Eu^{2+} change a lot as the firing temperature increases. As shown in Figure 4.4, the particle size increases from ~0.6 μm for the starting powder to ~3 μm for the sample fired at 1,950°C, and finally to ~5 μm at 2,050°C. The AlN particles showed a uniform and spherical morphology with smooth surfaces when the sample was synthesized at 2,050°C, whereas the particles had a plate-like shape and a broad size distribution when they were fired at 1,950°C.

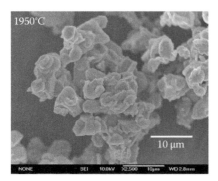

FIGURE 4.4
SEM images of AlN:Eu²⁺ fired at 1,950 and 2,050°C. (Reprinted from Inoue, K., Hirosaki, N., Xie, R.-J., and Takeda, T., *J. Phys. Chem. C*, 133, 9392–9397, 2009. With permission.)

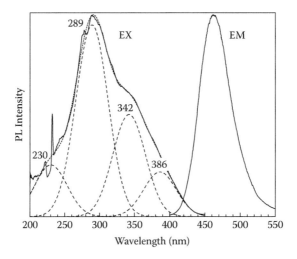

FIGURE 4.5
Excitation and emission spectra of AlN:Eu²⁺ (0.24 mol% Eu, 2.9 mol% Si) fired at 2,050°C. (Reprinted from Inoue, K., Hirosaki, N., Xie, R.-J., and Takeda, T., *J. Phys. Chem. C*, 133, 9392–9397, 2009. With permission.)

4.2.1.2.2 Photoluminescence

Figure 4.5 shows the excitation and emission spectra of the sample with 2.9 mol% Si and 0.24 mol% Eu²⁺. The emission spectrum of AlN:Eu²⁺ consists of a single and symmetric broad band centered at 465 nm when the sample was excited at 300 nm. The full width at half maximum (FWHM) of the band is about 52 nm. The blue emission can be ascribed to the allowed $4f^65d \rightarrow 4f^7$ transitions of Eu²⁺. The excitation spectrum, monitored at 465 nm, covers a broad range of 250–450 nm. Several peaks or shoulders at 280, 290, and 350 nm are observed. The structure in the excitation spectrum is due to the crystal field splitting of the 5d level of Eu²⁺ ions. The excitation band can be

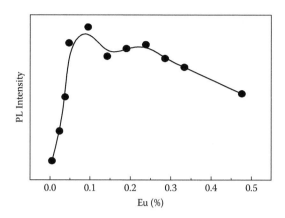

FIGURE 4.6
Concentration quenching of AlN:Eu^{2+} (0.24 mol% Eu, 2.9 mol% Si). (Reprinted from Inoue, K., Hirosaki, N., Xie, R.-J., and Takeda, T., *J. Phys. Chem. C*, 133, 9392–9397, 2009. With permission.)

fitted into four Gaussian subbands centered at 230, 289, 342, and 386 nm, respectively. The crystal field splitting is estimated to be 17,600 cm^{-1}, and the Stokes shift is calculated to be 4,400 cm^{-1}.

The effect of Eu^{2+} concentration on the luminescence of AlN:Eu^{2+} (2.9 mol% Si) is presented in Figure 4.6. The concentration quenching occurs at about 0.1 mol%. As the solubility of Eu^{2+} in the AlN lattice is quite low, the heavily doped samples will produce some Eu-containing impurity phases and decrease the crystallinity of AlN, which reduces the luminescence of AlN:Eu^{2+}.

The quantum efficiency and absorption as a function of the excitation wavelength are given in Figure 4.7. Under the 365 nm excitation, the internal quantum efficiency (η_i), external quantum efficiency (η_0), and absorption (α) of AlN:Eu^{2+} are 76, 46, and 63%, respectively. The high absorption and quantum efficiency of AlN:Eu^{2+} at the excitation wavelength of 365 nm indicate that it matches well with the emission wavelength of InGaN UV LED chips and is a promising blue-emitting downconversion phosphor in white LEDs.

4.2.1.3 Ln-Si-O-N:Ce^{3+} (Ln = Y, La)

4.2.1.3.1 Luminescence of Y-Si-O-N:Ce^{3+}

Van Krevel et al. (1998) investigated the luminescence of Ce^{3+} in several Y-Si-O-N compounds, such as Y$_5$(SiO$_4$)$_3$N, Y$_4$Si$_2$O$_7$N$_2$, YSiO$_2$N, and Y$_2$Si$_3$O$_3$N$_4$. The results showed that all samples had the blue-emitting color due to the luminescence of Ce^{3+}, and the excitation and emission spectra shifted toward the long-wavelength side when Ce^{3+} was coordinated to more N^{3-} vs. O^{2-} (Table 4.6) in the sequence of Y$_2$Si$_3$N$_3$N$_4$ > YSiO$_2$N > Y$_4$Si$_2$O$_7$N$_2$ > Y$_5$(SiO$_4$)$_3$N. Moreover, the structural nature of Si(O,N)$_4$ tetrahedral units has a great influence in the Ce^{3+} luminescence of Y-Si-O-N materials. The Stokes shift increases in the series of Y$_5$(SiO$_4$)$_3$N > Y$_4$Si$_2$O$_7$N$_2$ > YSiO$_2$N > Y$_2$Si$_3$N$_3$N$_4$, as the

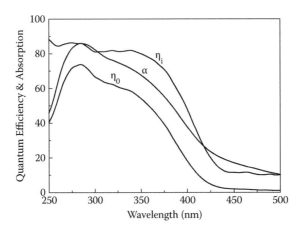

FIGURE 4.7
Quantum efficiency and absorption of AlN:Eu²⁺ (0.24 mol% Eu, 2.9 mol% Si) as a function of the excitation wavelength. (Reprinted from Inoue, K., Hirosaki, N., Xie, R.-J., and Takeda, T., *J. Phys. Chem. C*, 133, 9392–9397, 2009. With permission.)

TABLE 4.6

Emission Maxima, Excitation Maxima, the Crystal Field Splitting (CFS), the Stokes Shift (SS), and the Luminescence Intensity for Different Y-Si-O-N Materials

Materials	Emission/nm	Excitation/nm	CFS/cm⁻¹	SS/cm⁻¹
$Y_5(SiO_4)_3N$	475, 423	240, 290, 325, 355	13,500	7,100
$Y_4Si_2O_7N_2$	504	225, 280, 310, 325	16,100	5,800
$YSiO_2N$	405, 442	240, 290, 355, 390	24,300	3,300
$Y_2Si_3O_3N_4$	493	195, 260, 310, 390	25,700	3,200

Source: Data from van Krevel, J. W. H., Hintzen, H. T., Metselaar, R., and Meijerink, A., *J. Alloys Compd.*, 268, 272–277, 1998.

$Si(O,N)_4$ tetrahedra are cross-linked differently in these materials: there are isolated $Si(O,N)_4$ in $Y_5(SiO_4)_3N$, dimeric $Si_2O_5N_2$ units in $Y_4Si_2O_7N_2$, trimeric $Si_3O_6N_3$ rings in $YSiO_2N$, and sheets of $(Si_3O_3N_4)_n$ in $Y_2Si_3O_3N_4$.

The luminescence of $YSiO_2N:Ce^{3+}$ and $Y_2Si_3O_3N_4:Ce^{3+}$ is higher than that of the other two, which have large Stokes shifts, making them interesting blue phosphors for UV or near-UV LEDs.

4.2.1.3.2 Structure of La-Si-O-N:Ce³⁺

Table 4.7 presents the structural parameters of several La-Si-O-N compounds: $La_5Si_3O_{12}N$, $La_4Si_2O_7N_2$, $LaSiO_2N$, and $La_3Si_8O_4N_{11}$.

$La_5Si_3O_{12}N$ contains oxygen in O_{nb} nitridosilic (in SiO_4 and SiO_3N units in the ratio 16:6) and an ionic environment, in which oxygen is located on a C3-axis and is coordinated to three identical lanthanums (Titeux et al. 2000). The La1 atom has a sevenfold coordination with three free oxygen and four

TABLE 4.7

Crystal Structure of La-Si-O-N Compounds

Compounds	$La_5Si_3O_{12}N$	$La_4Si_2O_7N_2$	$LaSiO_2N$	$La_3Si_8O_4N_{11}$
Crystal structure	Hexagonal	Monoclinic	Hexagonal	Orthorhombic
Space group	P 63/m	P 1 21/c 1	P -6 c 2	C2/c
a-Axis (Å)	9.7051	8.036	7.310	15.850
b-Axis (Å)	9.7051	10.992	7.310	4.9029
c-Axis (Å)	7.2546	11.109	9.550	18.039
β (°)	120	110.92	120	114.849
Cell volume (Å³)	591.758	916.754	441.95	1,272.05
Z	1	4	6	4

Source: Data from Dierre, B., Xie, R.-J., Hirosaki, N., and Sekiguchi, T., *J. Mater. Res.*, 22, 1933–1941, 2007.

O/N atoms that belong to the silicon tetrahedra at an average distance of 2.576 Å. The La2 atom is coordinated to three O/N and six O atoms at an average distance of 2.624 Å, in which the oxygen and nitrogen are all bonded to silicon. There are isolated $Si(O,N)_4$ tetrahedra in $La_5Si_3O_{12}N$.

The $La_4Si_2O_7N_2$ compound, isostructural with the $Y_4Al_2O_9$ high-temperature phase, crystallizes in a monoclinic structure with a space group of $P2_1/C$ (Takahashi et al. 2003). All nitrogen atoms occupy the bridging site and the terminal sites of $Si_2(O_5N_2)$ ditetrahedra, and oxygen atoms occupy the non-metal sites surrounded by La atoms in $La_4Si_2O_7N_2$. There are four different sites for La: La1 and La3 atoms are coordinated to seven O/N at an average distance of 2.491 and 2.572 Å, respectively, the La2 atom is surrounded by eight O/N atoms with an average distance of 2.604 Å, and the La4 atom is coordinated to six O/N atoms at an average distance of 2.893 Å.

$LaSiO_2N$ has the α-wollastonite structure with the nitrogen atoms present in three-membered $[Si_3O_6N_3]$ rings, occupying bridging sites between pairs of Si-centered nitridosili and also linked to two lanthanum atoms (Harris et al. 1992). The nitrogen environment is therefore more ionic than Si_3N_4 or Si_2N_2O. All the oxygen atoms in the structure occupy nonbridging sites in the $Si_3O_6N_3$ three-membered rings.

The structure of $La_3Si_8O_4N_{11}$ contains ribbons as structural units with a composition of $Si_6(O,N)_{14}$. The ribbons extend along the (010) direction, and are formed by corner-sharing $Si(O,N)_4$ tetrahedra (Grins et al. 2001). The La1 atom is octahedrally coordinated by four O/N and two O atoms at distances of 2.45–2.75 Å, and the La2 atom is coordinated by five O/N, two O, and one N atom at distances of 2.48–3.04 Å, which approximately form a cubic antiprism.

4.2.1.3.3 Photoluminescence of La-Si-O-N:Ce³⁺

The photoluminescence of Ce^{3+}-doped La-Si-O-N was investigated by Dierre et al. (2007). The photoluminescence spectra of La-Si-O-N:Ce^{3+} are given in Figure 4.8.

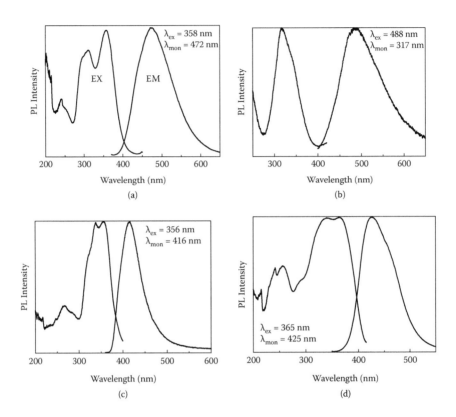

FIGURE 4.8
Excitation and emission spectra of (a) $La_5Si_3O_{12}N$, (b) $La_4Si_2O_7N_2$, (c) $LaSiO_2N$, and (d) $La_3Si_8O_4N_{11}$. (Reprinted from Dierre, B., Xie, R.-J., Hirosaki, N., and Sekiguchi, T., *J. Mater. Res.* 22, 1933–1941, 2007. With permission.)

As seen in Figure 4.8a, the emission spectrum of $La_5Si_3O_{12}N$ doped with 2 mol% Ce^{3+} extends from 400 to 650 nm upon the excitation at λ_{ex} = 358 nm, showing a broad structureless band. It has a maximum at about 472 nm with a full width at half maximum (FWHM) of intensity of 108 nm. The excitation spectrum monitored at λ_{mon} = 472 nm consists of five bands in the range of 220–400 nm, and the crystal field splitting is calculated as 13,800 cm^{-1} (see Table 4.8). The lowest excitation band is situated at 361 nm, so that the Stokes shift is relatively large, at 6,800 cm^{-1}.

Upon excitation at 317 nm, the emission spectrum of $La_4Si_2O_7N_2$ doped with 6 mol% Ce^{3+} shows a broad structureless band with a maximum at 488 nm and a FWHM of 117 nm (Figure 4.8b). The excitation spectrum, monitored at 488 nm, starts at 400 nm and extends to 275 nm. The splitting of the excitation spectrum is not distinct, which has a maximum at 317 nm and a shoulder at 345 nm.

A single-emission band centered at 416 nm is observed upon excitation at 356 nm for $LaSiO_2N$ doped with 4 mol% Ce^{3+} (Figure 4.8c). The excitation

TABLE 4.8

Excitation, Emission, Stokes Shift, Crystal Field Splitting, and Center of Gravity of Lanthanide Silicon Oxynitrides Doped with 6 mol% Ce^{3+}

Samples	Excitation (nm)	Emission (nm)	Stokes Shift (cm^{-1})	Crystal Field Splitting (cm^{-1})	Center of Gravity (cm^{-1})
$La_5Si_3O_{12}N$	361, 312, 296, 254, 241	478	6,800	13,800	35,000
$La_4Si_2O_7N_2$	345, 317	488	8,500	—	—
$LaSiO_2N$	356, 338, 322, 285, 266	416	4,100	9,500	32,000
$La_3Si_8O_4N_{11}$	365, 339, 290, 257	424, 458	3,800	14,000	34,000

Source: Data from Dierre, B., Xie, R.-J., Hirosaki, N., and Sekiguchi, T., *J. Mater. Res.*, 22, 1933–1941, 2007.

spectrum consists of three bands situated at 356, 338, and 266 nm and two shoulders at 285 and 322 nm, monitored at 416 nm. The Stokes shift and the crystal field splitting were calculated as 4,100 and 9,500 cm^{-1}, respectively.

$La_3Si_8O_4N_{11}$ doped with 6 mol% Ce^{3+} shows an emission spectrum consisting of a broad band situated at 424 nm and a shoulder at 458 nm, when it is excited at 365 nm (Figure 4.8d). The FWHM of the emission band is about 80 nm. It is clearly observed that the excitation spectrum is composed of five peaks or bands situated at 365, 339, 290, 257, and 242 nm, respectively. The crystal field splitting and the Stokes shift are then roughly estimated from the spectra, which are 14,000 and 3,800 cm^{-1}, respectively.

4.2.1.3.4 Thermal Quenching of La-Si-O-N:Ce^{3+}

Figure 4.9 presents the emission intensity of Ce^{3+}-doped La-Si-O-N phosphors as a function of measuring temperature. The temperature-dependent emission intensity of $La_4Si_2O_2N_7$ is not given due to its rather low emission intensity at room temperature. It is clearly seen that the emission intensity of all samples reduces as the temperature increases as a result of thermal vibration of the lattice. The quenching temperature, defined as the temperature at which the emission intensity drops to 50% of that measured at room temperature, is 120, 60, and 220°C for $La_5Si_3O_{12}N$:2%Ce, $LaSiO_2N$:4%Ce, and $La_3Si_8O_4N_{11}$:6%Ce, respectively.

Among these four La-Si-O-N:Ce^{3+} phosphors, $La_3Si_8O_4N_{11}$:Ce^{3+} shows the most interesting luminescent properties and small thermal quenching. This blue phosphor thus has the greatest potential to be used in white LED applications.

4.2.1.4 LaSi$_3$N$_5$:Ce^{3+}

4.2.1.4.1 Crystal Structure

The crystal structure of $LaSi_3N_5$ was characterized by Inoue et al. (1980) and Hatfield et al. (1990). $LaSi_3N_5$ crystallizes in the orthorhombic crystal system with the space group of $P2_12_12_1$. Its structure is built up from [SiN4]

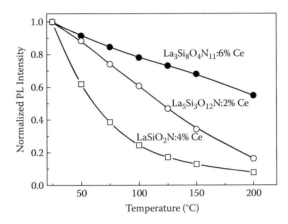

FIGURE 4.9
Thermal quenching of La-Si-O-N:Ce^{3+} phosphors. (Reprinted from Dierre, B., Xie, R.-J., Hirosaki, N., and Sekiguchi, T., *J. Mater. Res.* 22, 1933–1941, 2007. With permission.)

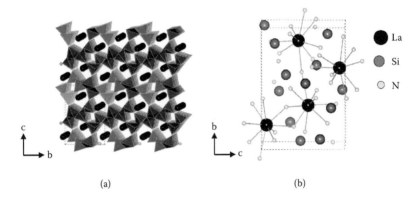

FIGURE 4.10
(a) Crystal structure of $LaSi_3N_5$ projected on the (100) plane, and (b) the coordination of La with N (c), showing that each La atom is coordinated to nine nitrogen atoms.

tetrahedra, which are linked by sharing corners nearly parallel to the (001) plane to form rings of five tetrahedral, and the La atoms are centrally located between the pentagonal holes along the c-axis (Figure 4.10a). The La atoms are coordinated by nine nitrogen atoms with an average distance of 2.78 Å (Figure 4.10b). There are two types of nitrogen environments. In the first, each La atom is connected to three silicon atoms in a way similar to the Si_3N_4 structure; in the second, each La atom is coordinated to two silicon and two lanthanum atoms.

Table 4.9 shows the atomic coordinates of $La_{0.9}Ce_{0.1}Si_3N_5$ powders (Suehiro et al. 2009), which are in agreement with those reported by Hatfield et al. (1990). In addition, both $LaSi_3N_5$ and $CeSi_3N_5$ share the same crystal structure, so that they can form complete solid solutions.

TABLE 4.9

Atomic Coordinates, Site Occupancy Fraction, and Crystallographic Data for $La_{0.9}Ce_{0.1}Si_3N_5$

Atom	Wyckoff	x	y	z	SOF
La/Ce	4a	0.5088(5)	0.0595(1)	0.16821(8)	1.00
Si(1)	4a	0.021(2)	0.3386(5)	0.1591(5)	1.00
Si(2)	4a	0.017(2)	0.2051(6)	0.4170(4)	1.00
Si(3)	4a	0.523(2)	0.4689(6)	0.0414(4)	1.00
N(1)	4a	0.185(3)	0.516(2)	0.094(1)	1.00
N(2)	4a	0.674(3)	0.389(1)	0.155(1)	1.00
N(3)	4a	0.306(3)	0.157(2)	0.505(1)	1.00
N(4)	4a	0.026(4)	0.159(1)	0.0745(9)	1.00
N(5)	4a	0.140(3)	0.289(2)	0.287(1)	1.00

Source: Data from Suehiro, T., private communication.

Note: Crystal system: Orthorhombic.

Space group: $P2_12_12_1$.

Lattice constant: a = 4.80781(8) Å, b = 7.8369(2) Å, c = 11.2572(2) Å, V = 424.15(1) Å³.

Z = 4.

Bond lengths: La – N(1): 3.0654 Å, La – N(1): 3.1335 Å, La – N(2): 2.6963 Å, La – N(2): 2.4199 Å, La – N(3): 2.8305 Å, La – N(4): 2.5496 Å, La – N(4): 2.8804 Å, La – N(5): 2.8275 Å, La – N(5): 2.6432 Å.

4.2.1.4.2 Photoluminescence

Suehiro et al. (2009) investigated the photoluminescence of Ce^{3+}-doped $LaSi_3N_5$ phosphors that were prepared by gas reduction and nitridation of the powder mixture of La_2O_3, SiO_2, and CeO_2. As seen in Figure 4.11, the excitation spectrum covers the spectral range of 240–400 nm with two distinct bands centered at 255 and 355 nm, respectively. These excitation bands can be fitted into five subbands centered at 234, 258, 297, 332, and 361 nm by using the Gaussian simulation. This gives the estimated crystal field splitting and center of gravity of 15,000 and 34,600 cm⁻¹, respectively. The emission spectrum of $La_{0.9}Ce_{0.1}Si_3N_5$ showed a broad band centered at 440 nm and the full width at half maximum (FWHM) of 95. The emission band can be decomposed into two Gaussian components at 421 and 458 nm, resulting in the energy difference of 1,920 cm⁻¹, which is close to that for the doublet of the 4f f¹ ground state ($^2F_{7/2}$ and $^2F_{5/2}$). The Stokes shift is estimated as 3,900 cm⁻¹.

As shown in Figure 4.12, the emission band of $La_{1-x}Ce_xSi_3N_5$ is red-shifted when the Ce^{3+} concentration increases. This red-shift can be ascribed to the enhanced Stokes shift since the excitation spectra of these samples basically do not alter in shape with the Ce^{3+} concentration.

The external quantum efficiency of the optimal composition (x = 0.10) is 67.0 and 34.3% under the excitation of 355 and 380 nm, respectively. This sample also shows a small thermal quenching, and its emission intensity measured at 150°C under the 355 nm excitation decreases about 15% of the

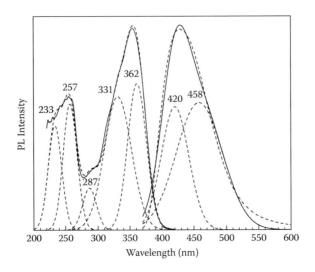

FIGURE 4.11
Excitation and emission spectra of $La_{0.9}Ce_{0.1}Si_3N_5$. (Reprinted from Suehiro, T., Hirosaki, N., Xie, R.-J., and Sato, T., *Appl. Phys. Lett.*, 95, 051903, 2009. With permission.)

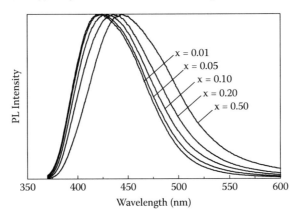

FIGURE 4.12
Emission spectra of $La_{1-x}Ce_xSi_3N_5$ (x = 0.01, 0.05, 0.10, 0.20, 0.50). (Reprinted from Suehiro, T., Hirosaki, N., Xie, R.-J., and Sato, T., *Appl. Phys. Lett.*, 95, 051903, 2009. With permission.)

initial intensity. The excellent photoluminescent properties of $LaSi_3N_5:Ce^{3+}$ make it a promising blue-emitting luminescent material for UV LEDs.

4.2.1.5 *$SrSi_9Al_{19}ON_{31}:Eu^{2+}$*

4.2.1.5.1 *Crystal Structure*

$SrSi_9Al_{19}ON_{31}$ crystallizes in a rhombohedral structure with the group space of R-3 and lattice parameters of a = 5.335Å and c = 79.1Å (Grins et al.

TABLE 4.10

Atomic Coordinates, Site Occupancy Fraction, and Crystallographic Data for $SrSi_{10-x}Al_{18+x}N_{32-x}O_x$ (x ~ 1); M = (Si,Al) and X = (O,N)

Atom	Wyckoff	x	y	z	SOF
Sr		0	0	0	1.00
M(1)		0	0.427	0.25	1.00
M(1)		0.434	0.185	0.057	1.00
M(2)		0.270	0.082	0.520	1.00
M(3)		0.293	0.333	0.337	1.00
X(1)		0.344	0.320	0.140	1.00
X(2)		0.383	0.210	0.438	1.00
X(3)		0.340	0.485	0.410	1.00
X(4)		0.110	0.314	0.363	1.00
X(5)		0.119	0.523	0.127	1.00

Source: Data from Grins, J., Shen, Z., Esmaeilzadeh, S., and Berastegui, P., *J. Mater. Chem.*, 11, 2358–2362, 2001.

Note: Crystal system: Orthorhombic.
Space group: Pbcn.
Lattice constant: a = 9.4303(7) Å, b = 9.7689(8) Å, c = 8.9386(6) Å.
Bond lengths: La – X(1): 2.88 Å, La – X(2): 2.95 Å, La – X(3): 2.45 Å, La – X(3): 2.51 Å, La – X(3): 2.74 Å, La – X(4): 2.67 Å, La – X(4): 2.72 Å.

1999). The atomic coordinates and some crystallographic data are given in Table 4.10. $SrSi_9Al_{19}ON_{31}$ shows a layered structure (see Figure 4.13), sharing a common chemical formula of $(SrM_4X_5)(M_6X_9)(M_3X_3)_{n-2}$, with n being the number of layers between the SrM_4X_5 regions (n = 8 for $SrSi_9Al_{19}ON_{31}$) and M = (Si/Al) and X = (O/N). The Sr atoms enter the structure by forming layers containing $Sr(O,N)_{12}$ cubo-octahedra and $(Si,Al)(O,N)_4$ tetrahedra, and these layers alternate with AlN type blocks into which $(Si,Al)_3(O,N)_{4.5}$ layers are incorporated. The Sr atoms are coordinated to 12 nearest (O,N) atoms at approximately equal distances of ~3 Å (Grins et al. 1999). The neighboring Sr atoms are separated by a distance of 5.3350 Å.

4.2.1.5.2 Photoluminescence

The photoluminescence spectra of Eu^{2+}-doped $SrSi_9Al_{19}ON_{31}$ were first reported by Fukuda (2007). We prepared the sample by gas-pressure firing of the powder mixture of $SrCO_3$, Si_3N_4, AlN, Eu_2O_3, and Al_2O_3 at 1,700°C for 2 h under the 0.5 MPa nitrogen atmosphere and investigated the photoluminescent properties in detail. Figure 4.14 presents the photoluminescence spectra of $(Sr_{0.80}Eu_{0.20})Si_9Al_{19}N_{31}$. A broad excitation band covering the spectral range of 250–400 nm is observed, which has maximal intensities at 285–350 nm. It thus indicates that the excitation of this phosphor

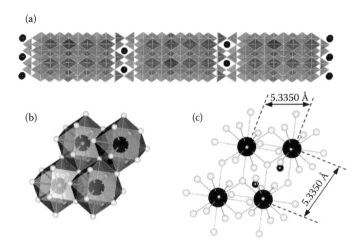

FIGURE 4.13
(a) Crystal structure of $SrSi_9Al_{19}ON_{31}$ projected on a (100) plan. (b) Illustration of Sr(O,N)12cubo-octahedra. (c) Distance between two neighboring Sr atoms.

matches well with the emission wavelength of near-UV LEDs (365–405 nm). The crystal field splitting is roughly calculated as 12,000 cm^{-1}. The emission spectrum is a single and symmetrical band centered at 470 nm, with the full width at half maximum (FWHM) value of 90 nm. The Stokes shift of this sample is about 1,500 cm^{-1}.

Concentration quenching is not seen for the $SrSi_9Al_{19}ON_{31}:Eu^{2+}$ phosphor, which can be due to the long distance between two neighboring Sr(Eu) atoms that leads to the lower possibility of energy transfer between Eu atoms. Furthermore, the reabsorption is quite small, as the overlap between the excitation and emission spectra is not so large.

The temperature-dependent emission intensity of $Sr_{0.80}Eu_{0.20}Si_9Al_{19}ON_{31}$ is illustrated in Figure 4.15. The PL intensity at 150°C reduces by 15% of the initial intensity, which is comparable to that of $LaSi_3N_5:Ce^{3+}$. As seen in Figure 4.16, under the 380 excitation, the blue $Sr_{0.80}Eu_{0.20}Si_9Al_{19}ON_{31}$ phosphor shows the absorption of 80% and the external quantum efficiency of 54%. These promising photoluminescent properties of $SrSi_9Al_{19}ON_{31}:Eu^{2+}$ indicate that it is an interesting blue phosphor for UV LEDs (350–400 nm).

4.2.1.6 Ca-α-sialon:Ce³⁺

4.2.1.6.1 Crystal Chemistry of Ca-α-sialon

Silicon nitride, Si_3N_4, occurs in two different crystallographic modifications, denoted α and β. The Si_3N_4 structure has been overviewed by several researchers (Grun 1979; Kohatsu and McCauley 1974; Cao and Metselaar 1991; Ekstrom and Nygen 1992). It is generally accepted that α-Si_3N_4 is the low-temperature modification, and β-Si_3N_4 the stable high-temperature

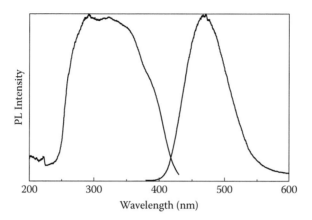

FIGURE 4.14

Excitation and emission spectra of $(Sr_{0.80}Eu_{0.20})Si_9Al_{19}ON_{31}$. The emission spectrum was measured under the 292 nm excitation, and the excitation spectrum was monitored at 470 nm.

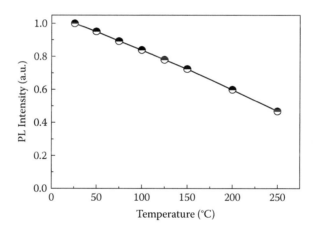

FIGURE 4.15

Thermal quenching of $(Sr_{0.80}Eu_{0.20})Si_9Al_{19}ON_{31}$.

modification at normal pressure. The space group is *P31c* of trigonal symmetry, with each unit cell containing four Si_3N_4 units. Both structures are built up of SiN4 tetrahedra, which are joined by sharing corners in such a way that each N is common to three tetrahedral; thus, each Si atom has four N atoms as nearest neighbors. The structures can be considered as consisting of stacking layers of silicon and nitrogen atoms in the sequence of ABAB … or ABCDABCD … for β- and α-Si_3N_4. α-Si_3N_4 consists essentially of alternate basal layers of β-Si_3N_4 and a mirror image of β, so that the c-spacing of α-Si_3N_4 is almost twice that of β-Si_3N_4. The β-modification has open channels along the c-axis, while the α-modification creates caves as a consequence of the mirror image of the A sequence as seen in Figure 4.17.

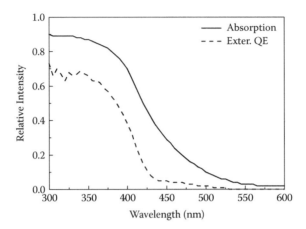

FIGURE 4.16
Absorption and external quantum efficiency of $Sr_{0.80}Eu_{0.20}Si_9Al_{19}ON_{31}$ with varying excitation wavelengths.

α-sialon is the solid solution of α-Si_3N_4, which is formed by partial replacement of Si^{4+} by Al^{3+} and of N^{4-} by O^{2-} and stabilized by trapping "modifying" cations, such as Li, Mg, Ca, Y, and some lanthanide elements, except La and Ce, in the voids of the $(Si,Al)(O,N)_4$ tetrahedral network. The general chemical formula of α-sialon can be given as $Me_xSi_{12-m-n}Al_{m+n}O_nN_{16-n}$, where x is the solubility of metal Me in the lattice of α-sialon, and x = m/v (v is the valence of Me); m and n denote the number of Si-N bonds replaced by Al-N and Al-O bonds, respectively. Since there are only two cages in one unit cell, the solubility of Me is no more than 2 (x ≤ 2). Furthermore, the bond length of Si-N (1.74 Å) is shorter than that of Al-O (1.75 Å) and of Al-N (1.78 Å), so that the lattice expansion occurs in α-sialon with the substitution. The increment in unit cell dimensions of α-sialon can be related to m and n values by the following equations:

$$\Delta a = 0.045m + 0.009n$$

$$\Delta c = 0.0040m + 0.008n \qquad (4.1)$$

Both x-ray and neutron diffraction studies of α-sialon have shown that the Me cations occupy the large interstitial sites (2b) at positions (1/3, 2/3, z) and (2/3, 1/3, z + 1/2) (Izumi et al. 1982; Cao et al. 1993). Si and Al are distributed over 6c sites, whereas O and N occupy 2a, 2b, and 6c sites. The NMR results indicate that silicon prefers to bond to nitrogen, whereas aluminum prefers to coordinate to oxygen (Kempgens et al. 2001). Furthermore, the extended x-ray absorption fine structure (EXAFS) study of α-sialon has shown that the general preference for the modifying Me cations is to locate along the c-axis (Cole et al. 1991). α-sialon is a nitrogen-rich compound with some

TABLE 4.11

Atomic Coordinates and Crystallographic Data for
Ca-α-sialon; M = (Si,Al) and X = (O,N)

Atom	Wyckoff	x	y	z	SOF
Ca	2b	1/3	2/3	0.228	0.339
M(1)	6c	0.5090	0.0820	0.207	1.00
M(2)	6c	0.1681	0.2512	−0.003	1.00
X(1)	2a	0	0	0	1.00
X(2)	2b	1/3	2/3	0.642	1.00
X(3)	6c	0.3477	−0.0441	−0.012	1.00
X(4)	6c	0.3164	0.3150	0.2460	1.00

Source: Data from Izumi, F., Mitomo, M., and Suzuki, J., *J. Mater. Sci. Lett.*, 1, 533–535, 1982.

Note: Crystal system: Trigonal.
Space group: P31c.
Lattice constant: a = 7.8383(7) Å, c = 5.7033(6) Å.
Bond lengths: Ca – X(2): 2.36 Å, Ca – X(3): 2.601 Å × 3,
Ca – X(4): 2.694 Å × 3.

oxygen dissolved in the structure. Izumi et al. (1982) reported that the N/O ratio in the sevenfold coordination was 6:1, and later Cole et al. (1991) suggested that the value was 5:2. These results indicate that most of the chemical bonds (71 or 86%) between the modifying metal Me and the anions are covalent. The bond length of Me-(O,N) is in the range of 2.360–2.694 Å for Ca-α-sialon (Izumi et al. 1982), with the bond parallel to the c-axis being much shorter than the other six ones. Table 4.11 gives the atomic coordinates and crystallographic data for Ca-α-sialon analyzed by Izumi et al. (1982). The crystal structure and sevenfold coordination of Ca atoms are illustrated in Figure 4.18.

4.2.1.6.2 Photoluminescence of Ca-α-sialon:Ce^{3+}

The photoluminescence of Ce^{3+} in Y- or Ca-α-sialon was reported by Xie et al. (2004b, 2005a). As shown in Figure 4.19, the excitation spectrum shows two distinct bands centered at 287 and 390 nm, and a shoulder at 332 nm, respectively. These structures are ascribed to the splitting of 5d energy levels of Ce^{3+} under the crystal field. The crystal field splitting is thus estimated as 9,200 cm^{-1} for this sample. The emission spectrum displays a broad band centered at 490 nm, and this band consists of two subbands centered at 482 and 527 nm, analyzed by the Gaussian simulation. These two bands are separated by the energy of 1,770 cm^{-1}, which corresponds to an energy difference of about 2,000 cm^{-1} for the doublet of the $4f^1$ ground state ($^2F_{7/2}$ and $^2F_{5/2}$). The Stokes shift of Ca-α-sialon:Ce^{3+} is calculated to be 5,200 cm^{-1}.

The chemical composition (i.e., m and n) has a great influence in the luminescence of *Ca-α-sialon:Ce^{3+}*. As seen in Figure 4.20, the sample with m = 2.0 exhibits the maximal luminescence. Moreover, the emission band

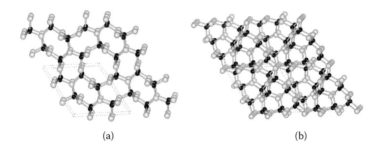

FIGURE 4.17
Crystal structure of (a) β-Si$_3$N$_4$ and (b) α-Si$_3$N$_4$ projected on the (001) plane.

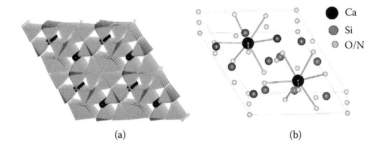

FIGURE 4.18
(a) Crystal structure of Ca-α-sialon projected on the (001) plane. (b) The sevenfold coordination of Ca atoms with O/N atoms.

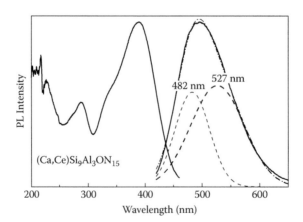

FIGURE 4.19
Excitation and emission spectra of Ca-α-sialon:Ce^{3+} (m = 2, n = 1, 5 mol% Ce). The emission spectrum was measured under the 390 nm excitation, and the excitation spectrum was monitored by 495 nm. (Reprinted from Xie, R. J., Hirosaki, N., Mitomo, M., Yamamoto, Y., Suehiro, T., and Ohashi, N., *J. Am. Ceram. Soc.*, 87, 1368–1370, 2004; Xie, R.-J., Hirosaki, N., Mitomo, M., Suehiro, T., Xu, X., and Tanaka, H., *J. Am. Ceram. Soc.*, 88, 2883–2888, 2005a. With permission.)

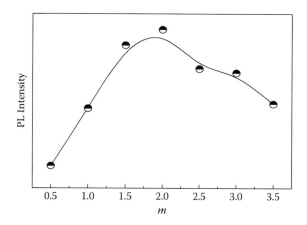

FIGURE 4.20
Effect of the chemical composition (m) on the luminescence of Ca-α-sialon:Ce³⁺ (n = 1/2m, 5 mol% Ce). (Reprinted from Xie, R.-J., Hirosaki, N., Mitomo, M., Suehiro, T., Xu, X., and Tanaka, H., *J. Am. Ceram. Soc.*, 88, 2883–2888, 2005a. With permission.)

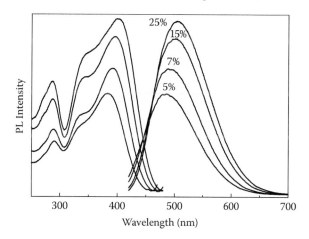

FIGURE 4.21
Excitation and emission spectra of Ca-α-sialon:Ce³⁺ (m = 2, n = 1) with varying Ce³⁺ concentration. (Reprinted from Xie, R.-J., Hirosaki, N., Mitomo, M., Suehiro, T., Xu, X., and Tanaka, H., *J. Am. Ceram. Soc.*, 88, 2883–2888, 2005a. With permission.)

is red-shifted as the m value increases, which is due to the increases in the absolute concentration of Ce^{3+} and in the Stokes shift as a result of lattice expansion. In addition, the concentration quenching of Ca-α-sialon:Ce^{3+} occurs at 25 mol%, and the emission band is red-shifted with increasing the Ce^{3+} concentration (see Figure 4.21).

Ca-α-sialon:Ce^{3+} shows strong absorptions of UV or near-UV light, and emits intense blue light, indicating that it is a promising blue phosphor for white LEDs when UV or near-UV LEDs are used as the primary light source.

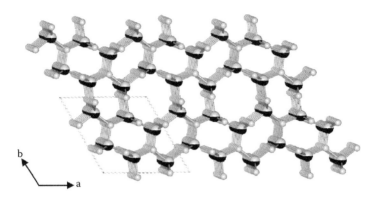

FIGURE 4.22
Crystal structure of β-sialon projected on the (001) plane.

4.2.2 Green-Emitting Phosphors

4.2.2.1 β-sialon:Eu²⁺

4.2.2.1.1 Crystal Chemistry of β-sialon

In the early 1970s, Oyama and Kamigaito (1971) and Jack (1976) observed that Al^{3+} can be dissolved in the silicon nitride lattice without changing its crystal structure by substituting Si^{4+} if N^{3-} is replaced by O^{2-} simultaneously. Such a solid solution is then named β-sialon, which is derived from $\beta\text{-}Si_3N_4$ by simultaneous equivalent substitution of Al-O for Si-N. β-sialon is commonly described by the formula $Si_{6-z}Al_zO_zN_{8-z}$. In this formula, the z value denotes the solubility of Al^{3+} in the $\beta\text{-}Si_3N_4$ lattice, which varies in the range of 0–4.2 (Jack 1981; Ekstrom 1992).

Like $\beta\text{-}Si_3N_4$, β-sialon also has open channels that are parallel to the [001] direction, as seen in Figure 4.22. It is accepted that the several types of modifying cations can be accommodated in the α-sialon structure for achieving charge neutralization, whereas cations except Al^{3+} cannot enter into the β-sialon lattice because the charge balance is not required. However, we have reported very intense luminescence of Eu^{2+} in β-sialon, which obviously indicates that Eu^{2+} is indeed dissolved in the lattice of β-sialon (Hirosaki et al. 2005; Xie et al. 2007b). The location of Eu^{2+} will be discussed later in this section.

4.2.2.1.2 Photoluminescence of β-sialon:Eu²⁺

Hirosaki et al. (2005) firstly report the green photoluminescence of Eu^{2+} in β-sialon, and later Xie et al. (2007b) made a comprehensive investigation of the Eu^{2+}-doped β-sialon phosphor. Figure 4.23 shows typical photoluminescence spectra of β-sialon:Eu²⁺. It is seen that the excitation spectrum covers the spectral range of 250–550 nm, showing two distinct bands centered at 293 and 403 nm as well as two shoulders at 336 and 471 nm, respectively. The crystal field splitting of β-sialon:Eu²⁺ is estimated as 13,000 cm^{-1}. The emission spectrum shows a narrow and symmetric band centered at 538 nm, which is

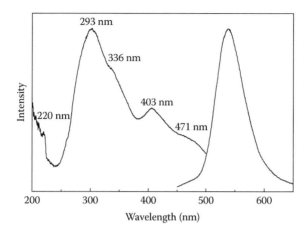

FIGURE 4.23
Excitation and emission spectra of β-sialon:Eu²⁺ (z = 0.3, 0.3 mol% Eu). The emission spectrum was recorded upon the 293 nm excitation, and the excitation spectrum was monitored at 538 nm. (Reprinted from Xie, R.-J., Hirosaki, N., Li, H. L., Li, Y. Q., and Mitomo, M., *J. Electrochem. Soc.*, 154, J314–J319, 2007. With permission.)

indicative of high symmetry of the Eu²⁺ sites. The full width at half maximum (FWHM) is 58 nm, which is extremely small for Eu²⁺. The Stokes shift is roughly computed as 2,700 cm⁻¹. Such a small Stokes shift implies that β-sialon:Eu²⁺ has high luminescence that is hardly quenched under thermal attacks.

4.2.2.1.3 Concentration Quenching and Composition-Dependent Luminescence

Figure 4.24 shows the luminescence intensity as a function of the Eu²⁺ concentration under the 305 nm excitation, for samples with varying z values. It is seen that the concentration quenching occurs at 0.5–0.7 mol% for z = 0.1 and 0.5, and at 0.3 mol% for z = 1.0–2.0. The concentration quenching occurring at higher Eu²⁺ concentration in samples with low z values could be due to the high solubility of Eu²⁺ in β-sialon with smaller z values (Xie et al. 2007b).

It is observed in Figure 4.24 that β-sialon:Eu²⁺ shows higher luminescence for compositions with smaller z values. This is related to the changes in particle size, particle size distribution, and phase purity with varying z values. The compositions with lower z values usually exhibit fine particle size and narrow particle size distribution, and also show higher phase purity (Xie et al. 2007b).

The red-shift of the emission color is enhanced when the z value increases, as seen in Figure 4.25. Usually a high z value yields lattice expansion because the bond length of Al-O (1.75Å) is longer than that of Si-N (1.74Å), which leads to less rigidity of β-sialon. Consequently, the Stokes shift increases with increasing the z value, and finally results in the shift of the emission band toward the long-wavelength side. On the other hand, the change of the Eu²⁺ concentration does not shift the emission band unless the concentration is quite low (<0.3 mol%).

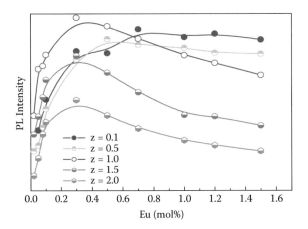

FIGURE 4.24
Concentration quenching for β-sialon:Eu²⁺ with varying z values. (Reprinted from Xie, R.-J., Hirosaki, N., Li, H. L., Li, Y. Q., and Mitomo, M., *J. Electrochem. Soc.*, 154, J314–J319, 2007. With permission.)

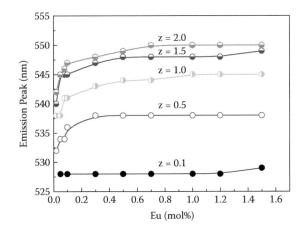

FIGURE 4.25
Effect of the Eu²⁺ concentration on the emission color of β-sialon:Eu²⁺ with varying z values. (Reprinted from Xie, R.-J., Hirosaki, N., Li, H. L., Li, Y. Q., and Mitomo, M., *J. Electrochem. Soc.*, 154, J314–J319, 2007. With permission.)

The temperature-dependent luminescence intensity of β-sialon:Eu²⁺ is shown in Figure 4.26. At 150°C the luminescence of β-sialon:Eu²⁺ remains 83% of the initial intensity measured at room temperature, indicative of small thermal quenching. In addition, the optimized β-sialon:Eu²⁺ has the absorption and external quantum efficiency of 68.2 and 49.8% when it is excited at 455 nm. Furthermore, there are no changes in chromaticity coordinates when β-sialon:Eu²⁺ experiences the high temperature and high-humidity test (85°C, 86% RH) for 6,000 h, or is irradiated by 365 nm light with a powder of

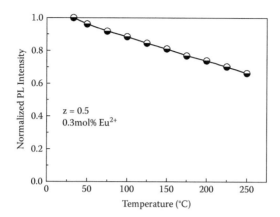

FIGURE 4.26
Thermal quenching of β-sialon:Eu²⁺. (Reprinted from Xie, R.-J., Hirosaki, N., Li, H. L., Li, Y. Q., and Mitomo, M., *J. Electrochem. Soc.*, 154, J314–J319, 2007. With permission.)

120 mW/cm² for 10,000 h, or heated up to 150°C for 2,500 h (Yamada 2010). These excellent photoluminescent properties and high reliability enable β-sialon:Eu²⁺ to be one of the most important green phosphors for white LEDs.

4.2.2.1.4 Location of Eu²⁺ in β-sialon

As described previously, metal ions cannot be accommodated in the lattice of β-sialon. However, the authors do observe the green emission of Eu²⁺ in β-sialon. Therefore, questions arise about the location of Eu²⁺ ions in β-sialon. By SEM and TEM analysis of β-sialon particles, Hirosaki et al. (2005) concluded that the distribution of Eu²⁺ was homogenous in a whole particle, and Eu²⁺ was not segregated in the amorphous layers on the surface of the β-sialon particle, but resided in special sites in the lattice. The uniformity of the Eu²⁺ luminescence was further investigated by means of cathodoluminescence (CL), which can give the luminescence spectrum of a single particle. Figure 4.27 shows the cathodoluminescence of a single β-sialon:Eu²⁺ particle and the monochromatic cathodoluminescence image (λ = 530 nm) of the whole sample. It is seen that the CL spectrum of β-sialon:Eu²⁺ is comparable to its photoluminescence spectrum, and each β-sialon:Eu²⁺ particle shows the green emission with equivalent luminescence intensity. These results indicate that it is the β-sialon:Eu²⁺ particle that emits the green color.

Y. Q. Li et al. (2008a) proposed that Eu²⁺ ions were located in the large channels along the c-axis, which are formed by six (Si,Al)(O,N)₄ tetrahedra (see Figure 4.22), and gave the formula $Eu_xSi_{6-z}Al_{z-x}N_{z+x}N_{8-z-x}$ for Eu²⁺-doped β-sialon. By the Reitveldt refinement analysis, Y. Q. Li et al. (2008a) addressed that the Eu²⁺ ions were coordinated to the six nearest (O,N) atoms at an equivalent distance of 2.4932 Å. Moreover, by using scanning transmission electron microscopy (STEM), the direct observation of Eu atoms in the β-sialon lattice was reported by Kimoto et al. (2009). The STEM annular dark-field

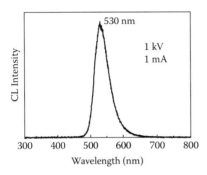

FIGURE 4.27
Cathodoluminescence spectrum of β-sialon:Eu²⁺ (left) and the monochromatic cathodo-luminescence image of β-sialon:Eu²⁺ (right) taken at the wavelength of 530 nm. (Reprinted from Hirosaki, H., Xie, R-J., Kimoto, K., Sekiguchi, T., Yamamoto, Y., Suehiro, T., and Mitomo, M., *Appl. Phys. Lett.*, 86, 211905, 2005. With permission.)

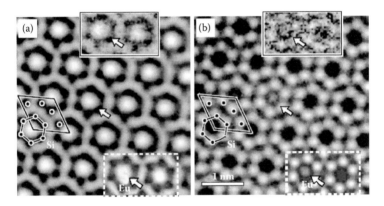

FIGURE 4.28
(a) STEM-BF (black field) and (b) STEM-ADF (asymptotically distribution free) images of β-sialon:Eu²⁺ phosphor. Upper insets in solid-line rectangles show single-scanning images obtained in a conventional manner, resulting in severe quantum noise. Lower insets in dotted rectangles show simulation results. White arrows show Eu atom positions. (Reprinted from Kimoto, K., Xie, R.-J., Matsui, Y., Ishizuka, K., and Hirosaki, N., *Appl. Phys. Lett.*, 94, 041908, 2009. With permission.)

image shows that Eu atoms are located in the continuous channels parallel to the c-axis (see Figure 4.28), which is in good agreement with the results reported by Y. Q. Li et al. (2008a).

4.2.2.2 $MSi_2O_2N_2$:Eu^{2+} (M = Ca, Sr, Ba)

4.2.2.2.1 Crystal Chemistry of $MSi_2O_2N_2$ (M = Ca, Sr, Ba)

The crystal structure of alkaline earth oxonitridosilicates, $MSi_2O_2N_2$ (M = Ca, Sr, Ba), was reported by Schnick's group (Hoppe et al. 2004; Oeckler et al. 2007;

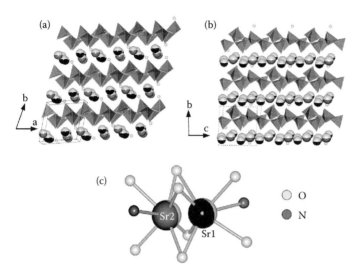

FIGURE 4.29
Crystal structure of $SrSi_2O_2N_2$ projected on (a) the (001) plane and (b) the (100) plane, showing the layered $SiON_3$ tetrahedra separated by Sr atoms. (c) The Sr atoms are surrounded by six oxygen atoms and one nitrogen atom.

Kechele et al. 2009b). The structures of these oxonitridosilicates are closely related, consisting of highly dense and corner-sharing $SiON_3$ tetrahedra (Q^3 type) layers that are separated by alkaline earth metal layers, as seen in Figures 4.29 and 4.30. The crystallographic data for $MSi_2O_2N_2$ (M = Ca, Sr, Ba) are summarized in Table 4.12. $CaSi_2O_2N_2$ crystallizes in the monoclinic structure with the space group of $P2_1$ (No. 4), $SrSi_2O_2N_2$ in the triclinic system with the space group of P1 (No. 1), and $BaSi_2O_2N_2$ in the orthorhombic system with the Pbcn (No. 60) space group.

In the anion $[Si2O2N2]^{2-}$ layer of $MSi_2O_2N_2$ (M = Ca, Sr, Ba), every N atom is connected to three neighboring Si tetrahedron centers ($N^{[3]}$), and the O atoms are exclusively bonded terminally to Si atoms. The configuration of the Q3 type $SiON_3$ tetrahedra in $SrSi_2O_2N_2$ (up-down sequence) is quite analogous to $BaSi_2O_2N_2$, but the silicate layers are shifted against each other in the Sr compound. In $CaSi_2O_2N_2$, however, the $SiON_3$ tetrahedra are connected with a different up-down sequence (Kechele et al. 2009b). In $CaSi_2O_2N_2$ and $SrSi_2O_2N_2$, Ca/Sr atoms are coordinated to six O atoms, forming a distorted trigonal prism that is capped by a single N atom (coordination number 6 + 1), whereas in $BaSi_2O_2N_2$, Ba atoms are surrounded by six atoms, forming a cuboid that is additionally capped by two N atoms (coordination number 6 + 2). The atomic coordinate data for $MSi_2O_2N_2$ (M = Sr, Ba) are given in Tables 4.13 and 4.14.

4.2.2.2.2 Photoluminescence of $MSi_2O_2N_2$:Eu^{2+} (M = Ca, Sr, Ba)

The photoluminescence of Eu^{2+}-doped $MSi_2O_2N_2$ (M = Ca, Sr, Ba) was reported by Y. Q. Li et al. (2006). Generally, $SrSi_2O_2N_2$:Eu^{2+} is a green-emitting

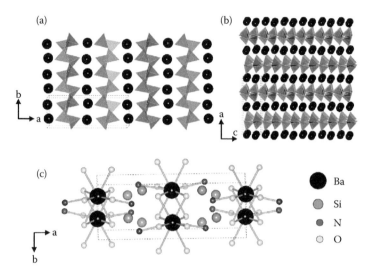

FIGURE 4.30
Crystal structure of $BaSi_2O_2N_2$ projected on (a) the (001) plane and (b) the (010) plane, showing the layered $SiON_3$ tetrahedra separated by Ba atoms. (c) The Ba atoms are surrounded by six oxygen atoms and two nitrogen atoms.

TABLE 4.12

Crystallographic Data of $MSi_2O_2N_2$ (M = Ca, Sr, Ba)

Formula	$CaSi_2O_2N_2$	$SrSi_2O_2N_2$	$BaSi_2O_2N_2$
Crystal system	Monoclinic	Triclinic	Orthorhombic
Space group	$P2_1$ (No. 40)	P1 (No. 1)	Pbcn (No. 60)
a(Å)	7.344 (2)	7.0802(2)	14.3902(3)
b(Å)	12.656(3)	7.2306(2)	5.34330(10)
c(Å)	10.483(2)	7.2554(2)	4.83254(6)
$\alpha(°)$	90	88.767(3)	90
$\beta(°)$	102.04(3)	84.733(2)	90
$\gamma(°)$	90	75.905(2)	90
V(Å³)	1028.3(4)	358.73(2)	371.58
Z	12	4	4

Source: Data from Kechele, J. A., Oeckler, O., Stadler, F., and Schnick, W., *Solid State Sci.*, 11, 537–543, 2009b; Oeckler, O., Stadler, F., Rosenthal, T., and Schnick, W., *Solid State Sci.*, 9, 205–212, 2007.

phosphor, $CaSi_2O_2N_2{:}Eu^{2+}$ a yellow-green phosphor, and $BaSi_2O_2N_2{:}Eu^{2+}$ a blue-green phosphor. Figure 4.31 presents the photoluminescence spectra of $MSi_2O_2N_2{:}Eu^{2+}$ (M = Ca, Sr, Ba). The photoluminescence data are summarized in Table 4.15.

The series of $MSi_2O_2N_2{:}Eu^{2+}$ (M = Ca, Sr, Ba) phosphors show broad excitation spectra extending from 230 to 500 nm, and all have strong absorptions

TABLE 4.13

Atomic Coordinate and Site Occupancy Fraction Data for $SrSi_2O_2N_2$

Atom	Wyckoff	x	y	z	SOF
Sr(1)		0.175(2)	0.1238(2)	0.068(1)	0.801
Sr(2)		0.631(2)	0.225(2)	0.3227(9)	0.801
Sr(3)		0.178	0.104	-0.4271	0.801
Sr(4)		0.631(1)	0.221(2)	0.8217(8)	0.801
Sr(5)		0.050(3)	0.208(3)	-0.171(3)	0.228
Sr(6)		0.661(4)	0.121(4)	0.012(3)	0.228
Sr(7)		0.089(3)	0.218(4)	0.313(3)	0.228
Sr(8)		0.676(4)	0.117(4)	0.557(4)	0.228
Si(1)		0.4751(2)	-0.2101(1)	0.1774(1)	1.00
Si(2)		0.3651(1)	0.5713(1)	0.0771(1)	1.00
Si(3)		-0.224(1)	0.5642(1)	-0.0027(1)	1.00
Si(4)		0.9769(1)	-0.2426(1)	0.2454(1)	1.00
Si(5)		0.0615(1)	-0.2276(1)	0.8310(1)	1.00
Si(6)		0.5462(1)	-0.2240(1)	0.3338(1)	1.00
Si(7)		0.2818(2)	-0.4426(1)	0.5050(1)	1.00
Si(8)		0.8578(1)	0.5625(1)	0.5816(1)	1.00
O(1)		0.3229	0.3637	0.0701	1.00
O(2)		0.4427	0.0143	0.7784	1.00
O(3)		0.9677	-0.0402	0.3373	1.00
O(4)		-0.1766	0.3596	0.0968	1.00
O(5)		-0.0288	-0.0180	0.7513	1.00
O(6)		0.4330	-0.0033	0.3499	1.00
O(7)		0.8311	0.3499	0.5574	1.00
O(8)		0.3362	-0.6671	0.5465	1.00
N(1)		0.5445	-0.3740	0.9196	1.00
N(2)		0.6393	-0.2711	0.5485	1.00
N(3)		0.1450	-0.2636	0.0508	1.00
N(4)		0.0484	-0.3953	-0.5709	1.00
N(5)		0.9173	-0.3864	0.800	1.00
N(6)		0.757	-0.2791	0.1818	1.00
N(7)		0.2696	-0.2723	0.6747	1.00
N(8)		0.4134	0.6072	0.3030	1.00

Source: Data from Oeckler, O., Stadler, F., Rosenthal, T., and Schnick, W., *Solid State Sci.*, 9, 205–212, 2007.

of blue light. With the Gaussian simulation, the excitation spectrum of $CaSi_2O_2N_2$:Eu^{2+} consists of five bands centered at 262, 315, 357, 396, and 446 nm, respectively. The crystal field splitting of $CaSi_2O_2N_2$:Eu^{2+} is estimated as 15,700 cm^{-1}. For $SrSi_2O_2N_2$:Eu^{2+}, the excitation spectrum is also composed of five bands centered at 262, 311, 365, 412, and 456 nm, leading to the crystal field splitting of 16,200 cm^{-1}. The excitation spectrum of $BaSi_2O_2N_2$:Eu^{2+}

TABLE 4.14

Atomic Coordinates, Equivalent Displacement Parameters (Å in A2), and Site Occupancy Fraction for the Average Structure of $BaSi_2O_2N_2$ in Pbcn

Atom	Wyckoff Position	x	y	z	U_{eq}	SOF
Ba	4c	0	0.2497(4)	1/4	0.0171(4)	1
Si	8d	0.7979(2)	0.3360(6)	0.709(1)	0.033(1)	1
O	8d	0.9057(2)	0.231(2)	0.740(2)	0.003(1)	1
N	8d	0.7244(4)	0.1224(8)	0.860(1)	0.003(1)	1

Source: Data from Kechele, J. A., Oeckler, O., Stadler, F., and Schnick, W., *Solid State Sci.*, 11, 537–543, 2009b.

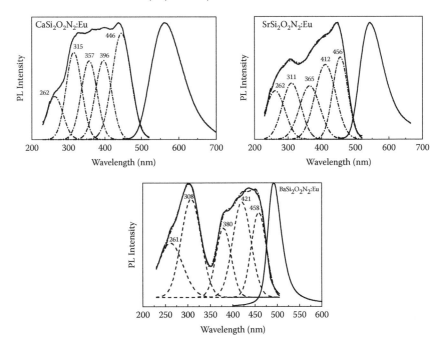

FIGURE 4.31
Excitation and emission spectra of $MSi_2O_2N_2:Eu^{2+}$ (M = Ca, Sr, Ba).

can be again fitted into five bands centered at 261, 308, 386, 421, and 458 nm, so that the crystal field splitting is roughly calculated as 18,000 cm^{-1}. The centers of gravity are 29,100, 28,800, and 28,500 cm^{-1} for $CaSi_2O_2N_2:Eu^{2+}$, $SrSi_2O_2N_2:Eu^{2+}$, and $BaSi_2O_2N_2:Eu^{2+}$, respectively. The large crystal field splitting of these oxonitrido-silicates contributes greatly to the significantly red-shifted excitation spectra.

The emission color is quite different for the three phosphors. The emission spectrum of $CaSi_2O_2N_2:Eu^{2+}$ shows a broad band with the maximum at 563 nm. The full width at half maximum is about 98 nm. With the Gaussian

TABLE 4.15

Photoluminescence Properties of $MSi_2O_2N_2:Eu^{2+}$ (M = Ca, Sr, Ba)

	$CaSi_2O_2N_2:Eu$	$SrSi_2O_2N_2:Eu$	$BaSi_2O_2N_2:Eu$
Excitation peak, nm	262, 315, 357, 396, 446	262, 311, 365, 412, 456	261, 308, 386, 421, 458
Emission peak, nm	563 (554, 601)	544 (541, 591)	492
FWHM, nm	98	83	36
Crystal field splitting, cm^{-1}	15,700	16,200	18,000
Stokes shift, cm^{-1}	4,370	3,450	1,500
Center of gravity, cm^{-1}	29,100	28,800	28,500
Absorption @ 450 nm	83%	80%	78%
External quantum efficiency @ 450 nm	72%	69%	41%

simulation, the emission spectrum can be fitted into two bands centered at 554 and 601 nm. This leads to a rough calculation of the Stokes shift of 4,370 cm^{-1}. The two bands are perhaps due to two crystallographic sites of Ca atoms. $SrSi_2O_2N_2:Eu^{2+}$ shows an emission band with the peak at 544 nm and the FWHM of 83 nm. This emission band can also be fitted into two bands centered at 541 and 591 nm, yielding the Stokes shift of 3,500 cm^{-1}. $BaSi_2O_2N_2:Eu^{2+}$ shows an extremely narrow emission band centered at 492 nm, with a FWHM of 36 nm. The small value of FWHM is explained by the high symmetry of the Ba sites, which is similar to that observed in β-sialon:Eu^{2+}. The Stokes shift of $BaSi_2O_2N_2:Eu^{2+}$ is also as small as 1,500 cm^{-1}.

Under the excitation of 450 nm, the absorptions of these alkaline earth oxonitridosilicates are 83%, 80%, and 78% for $CaSi_2O_2N_2:Eu^{2+}$, $SrSi_2O_2N_2:Eu^{2+}$, and $BaSi_2O_2N_2:Eu^{2+}$, respectively. The external quantum efficiencies for the corresponding phosphors are 72%, 79%, and 41%, respectively. The low quantum efficiency of $BaSi_2O_2N_2:Eu^{2+}$ can be ascribed to the large overlap between the excitation and emission spectra that results in the strong reabsorption.

The excellent photoluminescence properties of these alkaline earth oxonitridosilicates make them very suitable for use as yellow-green ($CaSi_2O_2N_2:Eu^{2+}$), green ($SrSi_2O_2N_2:Eu^{2+}$), and blue-green ($BaSi_2O_2N_2:Eu^{2+}$) phosphors in white LEDs.

4.2.2.3 $Ba_3Si_6O_{12}N_2:Eu^{2+}$

4.2.2.3.1 Crystal Chemistry of $Ba_3Si_6O_{12}N_2$

The crystal structure of $Ba_3Si_6O_{12}N_2$ was reported by Uheda et al. (2008) and Braun et al. (2010). It crystallizes in the trigonal crystal system with the space group of P-3 (No. 147). Table 4.12 shows the atomic coordinates and crystallographic data for $Ba_3Si_6O_{12}N_2$. The crystal structure of $Ba_3Si_6O_{12}N_2$ has fused rings-sheets, $2_\infty[(Si_6^{[4]}O_6^{[2]}N_2^{[3]})O_6^{[1]}]^{6-}$, which are composed of eight-membered Si-(O,N) and twelve-membered Si-O rings. It is built up of

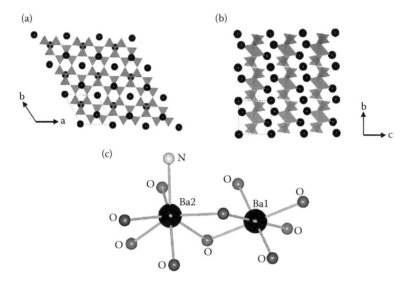

FIGURE 4.32
Crystal structure of $Ba_3Si_6O_{12}N_2$ projected on (a) the (001) plane and (b) the (100) plane. (c) The coordination of Ba atoms by (O,N) atoms.

corrugated corner-sharing SiO3N tetrahedron layers that were reseparated by Ba atoms (see Figure 4.32). The Ba^{2+} ions occupy two different crystallographic sites; one is trigonal antiprism (distorted octahedron) with six oxygen atoms, and the other is trigonal antiprism with six oxygen atoms, which is further capped with one nitrogen atom (see Figure 4.32c). The bond lengths of Ba-O/N are given in Table 4.16.

4.2.2.3.2 Photoluminescence of $Ba_3Si_6O_{12}N_2:Eu^{2+}$

Uheda et al. (2008) and Braun et al. (2010) reported the photoluminescence of Eu^{2+} in $Ba_3Si_6O_{12}N_2$. As seen in Figure 4.33, $Ba_3Si_6O_{12}N_2:Eu^{2+}$ has a broad excitation band that covers a spectral range of 250–500 nm, showing the strong absorption of UV blue light. The emission spectrum shows a narrow band centered at 525 nm, with a FWHM of 68 nm. The narrow green emission observed in $Ba_3Si_6O_{12}N_2:Eu^{2+}$ may be expected from the Ba2 site (coordinated with six oxygen atoms and one nitrogen atom) rather than the Ba1 site (with six oxygen atoms only).

This green phosphor exhibits high color purity with the International Commission on Illumination (CIE) chromaticity coordinates of x = 0.28 and y = 0.64) as well as small thermal quenching. At 150°C, the luminescence of $Ba_3Si_6O_{12}N_2:Eu^{2+}$ is quenched by 10% of the initial intensity at room temperature, whereas it is reduced by 25% for $(Sr,Ba)_2SiO_4:Eu^{2+}$. The useful photoluminescence and small thermal quenching enable $Ba_3Si_6O_{12}N_2:Eu^{2+}$ to be a promising green-emitting luminescent material in white LEDs for general lighting or liquid crystal display (LCD) backlights.

TABLE 4.16

Atomic Coordinates and Site Occupancy Fraction for $Ba_3Si_6O_{12}N_2$

Atom	Wyckoff	x	y	z	SOF
Ba(1)	1a	0	0	0	1.00
Ba(2)	2d	0.3333	0.6667	0.1039	1.00
Si	6g	0.2366	0.8310	0.6212	1.00
N	2d	0.3333	0.6667	0.5680	1.00
O(1)	6g	0.3560	0.2950	0.1730	1.00
O(2)	6g	0	0.6810	0.5890	1.00

Source: Data from Uheda, K., Shimooka, S., Mikami, M., Imura, H., and Kijima, N., Synthesis and characterization of new green oxo-nitridosilicate phosphor, $(Ba,Eu)_3Si_6O_{12}N_2$, for white LED, paper presented at the 214th ECS Meeting, Honolulu, Hawaii, 2008.

Note: Crystal system: Trigonal.
Space group: P-3.
Lattice constant: a = 7.5048(8) Å, c = 6.4703(5) Å.
Z = 1.
Bond lengths: Ba(1) – O(1): 2.7161 Å × 6, Ba(2) – O(1): 2.9126 Å × 3, Ba(2) – O(1): 2.8386 Å × 3, Ba(2) – N: 3.0029 Å.

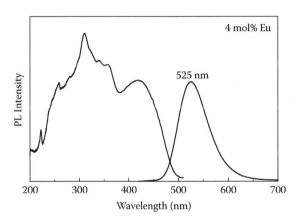

FIGURE 4.33
(a) Excitation and (b) emission spectra of $Ba_3Si_6O_{12}N_2:Eu^{2+}$.

4.2.2.4 $Sr_5Al_{5+x}Si_{21-x}N_{35-x}O_{2+x}:Eu^{2+}$ ($x \approx 0$)

4.2.2.4.1 Crystal Structure of $Sr_5Al_{5+x}Si_{21-x}N_{35-x}O_{2+x}$ ($x \approx 0$)

The investigation of the crystal structure of $Sr_5Al_{5+x}Si_{21-x}N_{35-x}O_{2+x}$ ($x \approx 0$) was made by Oeckler et al. (2009). As shown in Table 4.17 $Sr_5Al_5Si_{21}N_{35}O_2$ crystallizes in the orthorhombic crystal system with the space group of $Pmn2_1$ (No. 31). This oxonitridosilicate compound is considered one of the commensurate phases that belong to a series of compositely modulated structures (Michiue et al. 2009). It has a complex intergrowth structure that is composed of highly

TABLE 4.17

Crystallographic Data of $Sr_5Al_{5+x}Si_{21-x}N_{35-x}O_{2+x}$ ($x \approx 0$)

Formula	$Sr_5Al_5SiN_{35}O_2$
Crystal system	Orthorhombic
Space group	$Pmn2_1$ (No. 31)
a(Å)	23.614
b(Å)	7.487
c(Å)	9.059
V(Å³)	1601.5(6)
Z	2

Source: Data from Oeckler, O., Kechele, J. A., Koss, H., Schmidt, P. J., and Schnick, W., *Chem. Eur. J.*, 15, 5311–5319, 2009.

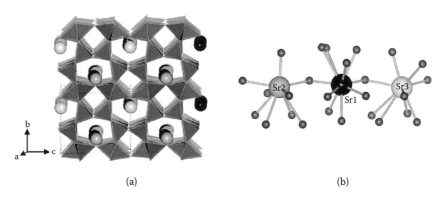

FIGURE 4.34

(a) Crystal structure of $Sr_5Al_5Si_{21}N_{35}O_2$ projected on the (100) plane and (b) coordination spheres of the Sr atoms.

condensed Dreier layers alternating with Sechser ring layers that consist of both vertex- and edge-sharing (Si,Al)-(O,N) nitridosili (see Figure 4.34a). The Sr atoms occupy three different crystallographic sites, which are all coordinated by nine (O,N) atoms (see Figure 4.34b). The bond length of Sr – (O,N) varies in the range of 2.45–3.44 Å. Table 4.18 gives the atomic coordinates for $Sr_5Al_5Si_{21}N_{35}O_2$.

4.2.2.4.2 Photoluminescence of $Sr_5Al_{5+x}Si_{21-x}N_{35-x}O_{2+x}$:$Eu^{2+}$ ($x \approx 0$)

The photoluminescence of Eu^{2+} in $Sr_5Al_{5+x}Si_{21-x}N_{35-x}O_{2+x}$ ($x \approx 0$) was reported by Oeckler et al. (2009). $Sr_5Al_{5+x}Si_{21-x}N_{35-x}O_{2+x}$:$Eu^{2+}$ is a green-emitting phosphor that shows a narrow emission band centered at 510 nm under the 450 nm excitation. The full width at half maximum is about 69 nm, which is similar to that of $SrSi_2O_2N_2$:Eu^{2+}. The rather narrow emission band is explained by the similarity in the coordination of different Sr sites in the structure. The chromaticity coordinates (x,y) of this green phosphor are (0.210, 0.539).

TABLE 4.18

Atomic Coordinates and Site Occupancy Fraction for $Sr_5Al_{15+x}Si_{21-x}N_{35-x}O_{2+x}$ ($x \approx 0$) (standard deviations are given in parentheses)

Atom	x	y	z	SOF
Sr(1)	0.20139(14)	0.2537(3)	0.48643(13)	1.00
Sr(2)	0	0.2341(4)	0.5143(3)	1.00
Sr(3)	0.09912(14)	0.7458(2)	0.0095(3)	1.00
(Si,Al)(1)	0	−0.0479(8)	1.1890(8)	0.808 Si / 0.192 Al
(Si,Al)(2)	0.3052(2)	−0.0146(5)	0.6607(5)	
(Si,Al)(3)	0.1963(2)	0.9380(5)	0.8132(4)	
(Si,Al)(4)	0	0.0103(11)	0.8372(9)	
(Si,Al)(5a)	0.1133(2)	−0.0086(11)	0.3358(9)	0.524 Si / 0.124 Al
(Si,Al)(6a)	0.1147(3)	1.0592(9)	0.6805(7)	
(Si,Al)(5b)	0.0755(5)	0.0425(13)	0.3119(12)	0.284 Si / 0.067Al
(Si,Al)(6b)	0.0796(5)	0.9872(16)	0.6656(14)	
(Si,Al)(7)	0	0.3645(9)	1.1608(8)	0.808 Si / 0.192Al Al
(Si,Al)(8)	0.1254(2)	0.6310(6)	0.6584(5)	
(Si,Al)(9)	0.1866(2)	0.3584(5)	0.8533(5)	
(Si,Al)(10)	0.0631(2)	0.3580(6)	0.8620(5)	
(Si,Al)(11)	0.2491(2)	0.3781(6)	0.1539(6)	
(Si,Al)(12)	0.3133(2)	0.3455(5)	0.8584(4)	
(Si,Al)(13)	0.1264(2)	0.3688(6)	0.1559(5)	
(Si,Al)(14)	0.0614(2)	0.6348(6)	1.3599(5)	
(Si,Al)(15)	0	0.6384(9)	0.6595(7)	
N(1)	0	0.377(2)	0.965(2)	1.00
N(2)	0.1894(6)	0.5154(12)	0.7114(10)	1.00
N(3)	0.0636(6)	0.4760(16)	1.2192(12)	1.00
N(4)	0.1239(5)	0.6175(17)	0.4646(14)	1.00
N(5)	0.1232(5)	0.3927(16)	−0.0359(14)	1.00
N(6)	0.2505(6)	0.3827(15)	−0.0419(13)	1.00
N(7)	0.1876(6)	0.4807(12)	0.2180(10)	1.00
N(8)	0.5	1.413(2)	0.9668(19)	1.00
N(9)	0.0605(6)	0.5381(15)	0.7332(11)	1.00
N(10)	0.2479(9)	0.154(2)	0.2167(16)	1.00
N(11)	0.0602(8)	0.148(2)	0.7859(17)	1.00
N(12)	0.0611(9)	−0.153(2)	0.2720(16)	1.00
N(13)	0.1333(7)	0.1477(18)	0.1973(15)	1.00
N(14)	0.1339(7)	0.843(2)	0.7345(16)	1.00
N(15)	0.3201(5)	−0.1521(17)	0.2530(14)	1.00
N(16)	0.3249(6)	0.1266(16)	0.7985(15)	1.00
(N,O)(17)	0	0.164(2)	1.2397(17)	
(N,O)(18)	0	−0.134(2)	0.6947(17)	
(N,O)(19)	0.1093(4)	0.1020(17)	0.5030(14)	0.714 N / 0.286 O
(N,O)(20)	0.2932(4)	0.0916(15)	0.4992(12)	
(N,O)(21)	0	−0.082(2)	1.005(2)	

Source: Data from Oeckler, O., Kechele, J. A., Koss, H., Schmidt, P. J., and Schnick, W., *Chem. Eur. J.*, 15, 5311–5319, 2009.

The excitation spectrum shows a maximum at ~400 nm, indicating that it can be excited by UV to blue light (350–450 nm) efficiently.

4.2.2.5 $Sr_{14}Si_{68-s}Al_{6+s}O_sN_{106-s}$:$Eu^{2+}$ (s ≈ 7)

4.2.2.5.1 Crystal Structure of $Sr_{14}Si_{68-s}Al_{6+s}O_sN_{106-s}$ (s ≈ 7)

Shioi et al. (2010) recently reported the structure and photoluminescence of Eu^{2+}-doped $Sr_{14}Si_{68-s}Al_{6+s}O_sN_{106-s}$ (s ≈ 7). This compound is also the commensurate phase belonging to the composite crystals, which is similar to $Eu_3Si_{15-x}Al_{1+x}O_xN_{23-x}$ and $Sr_5Si_{21-x}Al_{5+x}O_{2+x}N_{35-x}$. The composite crystal model generally has two substructures with different periodicities along the a-axis. The chemical formula is AM_2X (M = Si or Al, X = O or N) for the first substructure and M_2X_4 for the second one. If in the two substructures the ratio of dimensions for the a-axis is represented by $a_1/a_2=n/m$, the chemical composition of a whole crystal is then generally given by $(AM_2X)_m(M_2X_4)_n$. For $Eu_3Si_{15-x}Al_{1+x}O_xN_{23-x}$ and $Sr_5Si_{21-x}Al_{5+x}O_{2+x}N_{35-x}$ materials, the [n/m] is [5/3] and [8/5], respectively. Shioi et al. (2010) reported that $Sr_{14}Si_{68-s}Al_{6+s}O_sN_{106-s}$ (s ≈ 7) had the [n/m] value of [23/14]. It crystallizes in the monoclinic system with the space group of $P2_1$ (No. 4). The lattice parameters are a = 67.789(3)Å, b = 9.0416(2)Å, c = 7.4686(1)Å, and β = 90.121(4)°.

As seen in Figure 4.35, $Sr_{14}Si_{68-s}Al_{6+s}O_sN_{106-s}$ (s ≈ 7) has a crystal structure closely related to that of $Sr_5Al_{5+x}Si_{21-x}N_{35-x}O_{2+x}$, and the structure is built up of highly condensed Dreier layers alternating with Sechser ring layers that consist of both vertex- and edge-sharing M – X (M = Si, Al; X = O,N) nitridosili.

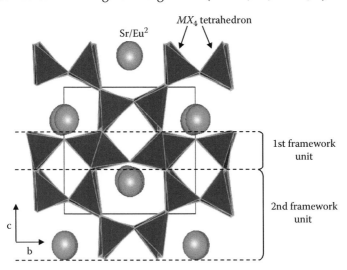

FIGURE 4.35
Crystal structure of $Sr_{14}Si_{68-s}Al_{6+s}O_sN_{106-s}$ (s ≈ 7) projected on the (100) plane. (Reprinted from Shioi, K., Michiue, Y., Hirosaki, N., Xie, R.-J., Takeda, T., Matsushita, Y., Tanaka, M., and Li, Y. Q., *J. Alloy Compds.*, 509, 332–337, 2011. With permission.)

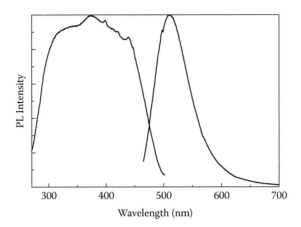

PL Intensity

Wavelength (nm)

FIGURE 4.36

Photoluminescence spectra of $Sr_{14}Si_{68-s}Al_{6+s}O_sN_{106-s}$:$Eu^{2+}$ (s ≈ 7). (Reprinted from Shioi, K., Michiue, Y., Hirosaki, N., Xie, R.-J., Takeda, T., Matsushita, Y., Tanaka, M., and Li, Y. Q., *J. Alloy Compds.*, 509, 332–337, 2011. With permission.)

The Sr atoms are located in the channel along the [100] direction, which are surrounded by six MX_4 tetrahedra.

4.2.2.5.2 *Photoluminescence of* $Sr_{14}Si_{68-s}Al_{6+s}O_sN_{106-s}$ *(s ≈ 7)*

The photoluminescence of Eu^{2+} in $Sr_{14}Si_{68-s}Al_{6+s}O_sN_{106-s}$ (s ≈ 7) was investigated by Shioi et al. (2010). As seen in Figure 4.36, the excitation spectrum shows a very broad band covering the spectral region of 250–500 nm, with the maximum at 300–450 nm. The emission band shows the maximum at 508 nm under the 376 nm excitation, and has a FWHM value of ~90 nm.

The green-emitting $Sr_{14}Si_{68-s}Al_{6+s}O_sN_{106-s}$:$Eu^{2+}$ (s ≈ 7) phosphor exhibits a smaller thermal quenching than the Ba_2SiO_4:Eu^{2+} green phosphor. The emission intensity of $Sr_{14}Si_{68-s}Al_{6+s}O_sN_{106-s}$:$Eu^{2+}$ (s ≈ 7) at 100°C remains 86% of the initial intensity at room temperature, whereas it is only 69% for Ba_2SiO_4:Eu^{2+}.

4.2.2.6 *Ca-α-sialon:Yb^{2+}*

The photoluminescence of Yb^{2+} in Ca-α-sialon was reported by Xie et al. (2005b). As seen in Figure 4.37, the excitation spectrum of Ca-α-sialon:Yb^{2+} consists of three bands centered at 299, 344, and 444 nm, and two shoulders at 258 and 285 nm, respectively. It can be fitted into five broad bands at 249, 295, 343, 401, and 452 nm with the Gaussian simulation. The structure in the excitation spectrum is due to the crystal field splitting of the 5d level of the Yb^{2+} ion. The crystal field splitting is estimated to be 18,000 cm^{-1}.

The emission spectrum consists of a single and symmetric broad band with the maximum at 549 nm, which is assigned to the transition between the allowed $4f^{13}5d$ and $4f^{14}$ configurations of the Yb^{2+} ion. A rough estimate

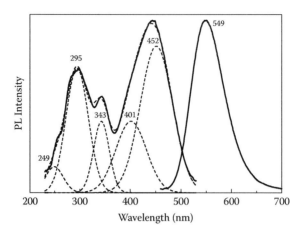

FIGURE 4.37

Excitation and emission spectra of $(Ca_{1-x}Yb_x)_{m/2}Si_{12-m-n}Al_{m+n}O_nN_{16-n}$ (m = 2, n = 1, x = 0.005). The emission spectrum was recorded under the 444 nm excitation, and the excitation spectrum was monitored at 549 nm. (Reprinted from Xie, R,-J., Hirosaki, N., Mitomo, M., Uheda, K., Suehiro, T., Xu, X., Yamamoto, Y., and Sekiguchi, T., *J. Phys. Chem. B,* 109, 9490–9494, 2005b. With permission.)

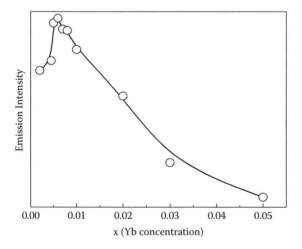

FIGURE 4.38

Concentration quenching of $(Ca_{1-x}Yb_x)_{m/2}Si_{12-m-n}Al_{m+n}O_nN_{16-n}$ (m = 2, n = 1). (Reprinted with permission from Xie, R,-J., Hirosaki, N., Mitomo, M., Uheda, K., Suehiro, T., Xu, X., Yamamoto, Y., and Sekiguchi, T., *J. Phys. Chem. B,* 109, 949–494, 2005. With permission.)

of the Stokes shift can be made by assuming that the excitation band is the mirror image of the emission band. In this way, the Stokes shift is calculated as 3,900 cm^{-1}.

The concentration-dependent luminescence of Ca-α-sialon:Yb^{2+} is shown in Figure 4.38. The concentration quenching occurs at a very low Yb^{2+}

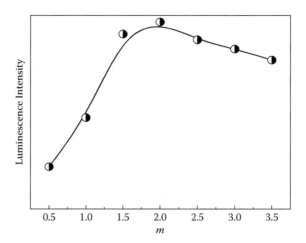

FIGURE 4.39
Effect of the m value on the luminescence of $(Ca_{1-x}Yb_x)_{m/2}Si_{12-m-n}Al_{m+n}O_nN_{16-n}$ ($n = 1/2m$, $x = 0.005$). (Reprinted from Xie, R,-J., Hirosaki, N., Mitomo, M., Uheda, K., Suehiro, T., Xu, X., Yamamoto, Y., and Sekiguchi, T., *J. Phys. Chem. B*, 109, 9490–9494, 2005b. With permission.)

concentration, which is as low as 0.5 mol%. The effect of the chemical composition of the host lattice on the luminescence of Yb^{2+} is given in Figure 4.39. It is seen that the composition of m = 2 and n = 1 shows the highest luminescence. Xie et al. (2005b) have stated that the compositional dependence of luminescence properties is dominantly attributed to the changes in crystallinity, phase purity, and particle morphology of α-sialon with varying m. With lower m values the crystallization of the powders is poor, and there are some defects in each particle that can trap or scatter the emitted light. With high m values large hard agglomerates form, and this results in a low packing density of the powder, which causes a strong light scattering.

4.2.2.7 γ-ALON:Mn²⁺

4.2.2.7.1 Crystal Chemistry of γ-ALON:Mn²⁺

Xie et al. (2008) reported the structure and photoluminescence of Mn^{2+}-doped γ-ALON. Aluminum oxynitride, or γ-ALON, is a solid solution in the binary system of Al_2O_3-AlN, with the aluminum nitride content between 27 and 40 mol%. It is a face-centered cubic phase with a spinel structure and Fd-3m space group. γ-ALON is a nonstoichiometric spinel because some cation sites are occupied by vacancies. A general formula of the defective γ-ALON was proposed by McCauley:

$$Al_{(64+x)/3}\boxtimes_{(8-x)/3}O_{32-x}N_x \quad (0 \leq x \leq 8) \tag{4.2}$$

The symbol ⊠ denotes cation vacancies. For γ-ALON, values of x range from approximately 3.5 to 5.8.

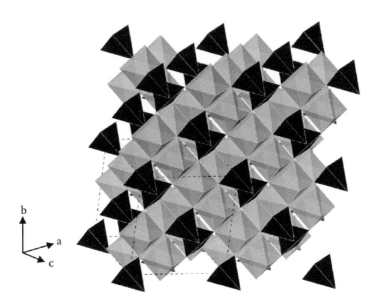

FIGURE 4.40
Crystal structure of spinel γ-ALON, showing the AlO4 tetrhedra (black) and AlO6 octahedra (grey).

TABLE 4.19

Atomic Coordinates, Site Occupancy Fraction (SOF), and Isotropic Displacement Parameters (B) of Mn^{2+}-Mg^{2+} Codoped γ-ALON

Atom	Site	SOF	x	y	z	B
Al1	8a	0.92	0	0	0	0.108
Mn	8a	0.02	0	0	0	0.108
Mg	8a	0.05	0	0	0	0.108
Al2	16d	0.87	0.625	0.625	0.625	0.023
O	32e	0.89	0.3839	0.3839	0.3839	0.744
N	32e	0.11	0.3839	0.3839	0.3839	0.744

Source: Data from Xie, R.-J., Hirosaki, N., Liu, X.-J., Takeda, T., and Li, H.-L., *Appl. Phys. Lett.*, 92, 201905, 2008.

In the structure of γ-ALON, the aluminum cations occupy both octahedral (AlO6) and tetrahedral (AlO4) sites, as depicted in Figure 4.40, and oxygen/nitrogen occupy the anion sites of a spinel lattice. This spinel phase is stable only when a disordered vacancy also exists on the octahedral sites. This is confirmed by the Rietveldt refinement of the structure of γ-ALON, as shown in Table 4.19 It is observed that Al atoms occupy both tetrahedral (8a) and octahedral (16d) sites. On the other hand, Mn and Mg substitute the Al atoms only

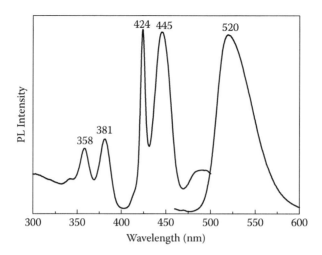

FIGURE 4.41
Excitation and emission spectra of Mn^{2+}-Mg^{2+} codoped γ-ALON (7 mol% Mn, 10 mol% Mg). The excitation spectrum was monitored at 520 nm, and the emission spectrum was measured by the 445 nm irradiation. (Reprinted from Xie, R.-J., Hirosaki, N., Liu, X.-J., Takeda, T., and Li, H.-L., *Appl. Phys. Lett.*, 92, 201905, 2008. With permission.)

in the tetrahedral sites. In addition, γ-ALON contains vacancies as a regular part of the crystal, which are distributed mainly in the octahedral site.

4.2.2.7.2 Photoluminescence of γ-ALON:Mn^{2+}

The Mn^{2+}-Mg^{2+} codoped γ-ALON ($Al_{1.7}O_{2.1}N_{0.3}$) powders were prepared by firing the powder mixture of Al_2O_3, AlN, MgO, and $MnCO_3$ at 1,800°C for 2 h under a 0.5 MPa N_2 atmosphere. The photoluminescence spectra of the sample are given in Figure 4.41. The excitation spectrum of γ-ALON:Mn^{2+} consists of five peaks at 340, 358, 381, 424, and 445 nm, which are due to the electronic transitions from the ground state of $^6A_1$ to the excited states of $^4T_2$ (^4P), ^4E (^4G), $^4T_2$, [^4E (^4G), ^4A (^4G)], and $^4T_2$ (^4G), respectively (Yen et al. 2006). The strong excitation peak at 445 nm makes γ-ALON:Mn^{2+} a suitable green phosphor for white LEDs using blue LED chips. The emission spectrum consists of a single band centered at 520 nm with a FWHM of 44 nm. This green emission is the characteristic emission of Mn^{2+} due to a $^4T_1$ (^4G) $\rightarrow$ $^6A_1$ transition. The chromaticity coordinates are x = 0.217, y = 706 for Mn^{2+}-Mg^{2+}-doped γ-ALON. The decay time of γ-ALON:Mn^{2+},Mg^{2+} is in the order of 3–4 ms. Although the internal quantum efficiency of this green phosphor is as high as 61% under the 450 nm excitation, the absorption is quite low (20%) due to spin-forbidden transitions. This finally leads to an external quantum efficiency of ~ 12%.

The temperature-dependent luminescence of γ-ALON:Mn^{2+},Mg^{2+} is shown in Figure 4.42. This green phosphor exhibits a small thermal quenching, with the luminescence quenched by 13% of the initial intensity at 150°C.

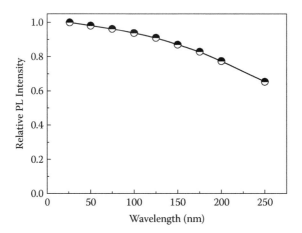

FIGURE 4.42
Thermal quenching of Mn^{2+}-Mg^{2+} codoped γ-ALON (7 mol% Mn, 10 mol% Mg). The sample was excited by the 450 nm blue light. (Reprinted from Xie, R.-J., Hirosaki, N., Liu, X.-J., Takeda, T., and Li, H.-L., *Appl. Phys. Lett.*, 92, 201905, 2008. With permission.)

4.2.3 Yellow-Emitting Phosphors

4.2.3.1 Ca-α-sialon:Eu^{2+}

Ce^{3+}-doped Ca-α-sialon emits blue/blue-green light under the UV or near-UV light irradiation. The Eu^{2+}-doped Ca-α-sialon was reported to be a promising yellow-emitting phosphor (Xie et al. 2002, 2004a, 2004c, 2005b). Ca-α-sialon:Eu^{2+} was prepared by firing the powder mixture of Si$_3$N$_4$, CaCO$_3$, AlN, Al$_2$O$_3$, and Eu$_2$O$_3$ at 1,700°C under the 0.5 MPa nitrogen atmosphere. Figure 4.43 shows the diffuse reflectance spectra of nondoped and Eu^{2+}-doped Ca-α-sialon. It is seen that the nondoped Ca-α-sialon shows an absorption edge at ~297 nm, and white body color. On the other hand, the Eu^{2+}-doped Ca-α-sialon shows strong absorptions in the UV-to-visible light spectral region, which is due to the absorption of Eu^{2+}.

The photoluminescence spectra of Ca-α-sialon:Eu^{2+} are given in Figure 4.44. The excitation spectrum consists of two broad bands peaking at 302 and 412 nm, respectively. It can be further fitted into four subbands centered at 300, 392, 444, and 487 nm by means of the Gaussian simulation. These subbands represent the splitting of the energy levels of Eu^{2+} ions under the crystal field. The crystal field splitting is estimated as to be 12,800 cm^{-1}, and the center of gravity is ~ 20,400 cm^{-1}. The emission spectrum shows a broad and symmetric band peaking at 580 nm, with a FWHM of 94 nm. A Stokes shift of ~ 3,300 cm^{-1} is then estimated for Ca-α-sialon:Eu^{2+}.

Figure 4.45 shows the luminescence of Eu^{2+} changing with the Eu^{2+} concentration and the chemical composition (i.e., m value). For Ca-α-sialon, it is seen that the luminescence has the maximum when the Eu^{2+} concentration is ~7.5 mol% or the m value is 2.0. Moreover, the emission band is red-shifted as

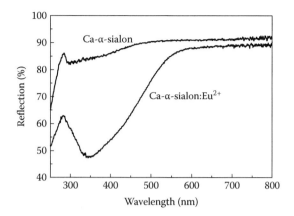

FIGURE 4.43
Diffuse reflectance spectra of non-doped and E^{u2+}-doped Ca-α-sialon. (Reprinted from Xie, R.-J., Hirosaki, N., Mitomo, M., Yamamoto, Y., and Suehiro, T., *J. Phys. Chem. B*, 108, 12027–12031, 2004a. With permission.)

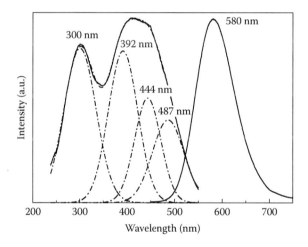

FIGURE 4.44
Excitation and emission spectra of Ca-α-sialon:Eu^{2+} (m = 2, n = 1, 7.5 mol% Eu). The excitation spectrum was monitored at 580 nm, and the emission spectrum was measured upon the 400 nm excitation. (Reprinted from Xie, R.-J., Hirosaki, N., Mitomo, M., Yamamoto, Y., and Suehiro, T., *J. Phys. Chem. B*, 108, 12027–12031, 2004a. With permission.)

the Eu^{2+} concentration or the m value increases. This is due to the reabsorption and enhanced Stokes shift.

In comparison with yellow YAG:Ce^{3+}, Ca-α-sialon:Eu^{2+} exhibits smaller thermal quenching, as seen in Figure 4.46. At 150°C, the luminescence is quenched by 23 and 14% of the initial intensity for YAG:Ce^{3+} and Ca-α-sialon:Eu^{2+}, respectively. The thermal quenching temperature is about 230°C for YAG:Ce^{3+}, whereas it is higher than 300°C for Ca-α-sialon:Eu^{2+}.

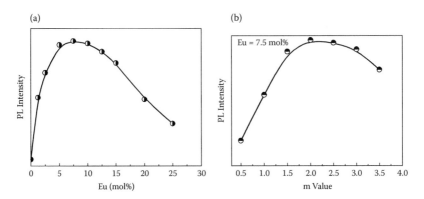

FIGURE 4.45

Effects of (a) the Eu^{2+} concentration and (b) the m value on the luminescence of Ca-α-sialon:Eu^{2+} (($Ca_{1-x}Eu_x)_{m/2}Si_{12-m-n}Al_{m+n}O_nN_{16-n}$). (Reprinted from Xie, R.-J., Hirosaki, N., Mitomo, M., Yamamoto, Y., and Suehiro, T., *J. Phys. Chem. B*, 108, 12027–12031, 2004a. With permission.)

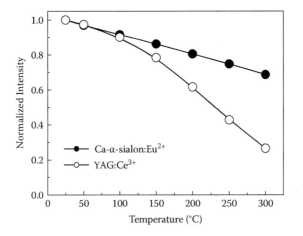

FIGURE 4.46

Thermal quenching of Ca-α-sialon:Eu^{2+} (m = 2, n = 1, 7.5 mol% Eu) and YAG:Ce^{3+}.

The absorption and external quantum efficiency of Ca-α-sialon:Eu^{2+} (m = 1.4, n = 0.7, 7.5 mol% Eu) are 75 and 56%, respectively. This indicates that Ca-α-sialon:Eu^{2+} is a very attractive yellow phosphor for white LEDs when combined with blue LEDs.

4.2.3.2 $CaAlSiN_3$:Ce^{3+}

4.2.3.2.1 Crystal Structure of $CaAlSiN_3$

The crystal structure of $CaAlSiN_3$ was reported by Uheda et al. (2006b). The atomic coordinates and crystallographic data are summarized in Table 4.20. $CaAlSiN_3$ crystallizes in the orthorhombic structure with the space group

TABLE 4.20

Atomic Coordinates and Crystallographic Data for CaAlSiN₃

Atom	Wyckoff	x	y	z	SOF
Ca	4a	0	0.3146	0.4952	1.00
Al/Si	8b	0.1741	0.1570	0.0291	1.00
N(1)	8b	0.21258	0.1300	0.3776	1.00
N(2)	4a	0	0.2446	0	1.00

Source: Data from Uheda, K., Hirosaki, N., Yamamoto, Y., Naoto, A., Nakajima, T., and Yamamoto, H., *Electrochem. Solid State Lett.*, 9, H22–H25, 2006b.

Note: Crystal system: Orthorhombic.
Space group: $Cmc2_1$ (No. 36).
Lattice constant: a = 9.8020(4) Å, b = 5.6506(2) Å, c = 5.0633(2) Å.
Bond lengths: Ca – N(1): 2.4304 Å × 2, Ca – N(2): 2.4197 Å, Ca – N(2): 2.4911 Å, Ca – N(2): 2.7029 Å.

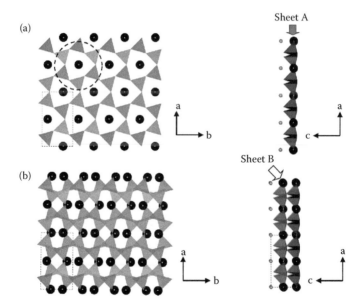

FIGURE 4.47

Crystal structure of $CaAlSiN_3$ projected on the (001) pane, showing (a) sheet A and a ring formed by corner sharing of six MN4 tetrahedra, and (b) sheet B.

of $Cmc2_1$ (No. 36), which is isotypic with $LiSi_2N_3$ and $NaSi_2N_3$. The Al and Si atoms are distributed randomly at the 8b site of the lattice.

The structure of $CaAlSiN_3$ consists of highly dense and corner-sharing MN4 (M = Al, Si) tetrahedral network. As seen in Figure 4.47, a ring is formed by linking six MN4 tetrahedra at their corners, which creates sheet A (see Figure 4.47a), which is parallel to the (001) plane. Sheet B, which is equivalent to sheet A rotated by 180°, is overlaid on sheet A to form a rigid three-dimensional framework (see Figure 4.47b). The Ca atoms are located in the large channels along the c-axis, which are coordinated by five nitrogen with an average distance of ~2.50 Å.

4.2.3.2.2 Photoluminescence Properties of CaAlSiN$_3$:Ce^{3+}

Uheda et al. (2006a, 2006b) first investigated the photoluminescence of Eu^{2+}-doped CaAlSiN$_3$, which is a very useful red phosphor for white LEDs and will be discussed later. The photoluminescence of Ce^{3+} in CaAlSiN$_3$ was reported by Y. Q. Li et al. (2008b). Figure 4.48 presents the photoluminescence spectra of CaAlSiN$_3$:Ce^{3+}.

The excitation spectrum covers the spectral range of 250–550 nm, and shows the strong absorption of blue light. It can be fitted into five bands centered at 259, 313, 370, 421, and 483 nm, respectively, by using the Gaussian simulation. Therefore, the crystal field splitting and the center of gravity are estimated as to be 13,900 and 19,700 cm^{-1}, respectively. These data clearly show that a large crystal field splitting and strong nephelauxetic effect are realized in CaAlSiN$_3$:Ce^{3+}, which significantly shifts downward the energy levels of Ce^{3+} ions.

The emission band of CaAlSiN$_3$:Ce^{3+} shows a very broad and asymmetrical band centered at 570 nm, with a FWHM of ~134 nm. The emission band of Ce^{3+} can be fitted into two subbands peaking at 556 and 618 nm, which roughly correspond to the transitions of the 5d excited states to the two ground-state configurations of $^2F_{7/2}$ and $^2F_{5/2}$ of Ce^{3+}, respectively, with a difference of about 1,800 cm^{-1} (in general ~2,000 cm^{-1}), as shown in Figure 4.48. The Stokes shift of this yellow-emitting phosphor is estimated to be 2,700 cm^{-1}.

The concentration quenching of CaAlSiN$_3$:Ce^{3+} occurs at about 10 mol%. Under the 460 nm excitation, the absorption and external quantum efficiency of CaAlSiN$_3$:Ce^{3+} are 70 and 50%, respectively. The yellow CaAlSiN$_3$:Ce^{3+} phosphor also exhibits a small thermal quenching, and its luminescence at 150°C is quenched by 10% of the initial intensity.

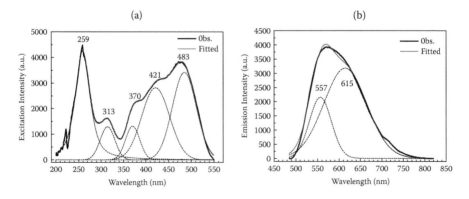

FIGURE 4.48
Excitation (a) and emission (b) spectra of CaAlSiN$_3$:Ce^{3+}. The emission spectrum was measured under 450 nm excitation, and the excitation spectrum was monitored at 570 nm.

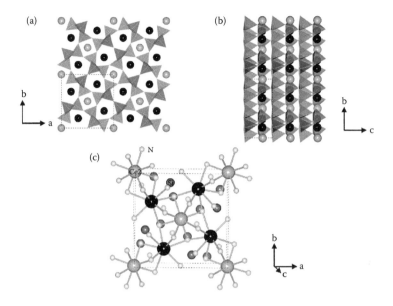

FIGURE 4.49
Crystal structure of La$_3$Si$_6$N$_{11}$ viewed from (a) the ab plane and (b) the bc plane. The black and grey balls are La(1) and La(2) atoms (c).

4.2.3.3 *La$_3$Si$_6$N$_{11}$:Ce^{3+}*

4.2.3.3.1 *Crystal Structure of La$_3$Si$_6$N$_{11}$*

The structure of La$_3$Si$_6$N$_{11}$ is equivalent to that of Ce$_3$Si$_6$N$_{11}$. La$_3$Si$_6$N$_{11}$ crystallizes in a tetrahedral structure with the space group of P4bm, and lattice constants of a = 10.189 Å and c = 4.837 Å (Schlieper and Schnick 1995b; Woile and Jeitschko 1995). As seen in Figure 4.49, La$_3$Si$_6$N$_{11}$ consists of corner-sharing SiN4 tetrahedra, forming highly dense and three-dimensional networks with large voids accommodating La atoms. The SiN4 tetrahedra are arranged in the ab plane. The SiN4 tetrahedra form two types of rings: Si4N4 and Si8N8. There are two kinds of La atoms: La(1) and La(2). La(1) is surrounded by the Si4N4 ring, and coordinated to nine neighboring nitrogen atoms, while La(2) is captured by the Si8N8 ring, and connected to the eight nearest nitrogen atoms. As seen from the ac plane, La atoms are not situated in the middle position of the layer of SiN4 tetrahedra arranged in the direction along the a-axis, but are situated near the position of the upper edge of the SiN4 tetrahedra.

4.2.3.3.2 *Photoluminescence of La$_3$Si$_6$N$_{11}$:Ce^{3+}*

La$_3$Si$_6$N$_{11}$:Ce^{3+} was synthesized by firing the powder mixture of SiO$_2$, La$_2$O$_3$, and CeO$_2$ by using the gas reduction and nitridation method (Suehiro et al. 2011). Seto et al. (2009) prepared the La$_3$Si$_6$N$_{11}$:Ce^{3+} phosphor by using the solid-state reaction method, i.e., firing the powder mixture of Si$_3$N$_4$, LaN, CeO$_2$, and CeN at 1,580–2,000°C under 1.0 Mpa N$_2$. The La/Si ratio has a

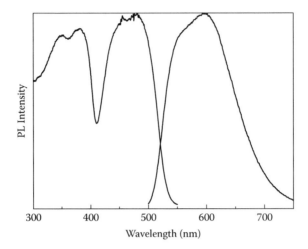

FIGURE 4.50
Excitation and emission spectra of La₃Si₆N₁₁:Ce³⁺. The emission spectrum was recorded under the 450 nm excitation, and the excitation spectrum was monitored at 600 nm. (Reprinted from Suehiro, T., Hirosaki, N., and Xie, R.-J., unpublished. With permission.)

great influence in the phase purity of $La_3Si_6N_{11}$. A phase-pure $La_3Si_6N_{11}$ could be synthesized when the La/Si ratio was 0.62. Below this value, the impurity phase $LaSi_3N_5$ was present.

The photoluminescence spectra of $La_3Si_6N_{11}$:Ce³⁺ are shown in Figure 4.50. The excitation spectrum of $La_3Si_6N_{11}$:Ce³⁺ shows four distinct bands centered at 347, 382, 457, and 480 nm, respectively. The emission spectrum is a broad band that consists of a peak at 595 nm and a shoulder at 552 nm, which is characteristic of the Ce³⁺ emission.

The optimal Ce³⁺ concentration occurs at about x = 0.1 in $La_{3-x}Ce_xSi_6N_{11}$. Moreover, the use of oxygen-less CeN leads to enhanced yellow emission in comparison with the use of CeO_2, indicating that the oxygen contamination suppresses the formation of the phase-pure $La_3Si_6N_{11}$ (Seto et al. 2009).

$La_3Si_6N_{11}$:Ce³⁺ has a smaller thermal quenching than YAG:Ce³⁺, and its emission intensity at 150°C decreases about 28% of the initial intensity. In comparison with yellow-emitting Ca-α-sialon:Eu²⁺ and CaAlSiN₃:Ce³⁺ phosphors, $La_3Si_6N_{11}$:Ce³⁺ exhibits lower thermal stability. The external quantum efficiency of $La_3Si_6N_{11}$:Ce³⁺ is 42% when it is excited at 450 nm.

4.2.4 Red-Emitting Phosphors

4.2.4.1 M₂Si₅N₈:Eu²⁺ (M = Ca, Sr, Ba)

4.2.4.1.1 Crystal Structure of M₂Si₅N₈ (M = Ca, Sr, Ba)

The crystal structure of ternary alkaline earth silicon nitride, M₂Si₅N₈ (M = Ca, Sr, Ba), was investigated by Schlieper et al. (1995a, 1995b). Their crystallographic

TABLE 4.21

Crystallographic Data of $M_2Si_5N_8$ (M = Ca, Sr, Ba)

Formula	$Ca_2Si_5N_8$	$Sr_2Si_5N_8$	$Ba_2Si_5N_8$
Crystal system	Monoclinic	Orthorhombic	Orthorhombic
Space group	Cc (No. 9)	$Pmn2_1$ (No. 31)	$Pmn2_1$ (No. 31)
a(Å)	1.4347(8)	5.712(3)	5.782(3)
b(Å)	5.606(3)	6.817(3)	6.954(3)
c(Å)	9.686(7)	9.336(1)	9.395(8)
α(°)	90	90	90
β(°)	112.03(8)	90	90
γ(°)	90	90	90
V(Å³)	723.03	363.9	378.0
Z	4	2	2
Coordination number of M	7	8,9	8,9

Source: Data from Schlieper, T., and Schnick, W., *Z. Anorg. Allg. Chem.*, 621, 1037–1041, 1995; Schlieper, T., Milius, W., and Schnick, W., *Z. Anorg. Allg. Chem.*, 621, 1380–1384, 1995.

data, bond length, and atomic coordinates are summarized in Tables 4.21–4.24. $Ca_2Si_5N_8$ has a monoclinic crystal system with the space group of Cc, whereas both $Sr_2Si_5N_8$ and $Ba_2Si_5N_8$ have an orthorhombic lattice with the space group of $Pmn2_1$. The Ca atoms are located in the channels along the [010] direction, whereas both the Sr and Ba atoms are introduced in the channels along the [100] direction, formed by a three-dimensional framework of corner-sharing SiN4 tetrahedra (see Figures 4.51 and 4.52). The local coordination in the structures is quite similar for these ternary alkaline earth silicon nitrides, with half of the nitrogen atoms connecting two Si neighbors ($N^{[2]}$) and the other half having three Si neighbors ($N^{[3]}$). There are two different crystallographic sites for alkaline earth metals. Each Ca atom in $Ca_2Si_5N_8$ is coordinated to seven nitrogen atoms, while Sr in $Sr_2Si_5N_8$ and Ba in $Ba_2Si_5N_8$ are coordinated to eight or nine nitrogen atoms. The average bond length between alkaline earth metals and nitrogen is about 2.880 Å.

4.2.4.1.2 Photoluminescence of $M_2Si_5N_8$:Eu^{2+} (M = Ca, Sr, Ba)

The photoluminescence of $Ba_2Si_5N_8$:Eu^{2+} and $M_2Si_5N_8$:Eu^{2+} (M = Ca, Sr, Ba) was investigated by Hoppe et al. (2000) and Y. Q. Li et al. (2006, 2008c). The photoluminescence properties of $M_2Si_5N_8$:Eu^{2+} (M = Ca, Sr, Ba) are presented in Table 4.25.

As seen Figure 4.53, the excitation spectra all show broad bands covering the spectral range of 250–600 nm, indicative of large crystal field splitting of Eu^{2+} energy levels. By using the Gaussian simulation, the excitation spectrum of each phosphor can be fitted into five subbands. Therefore, the crystal field splitting is estimated to be ~22,000 cm^{-1}, and the center of gravity for $M_2Si_5N_8$:Eu^{2+} (M = Ca, Sr, Ba), ~28,000 cm^{-2}.

TABLE 4.22

Bond Length Data for $M_2Si_5N_8$ (M = Ca, Sr, Ba)

Compounds	Bond Lengths of M-N
$Ca_2Si_5N_8$	Ca(1) – N(1): 2.343 Å Ca(2) – N(1): 2.645 Å
	Ca(1) – N(2): 2.315 Å Ca(2) – N(2): 2.471 Å
	Ca(1) – N(5): 2.595 Å Ca(2) – N(2): 3.128 Å
	Ca(1) – N(5): 2.956 Å Ca(2) – N(4): 2.848 Å
	Ca(1) – N(6): 2.627 Å Ca(2) – N(5): 2.398 Å
	Ca(1) – N(7): 2.676 Å Ca(2) – N(7): 2.397 Å
	Ca(1) – N(8): 3.059 Å Ca(2) – N(8): 3.070 Å
$Sr_2Si_5N_8$	Sr(1) – N(1): 2.891 Å × 2 Sr(2) – N(1): 2.542 Å
	Sr(1) – N(2): 2.627 Å × 2 Sr(2) – N(2): 2.720 Å × 2
	Sr(1) – N(3): 3.231 Å × 2 Sr(2) – N(2): 3.181 Å × 2
	Sr(1) – N(4): 2.861 Å Sr(2) – N(3): 2.959 Å × 2
	Sr(1) – N(5): 2.57 Å Sr(2) – N(5): 2.894 Å × 2
$Ba_2Si_5N_8$	Ba(1) – N(1): 2.925 Å × 2 Ba(2) – N(1): 2.677 Å
	Ba(1) – N(2): 2.753 Å × 2 Ba(2) – N(2): 2.823 Å × 2
	Ba(1) – N(3): 3.171 Å × 2 Ba(2) – N(2): 3.160 Å × 2
	Ba(1) – N(4): 2.931 Å Ba(2) – N(3): 3.004 Å × 2
	Ba(1) – N(5): 2.706 Å Ba(2) – N(5): 2.930 Å × 2

Source: Data from Schlieper, T., and Schnick, W., *Z. Anorg. Allg. Chem.*, 621, 1037–1041, 1995a; Schlieper, T., Milius, W., and Schnick, W., *Z. Anorg. Allg. Chem.*, 621, 1380–1384, 1995a.

TABLE 4.23

Atomic Coordinates for $Sr_2Si_5N_8$

Atom	Wyckoff	x	y	z
Sr(1)	2a	0	0.8695	0
Sr(2)	2b	0	0.8816	0.3686
Si(1)	4b	0.2518	0.6669	0.6836
Si(2)	2a	0	0.0549	0.6771
Si(3)	2a	0	0.4196	0.4619
Si(4)	2a	0	0.4014	0.9023
N(1)	2a	0	0.191	0.520
N(2)	4b	0.2478	0.9122	0.6728
N(3)	4b	0.2488	0.4443	0.0105
N(4)	2a	0	0.5871	0.7735
N(5)	2a	0	0.171	0.835
N(6)	2a	0	0.427	0.2722

Source: Data from Schlieper, T., Milius, W., and Schnick, W., *Z. Anorg. Allg. Chem.*, 621, 1380–1384, 1995.

TABLE 4.24

Atomic Coordinates for $Ca_2Si_5N_8$

Atom	x	y	z
Ca(1)	0	0.7637	0
Ca(2)	0.6112	0.7457	0.2000
Si(1)	0.0581	0.8055	0.3539
Si(2)	0.7557	0.2097	0.3182
Si(3)	0.7545	0.4966	0.0631
Si(4)	0.3627	0.2078	0.3681
Si(5)	0.8552	0.0027	0.1264
N(1)	0.9866	0.638	0.4289
N(2)	0.1286	0.009	0.9959
N(3)	0.7959	0.2424	0.1702
N(4)	0.8027	0.7484	0.1752
N(5)	0.9798	0.999	0.2178
N(6)	0.8335	0.0145	0.9349
N(7)	0.6309	0.157	0.2732
N(8)	0.7960	0.4826	0.4161

Source: Data from Schlieper, T., and Schnick, W., Z. *Anorg. Allg. Chem.*, 621, 1037–1041, 1995a.

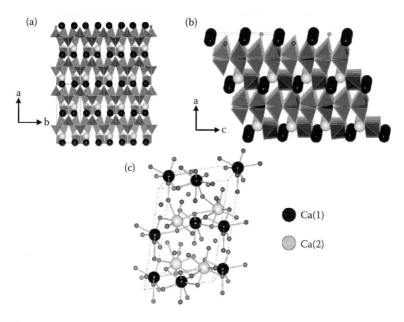

FIGURE 4.51

Crystal structure of $Ca_2Si_5N_8$ projected on (a) the (001) plane and (b) the (010) plane. The local coordination of Ca atoms that are connected to seven nitrogen atoms (c).

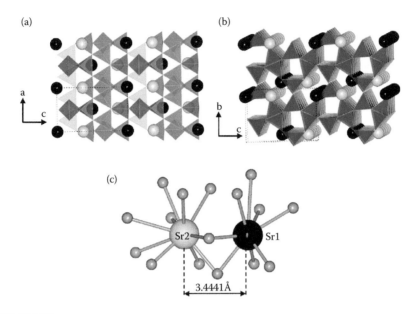

FIGURE 4.52
Crystal structure of $Sr_2Si_5N_8$ projected on (a) the (010) plane and (b) the (100) plane. The local coordination of Sr atoms that are connected to eight or nine nitrogen atoms (c).

TABLE 4.25
Photoluminescence Properties of $M_2Si_5N_8:Eu^{2+}$ (M = Ca, Sr, Ba)

	$Ca_2Si_5N_8$	$Sr_2Si_5N_8$	$Ba_2Si_5N_8$
Excitation peak, nm	238, 303, 372, 438, 511	237,308, 389, 449, 527	239, 318, 393, 441, 512
Emission peak, nm	618, 678	628, 687	638, 694
FWHM, nm	112	101	105
Crystal field splitting, cm^{-1}	22,500	23,000	22,000
Stokes shift, cm^{-1}	3,400	3,050	3,350
Center of gravity, cm^{-1}	28,900	28,300	28,700

The $M_2Si_5N_8:Eu^{2+}$ (M = Ca, Sr, Ba) phosphors show very intense red emission under blue light excitation. The emission color changes from orange to deep red when the ionic size of alkaline earth metals is increased. The emission spectra all exhibit a broad and asymmetrical band extending from 550 to 850 nm (see Figure 4.54). As the alkaline earth metal atoms occupy two different crystallographic sites, two emission bands are thus expected. In fact, the emission bands can be further decomposed into two Gaussian components, as seen in Figure 4.54. The Stokes shift is then roughly calculated to be 30,500–33,500 cm^{-1} for these red phosphors.

Upon the 450 nm excitation, $Sr_{1.96}Eu_{0.04}Si_5N_8$ shows an absorption and external quantum efficiency of 87 and 71%, respectively (see Figure 4.55).

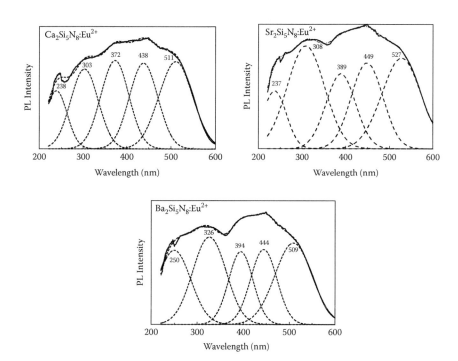

FIGURE 4.53
Excitation spectra of $M_2Si_5N_8$:Eu^{2+} (M = Ca, Sr, Ba). (Reprinted from Xie, R.-J., Hirosaki, N., Suehiro, T., Xu, F.-F., and Mitomo, M., *Chem. Mater.*, 18, 5578–5583, 2006c. With permission.)

In addition, the quantum efficiency is also high when the phosphor is excited by UV or near-UV light. The thermal quenching becomes larger upon increasing the ionic size of alkaline earth metals, with $Sr_2Si_5N_8$:Eu^{2+} having the smallest thermal quenching.

4.2.4.2 CaAlSiN₃:Eu²⁺

The red-emitting $CaAlSiN_3$:Eu^{2+} phosphor was reported by Uheda et al. (2006a, 2006b). Figure 4.56 presents the photoluminescence spectra of $CaAlSiN_3$:Eu^{2+}. This phosphor was prepared by firing the powder mixture of Si_3N_4, Ca_3N_2, and EuN at 1,600°C under the 0.5 Mpa nitrogen atmosphere. It is seen that the excitation spectrum covers a broad spectral range of 250 to 600 nm, showing an extremely broad band and strong absorption of visible light. This broad band can be decomposed into five Gaussian components at 258, 332, 426, 505, and 570 nm, respectively. The crystal field splitting and the center of gravity are estimated to be 21,000 and 26,000 cm^{-1}, respectively. The emission spectrum extends from 550 to 800 nm, showing a broad band centered at 655 nm with a FWHM value of 93 nm. The emission band can be fitted into two subbands centered at 651 and 697 nm by using the Gaussian simulation.

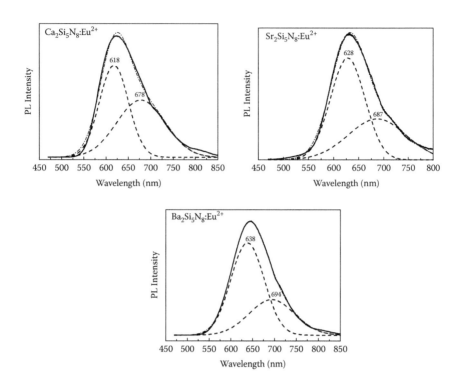

FIGURE 4.54
Emission spectra of $M_2Si_5N_8:Eu^{2+}$ (M = Ca, Sr, Ba). (Reprinted from Xie, R.-J., Hirosaki, N., Suehiro, T., Xu, F.-F., and Mitomo, M., *Chem. Mater.*, 18, 5578–5583, 2006c. With permission.)

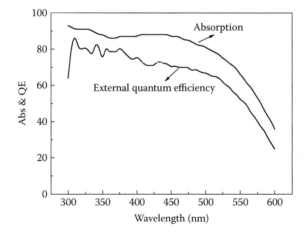

FIGURE 4.55
Absorption and quantum efficiency of $Sr_{1.96}Eu_{0.04}Si_5N_8$ at varying excitation wavelengths. (Reprinted from Xie, R.-J., Hirosaki, N., Suehiro, T., Xu, F.-F., and Mitomo, M., *Chem. Mater.*, 18, 5578–5583, 2006c. With permission.)

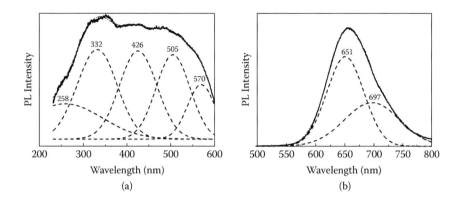

FIGURE 4.56
Excitation (a) and emission (b) spectra of $Ca_{0.99}Eu_{0.01}AlSiN_3$. The emission spectrum was measured under 450 nm excitation, and the excitation spectrum was monitored at 655 nm.

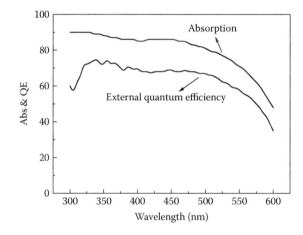

FIGURE 4.57
Absorption and quantum efficiency of $Ca_{0.99}Eu_{0.01}AlSiN_3$ measured upon excitation at varying wavelengths.

The Stokes shift is then calculated to be 2,200 cm^{-1}. The small Stokes shift is indicative of high conversion efficiency and small thermal quenching.

The concentration quenching occurs at about 1.6 mol% (Uheda et al. 2006b). As seen in Figure 4.57, the red-emitting $Ca_{0.99}Eu_{0.01}AlSiN_3$ strongly absorbs UV-to-visible light, and exhibits high quantum efficiency. When the phosphor is excited by the blue light (λ = 450 nm), the absorption is 86% and the external quantum efficiency is up to 70%. In addition, it has a small thermal quenching. The luminescence is quenched by 10% of the initial intensity when the phosphor is heated to 150°C and excited at 450 nm (see Figure 4.58). These excellent photoluminescence properties and small thermal quenching enable $CaAlSiN_3:Eu^{2+}$ to be a very attractive red phosphor for white LED applications.

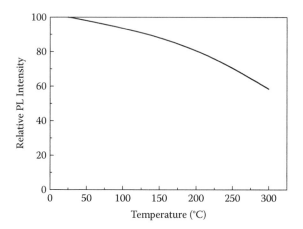

FIGURE 4.58
Thermal quenching of $Ca_{0.99}Eu_{0.01}AlSiN_3$ under 450 nm excitation.

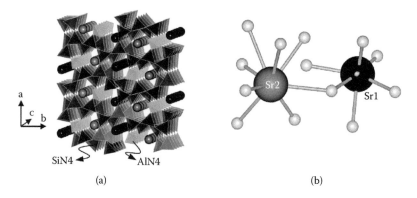

FIGURE 4.59
(a) Crystal structure of $SrAlSi_4N_7$ projected on the (001) plane, and (b) the coordination of Sr by N atoms.

4.2.4.3 $SrAlSi_4N_7{:}Eu^{2+}$

4.2.4.3.1 Crystal Structure of $SrAlSi_4N_7$

The crystal structure and photoluminescence of $SrAlSi_4N_7$ were reported by Hecht et al. (2009). The crystallographic parameters and atomic coordinates are summarized in Tables 4.26 and 4.27.

$SrAlSi_4N_7$ crystallizes in the orthorhombic structure with the space group of $Pna2_1$ (No. 33). It contains a highly condensed network structure that is built up of SiN4 and AlN4 tetrahedra. In $SrAlSi_4N_7$, the SiN4 tetrahedra are connected via common corners, which are usually observed in nitrido-silicates compounds. On the other hand, the structure of $SrAlSi_4N_7$ also contains edge-sharing AlN4 tetrahedra, forming infinite chains along the c-axis

TABLE 4.26

Crystallographic Data of $SrAlSi_4N_7$

Formula	$SrAlSi_4N_7$
Crystal system	Orthorhombic
Space group	$Pna2_1$ (No. 33)
a(Å)	11.742(2)
b(Å)	21.391(4)
c(Å)	4.966(1)
V(Å^3)	1247.2(49
Z	8

Bond lengths (Å) of Sr-N

Sr(1) – N(1): 2.504(5)	Sr(2) – N(12): 2.653(7)
Sr(1) – N(13): 2.572(8)	Sr(2) – N(8): 2.708(4)
Sr(1) – N(10): 2.634(5)	Sr(2) – N(7): 2.716(5)
Sr(1) – N(13): 2.700(8)	Sr(2) – N(11): 2.821(5)
Sr(1) – N(3): 2.723(5)	Sr(2) – N(14): 2.980(6)
Sr(1) – N(12): 3.143(5)	Sr(2) – N(12): 3.011(7)
	Sr(2) – N(5): 3.024(6)
	Sr(2) – N(6): 3.057(6)

Source: Data from Hecht, C., Stadler, F., Schmidt, P. J., auf der Gunne, S. J., Baumann, V., and Schnick, W., *Chem. Mater.*, 21, 1595–1601, 2009.

(see Figure 4.59). These trans-linked chains are connected to SiN4 tetrahedral networks through common corners. The Sr atoms are hosted in the channels along [001], which are formed by the corner-sharing SiN4 and AlN4 tetrahedra. There are two different crystallographic positions for Sr, with Sr1 coordinated by irregular polyhedra made up of six nitrogen atoms (coordination number 6) and Sr2 by eight nitrogen atoms (see Figure 4.59). The Sr-N distances range from 2.504 to 3.143 Å for Sr1 and from 2.653 to 3.057 Å for Sr2, as seen in Table 4.26.

4.2.4.3.2 Photoluminescence

Hecht et al. (2009) prepared the Eu^{2+}-$SrAlSi_4N_7$ by firing the powder mixture of Si_3N_4, AlN, Sr, and EuF_3 in a radiofrequency furnace at 1,630°C. Although the $SrAlSi_4N_7$ phase was present as the major phase in the resultant sample, $Sr_2Si_5N_8$ was also formed as a by-product. To suppress the formation of $Sr_2Si_5N_8$, Ruan et al. (2010) added more AlN into the composition, and obtained the phase-pure $SrAlSi_4N_7$ by using the gas-pressure sintering furnace.

Figure 4.60 presents the photoluminescence spectra of $SrAlSi_4N_7$:Eu^{2+}. The excitation spectrum shows a broad band covering the spectral range of 250–600 nm, indicative of the strong absorption of visible light. The excitation band can be fitted into five Gaussian components with maxima at 299, 352, 407, 466, and 524 nm, respectively. The crystal field splitting and the center of gravity are thus estimated to be 14,400 and 25,400 cm^{-1}, respectively.

TABLE 4.27

Atomic Position of $SrAlSi_4N_7$

Atom	x	y	z
Sr(1)	0.27919	0.50892	0.43863
Sr(2)	0.64517	0.69726	0.4560
Si(1)	0.35835	0.72290	−0.0308
Si(2)	0.92208	0.70013	0.4655
Si(3)	0.55782	0.53286	0.4579
Si(4)	0.00385	0.45635	0.4653
Si(5)	0.19254	0.36055	0.4514
Si(6)	0.16242	0.43487	−0.0497
Si(7)	0.47151	0.39055	0.4579
Si(8)	0.39464	0.65244	0.4686
Al(1)	0.1551	0.65081	0.7351
Al(2)	0.6557	0.84843	0.2295
N(1)	0.2529	0.6248	0.4745
N(2)	0.2150	0.7015	−0.0142
N(3)	0.4894	0.4658	0.3003
N(4)	−0.1004	0.3996	0.4937
N(5)	0.4406	0.6656	0.8000
N(6)	0.8631	0.7066	0.7903
N(7)	0.8421	0.6420	0.2959
N(8)	0.5666	0.8162	0.4782
N(9)	0.0541	0.4804	0.7828
N(10)	0.8761	0.5736	0.7971
N(11)	0.4928	0.6014	0.3060
N(12)	0.3349	0.3653	0.4148
N(13)	0.2964	0.4687	−0.0750
N(14)	0.4063	0.7273	0.3000

Source: Data from Hecht, C., Stadler, F., Schmidt, P. J., auf der Gunne, S. J., Baumann, V., and Schnick, W., *Chem. Mater.*, 21, 1595–1601, 2009.

The emission spectrum displays a broad band centered at 539 nm, having a FWHM value of 116 nm. Since Sr atoms occupy two different crystallographic sites with different coordination numbers, two emission bands are expected. As seen in Figure 4.60b, the emission band can be fitted into two Gaussian subbands centered at 629 and 690 nm, respectively. The Stokes shift is then calculated to be 3,700 cm^{-1} for $SrAlSi_4N_7:Eu^{2+}$.

The absorption and quantum efficiency of $(Sr_{0.95}Eu_{0.05})AlSi_4N_7$ were measured at varying excitation wavelengths. It is seen in Figure 4.61, when the sample is excited at 450 nm, that the absorption and external quantum efficiency are 76 and 53%, respectively. In addition, this red-emitting phosphor shows high absorption and quantum efficiency under the UV to near-UV irradiation, indicating that (Sr, Eu)$AlSi_4N_7$ is an attractive red phosphor candidate for white LEDs when coupled to blue, near-UV, or UV LEDs.

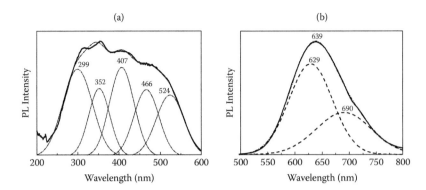

FIGURE 4.60
Excitation (a) and emission (b) of $(Sr_{0.95}Eu_{0.05})AlSi_4N_7$. The emission spectrum was measured under 450 nm excitation, and the excitation spectrum was monitored at 639 nm. (Reprinted from Ruan, J., Xie, R.-J., Hirosaki, N., and Takeda, T., *J. Am. Ceram. Soc.*, DOI: 10.1111/j.1551-2916.2010.04104.x, 2010. With permission.)

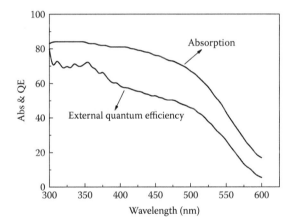

FIGURE 4.61
Absorption and quantum efficiency of $(Sr_{0.95}Eu_{0.05})AlSi_4N_7$. (Reprinted from Ruan, J., Xie, R.-J., Hirosaki, N., and Takeda, T., *J. Am. Ceram. Soc.*, DOI: 10.1111/j.1551–2916.2010.04104.x, 2010. With permission.)

4.2.4.4 $MSiN_2:Eu^{2+}/Ce^{3+}$

4.2.4.4.1 Crystal Structure

The structure of binary alkaline earth nitridosilicates, $MSiN_2$ (M = Ca, Sr, Ba), was investigated by Gal et al. (2004). The crystallographic data for $MSiN_2$ are summarized in Table 4.28, and the atomic coordinates and selected M – N (M = Ca, Sr, Ba) bond lengths are presented in Tables 4.29–4.31.

As seen in Figure 4.62, $CaSiN_2$ contains highly condense SiN4 tetrahedra that are joined only at their vertexes. The Ca atoms are hosted in the channels along [100] that are formed by the tetrahedral network. There

TABLE 4.28

Crystallographic Data of $MSiN_2$ (M = Ca, Sr, Ba)

Formula	$CaSiN_2$	$SrSiN_2$	$BaSiN_2$
Crystal system	Orthorhombic	Monoclinic	Orthorhombic
Space group	Pbca (No. 61)	$P2_1/c$ (No. 14)	Cmca (No.64)
a(Å)	5.1229(3)	5.9814(7)	5.6043(5)
b(Å)	10.2074(6)	7.3212(7)	11.3655(1)
c(Å)	14.8233(9)	5.5043(6)	7.5997(7)
$\alpha(°)$	90	90	90
$\beta(°)$	90	113.496	90
$\gamma(°)$	90	90	90
V(Å³)	775.13(8)	221.014(5)	484.06(1)
Z	16	4	8
Coordination number	6	8	8

Source: Data from Gal, Z. A., Mallinson, P. M., Orchard, H. J., and Clarke, S. J., *Inorg. Chem.*, 43, 3998–4006, 2004.

TABLE 4.29

Atomic Coordinate and Selected Bond Lengths for $CaSiN_2$

Atom	Wyckoff	x	y	z	SOF
Ca(1)	8c	0.2414	0.0097	0.0650	1.00
Ca(2)	8c	0.3462	0.2741	0.1876	1.00
Si(1)	8c	0.2265	0.0166	0.3121	1.00
Si(2)	8c	0.3172	0.2700	0.4372	1.00
N(1)	8c	0.0949	0.4774	0.2155	1.00
N(2)	8c	0.1495	0.2761	0.0354	1.00
N(3)	8c	0.1972	0.4268	0.4090	1.00
N(4)	8c	0.2762	0.1798	0.3382	1.00

Source: Data from Gal, Z. A., Mallinson, P. M., Orchard, H. J., and Clarke, S. J., *Inorg. Chem.*, 43, 3998–4006, 2004.

Note: Bond lengths: Ca(1) – N(1): 2.406 Å, Ca(1) – N(2): 2.794 Å, Ca(1) – N(2): 2.488 Å, Ca(1) – N(3): 3.022 Å, Ca(1) – N(3): 2.431 Å, Ca(1) – N(3): 2.412 Å, Ca(2) – N(1): 2.827 Å, Ca(2) – N(1): 2.447 Å, Ca(2) – N(2): 2.470 Å, Ca(2) – N(3): 2.778 Å, Ca(2) – N(4): 2.434 Å, Ca(2) – N(4): 2.458 Å.

are two different crystallographic positions for Ca atoms. The Ca1 atom is surrounded by four N atoms located between 2.40 and 2.49 Å and by two further N atoms with distances of 2.79 and 3.02 Å. The Ca2 atom is in a highly distorted octahedral environment, again connected to four N atoms with shorter distances (between 2.43 and 2.48 Å) and to two N atoms with longer distances (between 2.79 and 2.83 Å). Since all SiN4 tetrahedra are corner shared, all the N atoms are coordinated to two Si atoms.

Le Toquin and Cheetham (2006) reported a new structure type of $CaSiN_2$ that crystallizes in a *face-centered cubic* unit cell with a lattice parameter of a = 14.8822(5) Å. The difference in structure will lead to different photoluminescence, as will be discussed later.

TABLE 4.30

Atomic Coordinate and Selected Bond Lengths for $SrSiN_2$

Atom	Wyckoff	x	y	z	SOF
Sr	4e	0.3388	0.5738	0.1755	1.00
Si	4e	0.1049	0.1419	0.0682	1.00
N(1)	4e	0.2083	0.5877	0.5841	1.00
N(2)	4e	0.2154	0.2244	0.3850	1.00

Source: Data from Gal, Z. A., Mallinson, P. M., Orchard, H. J., and Clarke, S. J., *Inorg. Chem.*, 43, 3998–4006, 2004.

Note: Bond lengths: Sr – N(1): 2.564 Å, Sr – N(1): 2.658 Å, Sr – N(1): 2.753 Å, Sr – N(1): 3.027 Å, Sr – N(2): 2.618 Å, Sr – N(2): 2.980 Å, Sr – N(2): 3.001 Å, Sr – N(2): 3.145 Å.

TABLE 4.31

Atomic Coordinate and Selected Bond Lengths for $BaSiN_2$

Atom	Wyckoff	x	y	z	SOF
Ba	8f	0.3358	0.0651	0.49	1.00
Si	8f	0.0494	0.1400	0.29	1.00
N(1)	8f	0.3981	0.4200	0.60	1.00
N(2)	8f	0.0972	0.25	0.79	1.00

Source: Data from Gal, Z. A., Mallinson, P. M., Orchard, H. J., and Clarke, S. J., *Inorg. Chem.*, 43, 3998–4006, 2004.

Note: Bond lengths: Ba – N(1): 2.8923 Å × 2, Ba – N(1): 2.8805 Å, Ba – N(1): 2.7851 Å, Ba – N(2): 2.8800 Å × 2, Ba – N(2): 3.3615 Å × 2.

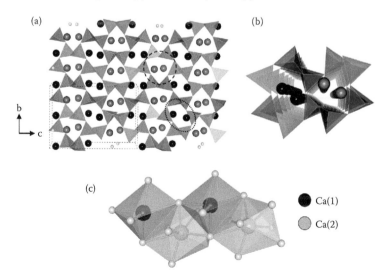

FIGURE 4.62

(a) Crystal structure of orthorhombic $CaSiN_2$ projected on the (100) plane. (b) Ca atoms are hosted in channels along [100]. (c) The coordination spheres for Ca.

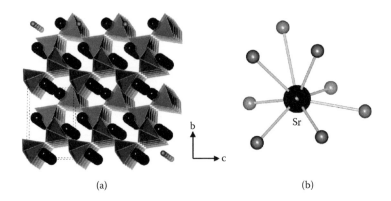

(a)   (b)

FIGURE 4.63
(a) Crystal structure of SrSiN$_2$ projected on the (100) plane. (b) The coordination of Sr by eight N atoms.

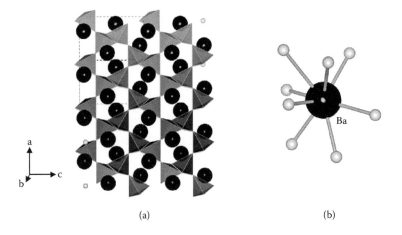

(a)   (b)

FIGURE 4.64
(a) Crystal structure of BaSiN$_2$ showing both edge- and corner-sharing SiN4 tetrahedra. (b) The coordination of Ba by eight N atoms.

The crystal structures of SrSiN$_2$ and BaSiN$_2$ are illustrated in Figures 4.63 and 4.64. The structure of SrSiN$_2$ is closely related to that of BaSiN$_2$, containing both edge-sharing and corner-sharing SiN4 tetrahedra. Pairs of SiN4 tetrahedra share edges to form "bow tie" units of Si2N6, and these units condense by vertex sharing of the four remaining unshared vertexes to form puckered two-dimensional sheets separated by Sr or Ba atoms. There is only one crystallographic site for Sr and Ba atoms, which are both connected to eight N atoms.

4.2.4.4.2 Photoluminescence

The photoluminescence of the Ce^{3+}-doped CaSiN$_2$ phosphor with the cubic structure was reported by Le Toquin and Cheetham (2006), and that of both

TABLE 4.32

Photoluminescence Properties of $MSiN_2:Eu^{2+}$ (M = Sr, Ba)

Properties	$SrSiN_2:Eu^{2+}$	$BaSiN_2:Eu^{2+}$
Absorption bands, nm	300–530	300–530
Excitation bands, nm	306, 336, 395, 466	312, 334, 395, 464
Emission bands, nm	670–685	600–630
Crystal field splitting, cm^{-1}	11,200	10,500
Stokes shift, cm^{-1}	6,500–6,900	4,850–5,550

Source: Data from Duan, C. J., Wang, X. J., Otten, W. M., Delsing, A. C. A., Zhao, J. T., and Hintzen, H. T., *Chem. Mater.*, 20, 1597–1605, 2008.

TABLE 4.33

Photoluminescence Properties of $MSiN_2:Ce^{3+}$ (M = Ca, Sr, Ba)

Properties	$CaSiN_2:Ce^{3+}$ (cubic)	$SrSiN_2:Ce^{3+}$	$BaSiN_2:Ce^{3+}$
Absorption bands, nm	450–550	370–420	370–420
Excitation bands, nm	365, 390, 440, 535	298, 330, 399	305, 403
Emission band, nm	625	535	485
Crystal field splitting, cm^{-1}	8,700	8,500	8,000
Stokes shift, cm^{-1}	2,700	6,350	4,300

Source: Data from Le Toquin, R., and Cheetham, A. K., *Chem. Phys. Lett.*, 423, 352–356, 2006; Duan, C. J., Wang, X. J., Otten, W. M., Delsing, A. C. A., Zhao, J. T., and Hintzen, H. T., *Chem. Mater.*, 20, 1597–1605, 2008; Li, Y. Q., Hirosaki, N., and Xie, X.-J., Structural and photoluminescence properties of $CaSiN_2:Ce^{3+}$, Li^+, paper presented at the Japan Society of Applied Physics, the 67th autumn meeting, Sapporo, Japan, 2007.

Eu^{2+}- and Ce^{3+}-doped $MsiN_2$ (M = Sr, Ba) was further investigated by Duan et al. (2008) and Y. Q. Li et al. (2007, 2009). The photoluminescence properties of $MsiN_2:Ce^{3+}$ and $MSiN_2:Eu^{2+}$ are summarized in Tables 4.32 and 4.33.

Both $SrSiN_2:Eu^{2+}$ and $BaSiN_2:Eu^{2+}$ show red emission under the blue light irradiation. The 5d energy levels were split into three dominant bands, resulting in a broad excitation spectrum for both samples. The emission spectrum shows a single broad band centered at 670–685 nm ($SrSiN_2:Eu^{2+}$) and 600–630 nm ($BaSiN_2:Eu^{2+}$), depending on the Eu concentration. The FWHM of $BaSiN_2:Eu^{2+}$ is smaller than that of $SrSiN_2:Eu^{2+}$, due to the higher symmetry of Ba atoms (see Tables 4.30 and 4.31). Duan et al. (2008) reported that quantum efficiency of $BaSiN_2:Eu^{2+}$ and $SrSiN_2:Eu^{2+}$ is 40 and 25% with respect to that of $YAG:Ce^{3+}$.

The Ce^{3+} luminescence is quite different for $MSiN_2$ with different alkaline earth metals. The cubic $CaSiN_2:Ce^{3+}$ emits red color; $SrSiN_2:Ce^{3+}$, green; and $BaSiN_2:Ce^{3+}$, blue. This difference can be related to the strength of the crystal field acting on the alkaline earth metal ions in the structure. Among these materials, $CaSiN_2:Ce^{3+}$ exhibits the largest crystal field splitting, leading to a

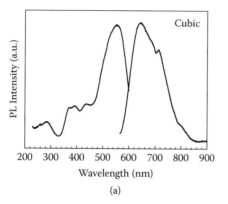
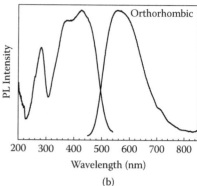

FIGURE 4.65
Excitation and emission spectra of (a) cubic and (b) orthorhombic $CaSiN_2:Ce^{3+}$. (Reprinted from Li, Y. Q., Hirosaki, N., and Xie, X.-J., paper presented at the Japan Society of Applied Physics, the 67th Autumn Meeting, Sapporo, Japan. With permission.)

broad excitation band. Furthermore, $CaSiN_2:Ce^{3+}$ can be excited either by UV to near-UV light, or by blue light, whereas both $SrSiN_2:Ce^{3+}$ and $BaSiN_2:Ce^{3+}$ can only be excited by near-UV light.

Figure 4.65 shows the photoluminescence of $CaSiN_2:Ce^{3+}$ with a different crystal structure. It is seen that the orthorhombic $CaSiN_2:Ce^{3+}$ emits a yellow color under the blue light irradiation, which is quite different from the cubic phase (Y. Q. Li et al. 2007).

4.3 Tunable Luminescence of Nitride Phosphors

As mentioned previously, the optical properties of white LEDs are strongly related to the photoluminescence properties of phosphors. White LEDs with desired color temperature and color point, as well as high luminous efficiency and long lifetime, can be achieved by selecting appropriate phosphors. To meet requirements for varying applications of white LEDs, one way is to develop new luminescent materials with suitable photoluminescence properties. Alternatively, another effective choice is to modify the photoluminescence properties of currently used phosphors. A notable example is Ce^{3+}-doped YAG: the emission band can be blue- or red-shifted by partial replacement of Al with Ga or of Y with Gd.

For traditional oxide or sulfide phosphors, cation substitution is the only choice to tune the emission color. On the other hand, there are two options for nitride phosphors to shift their emission bands: one is cation substitution, and the other is anion substitution (i.e., replacing nitrogen by oxygen).

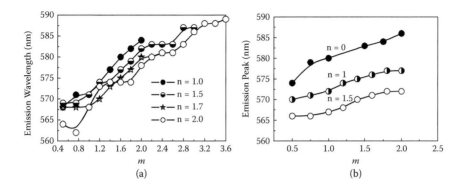

FIGURE 4.66
Variations of the emission maximum of (a) Ca-α-sialon:Eu²⁺ and (b) Li-α-sialon:Eu²⁺ with changes in m and n. The Eu²⁺ concentration is about 7.5 mol% with respect to Ca. (Reprinted from Xie, R. -J., Hirosaki, N., Mitomo, M., Sakuma, K., and Kimura, N., *Appl. Phys. Lett.*, 89, 241103, 2006a. With permission.)

4.3.1 Tuning of Luminescence by Varying the O/N Ratio

As we discussed in previous sections, α-sialon ($M_xSi_{12-m-n}Al_{m+n}O_nN_{16-n}$, x = m/v) is a solid solution of α-Si₃N₄, formed by substituting Si-N bonds with Al-O and Al-N bonds simultaneously. Therefore, it is able to modify the emission color of α-sialon:Eu²⁺ by compositional tailoring or varying the O/N ratio.

Figure 4.66 shows the changes in the emission maximum of α-sialon:Eu²⁺ with m and n values. As seen, the emission maximum red-shifts with increasing m, but it blue-shifts with increasing n. The red-shift can be explained by the enhanced Stokes shift, which is due to the decreased lattice stiffness and the increased absolute Eu²⁺ concentration in the lattice with increasing m. On the other hand, the blue-shift is due to the reduced covalence of the chemical bonding between M and (O,N) as the n value increases.

Figure 4.67 presents the emission bands for Ca-α-sialon:Eu²⁺ with varying compositions. Obviously, the emission maximum can be tuned between 569 and 603 nm. This indicates that the emission color of Ca-α-sialon:Eu²⁺ changes from yellow-green to orange if the m and n values (composition) are carefully controlled.

Since the bond length of Al-O is very close to that of Si-N, nitridosilicates can be changed to oxoaluminonitridosilicates (also known as sialon) by the partial substitution of Si-N by Al-O until the structure of nitridosilicates collapses. Therefore, the solubility of Al-O in the lattice of nitridosilicates is limited. Usually, the substitution increases the O/N ratio, and decreases the covalence of the chemical bonding between the dopant ions and anions, leading to the blue-shift of the emission band. On the other hand, the Al-O bonds in aluminates can be partially replaced by the Si-N bonds, which yield nitride-aluminates. This substitution increases the nephelauxetic effect due to

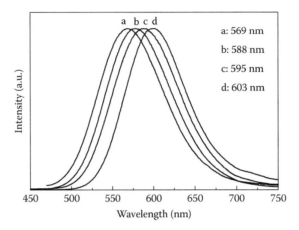

FIGURE 4.67
Emission spectra of Ca-α-sialon:Eu^{2+} phosphors with various m and n values. a, $(Ca_{0.925}Eu_{0.075})_{0.25}$ $Si_{11.25}Al_{0.75}O_{0.25}N_{15.75}$ (m = 0.5, n = 0.25); b, $(Ca_{0.925}Eu_{0.075})Si_9Al_3ON_{15}$ (m = 2, n = 1); c, $(Ca_{0.925}Eu_{0.075})$ $Si_{10}Al_2N_{16}$ (m = 2, n = 0). d, $(Ca_{0.925}Eu_{0.075})_2Si_8Al_4N_{16}$ (m = 4, n = 0).

the coordination of cations with nitrogen atoms, thus resulting in a red-shift of the emission band. Setlur et al. (2008) reported that the energy levels of Ce^{3+} were lowered with the incorporation of Si-N into the garnet lattice, which makes nitrido-YAG:Ce^{3+} possible for creating warm white LEDs.

4.3.2 Tuning of Luminescence by Cation Substitution

4.3.2.1 α-sialon:Eu^{2+}

The emission band of phosphors is usually shifted to the long-wavelength side when the dopant concentration increases. As seen in Figure 4.68, the emission maximum of Li-α-sialon:Eu^{2+} is red-shifted for all compositions by increasing the Eu^{2+} concentration. This red-shift is explained by the enhanced Stokes shifts and reabsorption as the Eu^{2+} concentration increases.

The α-sialon structure can host various kinds of metal atoms, such as Li, Mg, Ca, Y, and some lanthanides. This provides great possibilities to vary the emission band of α-sialon:Eu^{2+} by cation substitution. Figure 4.69 shows the variation of the emission spectra of (Ca,Li)-α-sialon:Eu^{2+}. With the substitution for Ca by Li, the emission band of Ca-α-sialon:Eu^{2+} is blue-shifted. The emission maximum varies from 588 nm for Ca-α-sialon to 577 nm for Li-α-sialon. The changes in chromaticity coordinates of α-sialon:Eu^{2+} with the substitution of Ca by Li are shown in Figure 4.70. Sakuma et al. (2007) reported that the replacement of Ca by Y led to the increase of the emission maximum. Therefore, with the cation substitution, the photoluminescence of α-sialon:Eu^{2+} can be tuned in a wide range, enabling it to produce white LEDs with varying color temperatures (Xie et al. 2006b).

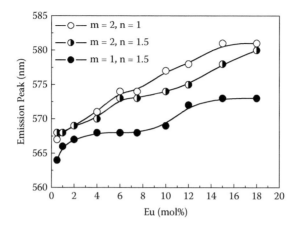

FIGURE 4.68
Variations of the emission maximum of Li-α-sialon:Eu²⁺, with varying Eu²⁺ concentrations. (Reprinted from Xie, R.-J., Hirosaki, N., Mitomo, M., Sakuma, K., and Kimura, N., *Appl. Phys. Lett.*, 89, 241103, 2006a. With permission.)

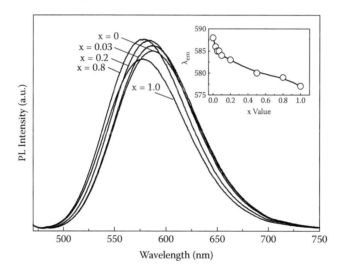

FIGURE 4.69
Emission spectra of $(Ca_{1-2x}Li_x)_{0.975}Eu_{0.075}Si_9Al_3ON_{15}$. (Reprinted from Xie, R.-J., Hirosaki, N., Mitomo, M., Sakuma, K., and Kimura, N., *Appl. Phys. Lett.*, 89, 241103, 2006a. With permission.)

4.3.2.2 $SrSi_2O_2N_2$:Eu^{2+}

The emission color of $SrSi_2O_2N_2$:Eu^{2+} can be tuned by the partial substitution for Sr by Ca or Ba (Bachmann et al. 2009). Although $SrSi_2O_2N_2$ has a chemical formula similar to those of $CaSi_2O_2N_2$ and $BaSi_2O_2N_2$, it has a different crystal structure (crystal system and space group), so that limited solid solutions can be formed among these materials. Table 4.34

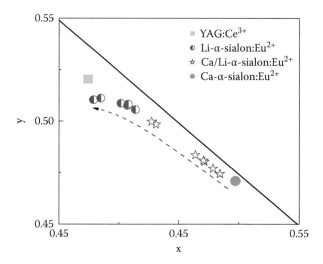

FIGURE 4.70
(See color insert.) Chromaticity coordinates of Ca-, Li-, and (Ca,Li)-α-sialon:Eu²⁺.

TABLE 4.34

Photoluminescence Properties of Ca- or Ba-Substituted
$SrSi_2O_2N_2$:Eu²⁺ Phosphors

Composition	(x,y)	QE (%)	λ_{em} (nm)	$T_{50\%}$ (K)
$Sr_{0.98}Eu_{0.02}Si_2O_2N_2$	(0.337, 0.619)	91	538	600
$Sr_{0.93}Ca_{0.05}Eu_{0.02}Si_2O_2N_2$	(0.342, 0.616)	89	538	
$Sr_{0.88}Ca_{0.10}Eu_{0.02}Si_2O_2N_2$	(0.345, 0.614)	93	538	
$Sr_{0.83}Ca_{0.15}Eu_{0.02}Si_2O_2N_2$	(0.349, 0.611)	93	536	
$Sr_{0.73}Ca_{0.25}Eu_{0.02}Si_2O_2N_2$	(0.351, 0.608)	93	542	557
$Sr_{0.48}Ca_{0.50}Eu_{0.02}Si_2O_2N_2$	(0.354, 0.601)	92	543	500
$Sr_{0.23}Ca_{0.75}Eu_{0.02}Si_2O_2N_2$	(0.378, 0.581)	81	546	
$Ca_{0.98}Eu_{0.02}Si_2O_2N_2$	(0.419, 0.556)	76	555	440
$Sr_{0.93}Ba_{0.05}Eu_{0.02}Si_2O_2N_2$	(0.343, 0.614)	91	538	
$Sr_{0.88}Ba_{0.10}Eu_{0.02}Si_2O_2N_2$	(0.351, 0.610)	90	539	
$Sr_{0.83}Ba_{0.15}Eu_{0.02}Si_2O_2N_2$	(0.358, 0.606)	89	544	
$Sr_{0.73}Ba_{0.25}Eu_{0.02}Si_2O_2N_2$	(0.361, 0.602)	79	544	560
$Sr_{0.48}Ba_{0.50}Eu_{0.02}Si_2O_2N_2$	(0.401, 0.573)	91	548	560
$Sr_{0.23}Ba_{0.75}Eu_{0.02}Si_2O_2N_2$	(0.434, 0.528)	66	564	
$Ba_{0.98}Eu_{0.02}Si_2O_2N_2$	(0.076, 0.440)	71	495	600

Source: Data from Bachmann, V., Ronda, C., Oeckler, O., Schnick, W., and
Meijerink, A., *Chem. Mater.*, 21, 316–325, 2009.

summarizes the photoluminescence properties (color point, quantum efficiency, emission maximum, and thermal quenching temperature) of Ca- or Ba-substituted $SrSi_2O_2N_2:Eu^{2+}$.

As seen, partial substitution of Sr by Ca or Ba always leads to a red-shifted Eu^{2+} emission. It is anticipated that in Ca-substituted samples, and the red-shift is explained by an increased crystal field splitting with the introduction of smaller Ca^{2+} ions into the $SrSi_2O_2N_2$ structure that will reduce the lattice volume, and then shorten the bonding distance between Eu and (O,N) anions. The red-shift of the Eu^{2+} emission in Ba-substituted samples is surprising because the ionic size of Ba^{2+} is larger than that of Sr^{2+}, and $BaSi_2O_2N_2:Eu^{2+}$ emits a blue-green color. Bachmann et al. (2009) addressed that this unusual behavior was also ascribed to an increased crystal field splitting. To preserve the crystal structure of $SrSi_2O_2N_2$ with the Ba substitution, the bonding distances between Eu and (O,N) anions will be unchanged or even reduced when larger Ba atoms are hosted in the lattice.

The thermal quenching temperature decreases with the Ca substitution, but increases with the Ba substitution. This indicates that the $(Sr,Ba)Si_2O_2N_2:Eu^{2+}$ phosphors are very attractive conversion materials in white LEDs.

4.3.2.3 *$CaAlSiN_3:Eu^{2+}$*

$CaAlSiN_3:Eu^{2+}$ is a deep red phosphor. It is required to blue-shift its emission band in order to improve the luminous efficiency of white LEDs by using it. Watanabe et al. (2008b) investigated the crystal structure and photoluminescence of $CaAlSiN_3$-$SrAlSiN_3$ solid solutions. They addressed that the structures of $CaAlSiN_3$ and $SrAlSiN_3$ are closely related, and complete solid solutions could be formed between the end members. The substitution of Ca by Sr leads to a blue-shift of the Eu^{2+} emission, the emission maximum shifting from 650 nm for $Ca_{0.99}Eu_{0.01}AlSiN_3$ to 625 nm for $Sr_{0.99}Eu_{0.01}AlSiN_3$. The blue-shift is explained by the decreasing crystal field splitting, which is caused by the lattice expansion when large Sr^{2+} ions are accommodated in the $CaAlSiN_3$ structure.

4.4 Synthesis of Nitride Phosphors

Oxosilicates are usually obtained by solid-state chemical reactions of silica (SiO_2) with metal oxides or carbonates at high temperatures. This route is also applicable to nitridosilicates, and is usually called a *solid-state reaction*. On the other hand, for the preparation of nitridosilicates, nitride starting powder or nitrogen-containing sources, such as silicon nitride (Si_3N_4), silicon diimide ($Si(NH)_2$), and metal nitride, are often utilized. In addition, in comparison with oxygen or halogens, the nitrogen molecule shows greater stability. The enthalpy of dissociation, i.e., 226 kcal/mol, is about two times higher than that of the oxygen molecule. This implies that nitriding reactions generally require higher temperature than that used for oxosilicates. At the same time, the chemical reactions need to take place in a nitriding atmosphere (N_2 or NH_3).

The use of different starting powder leads to a variety of synthetic approaches or routines for nitrides (Selvaduray and Sheet 1993). Table 4.35 summarizes the synthetic methods for silicon nitride, which is an important raw powder for nitridosilicates. These methods include direct nitridation, carbothermal reduction, ammonia reduction, vapor phase reaction, and the silicon diimide process. All the methods have their own advantages and disadvantages in terms of powder natures (impurity, particle size, particle shape, and particle size distribution), production cost, and production scale. With respect to binary silicon nitride, nitridosilicates are compounds in ternary, quaternary, or multinary systems. This leads to the fact that there are limited synthetic approaches to nitridosilicates. To date, the major methods reported for the synthesis of nitridosilicates are high-temperature solid-state reaction, gas reduction and nitridation, carbothermal reduction and nitridation, direct nitridation, supercritical ammonia nitridation, and self-propagating high-temperature synthesis (SHS). These methods, except the supercritical ammonia nitridation one, can also be applied to oxonitridosilicates and oxoaluminonitridosilicates.

In this section, we will give an overview of the synthetic methods for (oxo) nitridosilicates and (oxo)aluminonitridosilicate phosphors.

4.4.1 High-Temperature Solid-State Reaction

4.4.1.1 Introduction

The solid-state reaction method is a very common and simple way to synthesize nitride phosphors. A solid-state reaction, also called a *dry media reaction* or a *solventless reaction*, is a chemical reaction in which solvents are not used. The solid-state reaction initiates at interfaces between solid particles that are contacting each other. It is generally involved with four steps: (1) diffusions at solid interfaces, (2) chemical reactions at atomic levels, (3) nuclearation of new phases, and (4) mass transport of solid phases and growth of nuclei. It is clear that the reaction can be accelerated by using finer particles with larger specific surface areas, which result in larger contacting areas between particles. In addition, to encourage the reaction, high temperature is usually required to meet the thermodynamic condition.

4.4.1.2 Synthesis of Nitridosilicate Phosphors

The starting powders have a great impact on the synthesis temperature and thus the reaction procedure, because the reactivities of starting powders

TABLE 4.35

Summary of Si_3N_4 Powder Synthesis Methods

Method	Synthesis Reaction	Reaction Condition	Level of Development	Comments
Direct nitridation	$3Si(s) + 2N_2(g) = Si_3N_4(s)$	1,300–1,400°C	Industrial	Cheap, pulverization required
Carbothermal reduction	$3SiO_2(s) + 2N_2(g) + 6C(s) = Si_3N_4(s) + 6CO(g)$	1,400–1,500°C	Industrial	Cheap, C removal required
Ammonia reduction	$3SiO_2(s) + 4NH_3 = Si_3N_4(s) + 6H_2O$		Industrial	
Vapor phase reaction	$3SiCl_4(g) + 4NH_3(g) = Si_3N_4(s) + 12HCl(g)$	800–1,200°C	Laboratory scale	High quality, fine particle, expensive
	$3SiN_4(g) + 4NH_3(g) = Si_3N_4 + 12H_2(g)$	1,300–1,490°C		
Silicon diimide process	$SiCl_4(l) + 6NH_3(g) = Si(NH)_2(s) + 4NH_4Cl(s)$	0°C	Laboratory scale	High quality, fine particle, expensive
	$3Si(NH)_2(s) = Si_3N_4(s) + N_2(g) + 3H_2(g)$	1,200°C		

differ greatly from one to another. For nitridosilicates, they are usually syn-thesized through the chemical reactions (1) between silicon nitride and metal nitrides or (2) between silicon diimide and metals.

Uehda et al. (2006a) reported the preparation of red-emitting $CaAlSiN_3:Eu^{2+}$ by using all nitride starting powders. The powders were mixed in a nitro-gen-filled glove box with a concentration of oxygen and moisture less than 1 ppm. The powder mixtures were then launched into a gas-pressure sinter-ing furnace, and fired at 1,600°C under 1.0 MPa N_2. The chemical reaction can be expressed as

$$(1-x)Ca_3N_2 + Si_3N_4 + xEuN + 3AlN \rightarrow 3(Ca_{1-x}Eu_x)AlSiN_3 \qquad (4.3)$$

Y. Q. Li et al. (2008c) synthesized the $(Sr,Ca)_2Si_5N_8:5\%Eu^{2+}$ red phosphors by firing the powder mixtures of alkaline earth metal nitrides, europium nitride, and silicon nitride in an alumina horizontal tube furnace at 1,300–1,400°C for 12–16 h under flowing 90%N_2_10%H_2 atmosphere. The reaction is shown as

$$(1.90 - 1.90x)Sr_3N_2 + 1.90xCa_3N_2 + 3yEuN + 5Si_3N_4$$

$$\rightarrow 3(Sr_{1-x}Ca_x)_{1.90}Eu_{0.1}Si_5N_8 \qquad (4.4)$$

The chemical reaction between silicon nitride and metal nitrides is explained by the dissolution-diffusion-precipitation process. It is very similar to the liquid phase sintering used for the condensation of ceramic materials. In this liquid-assisted process, metal nitrides first melt during firing, forming some amount of transient liquid phase. Then, silicon nitride dissolves into the liquid phase with the reaction at interfaces. Moreover, the transient liquid phase enhances the diffusion rate by shortening the distance between reactants. The nitridosilicates product finally precipitates from the saturated liquid phase, and grows at the expense of dissolving silicon nitride and metal nitrides.

As seen above, silicon nitride is commonly used as the starting powder for the preparation of nitridosilicates, but the reaction temperature is usu-ally high (1,500 ~ 2,000°C) due to its chemical inertness (e.g., low diffusion coefficient). Moreover, to suppress the decomposition of silicon nitride at high firing temperatures (>1,820°C), high nitrogen gas pressure (>0.1 Mpa) is generally required. Alternatively, more reactive silicon diimide $(Si(NH)_2)$ was used as a silicon source to prepare nitridosilicates.

Schnick and coworkers (1999, 2001) developed a novel synthetic route to ternary and multinary nitridosilicates by high-temperature reactions of silicon diimide with pure metals instead of binary metal nitrides. This method has been proved to succeed in synthesizing many nitridosilicates, such as $M_2Si_5N_8$ (M = Ca, Sr, Ba) and $La_3Si_6N_{11}$. The nitridosilicates were prepared by firing the powder mixture of silicon diimide and alkaline earth/lanthanide metals in a radiofrequency furnace (type IG 10/200 Hy, frequency = 200 kHz, electrical output = 0 ± 12 kW; Huttinger, Freiburg) at 1,500–1,650°C. These reactions are shown as below:

$$2M + 5Si(NH)_2 \rightarrow M_2Si_5N_8 + N_2 + 5H_2 \tag{4.5}$$

$$3La + 6Si(NH)_2 \rightarrow La_3Si_6N_{11} + 1/2N_2 + 6H_2 \tag{4.6}$$

The chemical reaction between a metal and $Si(NH)_2$ is possibly interpreted as a dissolution of an electropositive metal in the nitride analogus, polymeric acid $Si(NH)_2$, accompanied by the evolution of hydrogen. Similar to the formation of Si_3N_4 whiskers, a vapor-solid (VS), vapor-liquid-solid (VLS), or liquid-solid (LS) mechanism is assumed for the synthesis of nitridosilicates. Basically, metals with melting points below 1,600°C can be used for such a procedure.

4.4.1.3 Synthesis of M-Si-Al-O-N Phosphors

The methods for nitridosilicates can be extended to synthesize oxonitridosilicates or oxoaluminonitridosilicates (M-Si-Al-O-N). For these oxygen-containing materials, metal oxides or metal carbonates, together with metal nitrides, are used as starting powders.

We have applied the solid-state reaction method to prepare a variety of oxynitride (oxonitridosilicate and oxoaluminonitridosilicate) phosphors, such as α-sialon:$Eu^{2+}$and β-sialon:Eu^{2+}. Ca-α-sialon, with the composition of $Ca_{m/2}Si_{12-m-n}Al_{m+n}O_nN_{16-n}$, can be synthesized by firing the powder mixture of Si_3N_4, Al_2O_3, $CaCO_3$, AlN, and Eu_2O_3 in a gas-pressure sintering furnace at 1,600–1,800°C under 0.5–1.0 MPa N_2. In addition, Eu_2O_3 is reduced to EuO in a nitriding atmosphere. The chemical reaction is described as

$$(12-m-n)/3Si_3N_4 + m/2CaCO_3 + (2n-m)/6Al_2O_3 + (4m+n)/3AlN$$

$$\rightarrow Ca_{m/2}Si_{12-m-n}Al_{m+n}O_nN_{16-n} + m/2CO_2 \tag{4.7}$$

$$6Eu^{3+} + 2N^{3-} \rightarrow 6Eu^{2+} + N_2 \tag{4.8}$$

For β-sialon ($Si_{6-z}Al_zO_zN_{8-z}$), it was synthesized by firing the powder mixture of Si_3N_4, Al_2O_3, and AlN at 1,800–2,000°C under 1.0 MPa N_2 atmosphere.

$$(2-z/3)Si_3N_4 + z/3AlN + z/3Al_2O_3 \rightarrow Si_{6-z}Al_zO_zN_{8-z} \tag{4.9}$$

4.4.2 Gas Reduction and Nitridation

Generally speaking, nitride phosphor powders that are prepared by a high-temperature solid-state reaction usually consist of hard agglomerates and exhibit large particle size and broad particle size distribution. Therefore, it is essential to pulverize the fired products to obtain fine and well-dispersed particles. This pulverization process, however, will inevitably damage the surface of the particles, which leads to surface defects/cracks, and finally reduces the luminescence. In addition, in case of using metal nitrides as starting powders, some of them are very sensitive to air and moisture. Moreover, metal nitrides are expensive, and even not commercially available. These make the solid-state reaction a complex and multistep processing method.

The gas reduction and nitridation (GRN) process is an effective and cheap way to synthesize oxynitride and nitride phosphors by using cheap and commercially available oxide starting powders. In this method, the reaction is generally performed in an alumina or quartz tube furnace through which NH_3 or NH_3-CH_4 gas flows. The NH_3 or NH_3-CH_4 gas acts as both the reducing and nitriding agents. Suehiro et al. (2002) used the GRN method to synthesize aluminum nitride (AlN) by using alumina and the NH_3-C_3H_8 mixture gas through the following reaction:

$$Al_2O_3 + 2NH_3(g) + C_3H_8(g) \rightarrow 2AlN + 7H_2(g) + 3CO(g) \qquad (4.10)$$

The GRN method was then extended to prepare multinary oxynitride and nitride phosphors by Suehiro et al. and Li et al. These phosphors include Ca-α-sialon:Eu^{2+} (Suehiro 2005; H. L. Li et al. 2008a), $Sr_2Si_5N_8$:Eu^{2+} (H. L. Li et al. 2008b), $LaSi_3N_5$:Ce^{3+} (Suehiro et al. 2009), Y-α-sialon:Eu^{2+} (Suehiro 2010), and $BaSi_7N_{10}$:Eu^{2+} (H. L. Li et al. 2010). They used multicomponent oxides as the starting powder and heated them in a flowing NH_3-CH_4 gas atmosphere mixture. Taking $Sr_2Si_5N_8$ and Ca-α-sialon as an example, the GRN reactions are shown as below:

$$2SrO + 5SiO_2 + 4NH_3 + 12CH_4 \rightarrow Sr_2Si_5N_8 + 25H_2 + 12CO \qquad (4.11)$$

$$mCaO + (24\text{--}2m\text{-}2n)SiO_2 + (m+n)Al_2O_3 + (32\text{--}2n)NH_3 + (48\text{-}n)CH_4$$

$$\rightarrow 2Ca_{m/2}Si_{12-m-n}Al_{m+n}O_nN_{16-n} + (144\text{--}5n)H_2 + (48\text{-}n)CO \qquad (4.12)$$

The phase purity of nitride phosphors after the GRN reaction is strongly related to processing parameters, such as the heating rate, firing temperature, holding time, gas flowing rate, and postannealing. By controlling these parameters carefully, one can achieve highly efficient phosphors with a small particle size (1 ~ 2 μm) and a narrow particle size distribution.

4.4.3 Carbothermal Reduction and Nitridation

The carbothermal reduction and nitridation (CRN) process was extensively used to prepare nitride ceramic powders, including silicon nitride, aluminum nitride, and β-sialons. Now it is currently one of two synthesis methods used to commercially produce AlN powders. It involves a solid-solid reaction between alumina and carbon powders, and the subsequent conversion of alumina in a nitrogen atmosphere, as shown below:

$$Al_2O_3(s) + C(s) + N_2(g) \rightarrow 2AlN(s) + 3CO(g) \tag{4.13}$$

When the nitridation process is complete, the residual carbon is removed through oxidation by heating to 600–800°C in an oxygen or ambient atmosphere. The prepared AlN powders have a uniform particle size distribution with a mean particle diameter of 0.01–0.5 µm, and a maximal total metallic impurity concentration of 0.5%.

The CRN method was also used to synthesize multinary nitride and oxynitride luminescent materials, such as Ca-α-sialon:Eu^{2+} (Zhang et al. 2007) and $M_2Si_5N_8$:Eu^{2+} (Piao et al. 2006). The reactants for this method include $CaCO_3$, Si_3N_4, Eu_2O_3, Al_2O_3, and carbon (graphite) for Ca-α-sialon:Eu^{2+}, and MCO_3 (M = Ca, Sr), Si_3N_4, Eu_2O_3, and carbon oxides, silicon nitride/silica, alumina, and carbon for $M_2Si_5N_8$:Eu^{2+}. For example, the overall CRN reaction for $Sr_2Si_5N_8$ is given by

$$6SrCO_3(s) + 5Si_3N_4(s) + 18C(s) + 12N_2(g) \rightarrow 3Sr_2Si_5N_8(s) + 18CO(g) \tag{4.14}$$

The phase purity of the nitride phosphor powders strongly relies on the starting composition (e.g., carbon/oxides ratio) and processing parameters, such as firing temperature, heating rate, holding time, and postannealing. Usually, the stoichiometric excess of carbon is essential for the success of the CRN process, because the excessive carbon serves to increase the reaction rate, facilitate the completion of the transformation, improve the dispersion of the powder, and control the powder aggregation. The solid-solid reaction between carbon and oxides is the rate-determining step in the process, and the availability of carbon is the rate-controlling mechanism.

The nitride phosphors prepared by the CRN method always contain some amount of residual carbon, which significantly reduces the absorption and luminescence of the phosphor powder itself. Therefore, it is necessary to remove the excess carbon after the carbothermal reduction process. The common way for the removal of carbon is to fire the powder in an oxidizing atmosphere at temperatures above 600°C. However, this process may also lead to the oxidation of the phosphor powders, which decreases the luminescence of phosphors. Annealing the phosphor powders in a carbon-free furnace in the N_2 or NH_3 atmosphere at high temperature is an alternative way to remove residual carbon in nitride phosphors.

4.4.4 Ammonothermal Synthesis

The synthesis of oxygen-based compounds by the hydrothermal method has been significantly investigated. Because ammonia is a closer match to the physical properties of water than any other known solvent, the ammonothermal technique is hence a promising method to synthesize nitrogen-based compounds. Moreover, crystal growth from solutions permits the low-temperature synthesis (600–800°C) and production of nano-sized materials.

The ammonothermal technique was originally used by Juza and Jacobs (1966) to prepare several amides and Be_3N_2 at 400°C and 272 bar. Later, a variety of metal amides, imides, and nitrides were synthesized in supercritical ammonia. These reactions were generally performed at fairly extreme conditions (400–600°C and 6 kbar).

The metal-ammonia solutions were first investigated by Weyl (1864). It is well known that alkali and alkaline earth metals (except Be), as well as lathanides, can dissolve in liquid ammonia under atmospheric pressure. Chemical reactions with ammonia have three major classes: (1) addition reactions, ammonolysis analogous by hydration; (2) substitution reactions, ammonolysis analogous to hydrolysis; and (3) oxidation-reduction reactions. Ammonia can react as a three-basic acid to form amides, imides, or nitrides with electropositive metals, depending on the temperature and ammonia partial pressure of the system (Wang and Callahan 2006). The reaction is accelerated by employing supercritical ammonia as the solvent and reactant at high temperatures and pressures.

J. W. Li et al. (2007) summarized the difference and similarity between the hydrothermal and ammonothermal reduction in Table 4.36, and also first investigated the synthesis of nitride phosphors by using the ammonothermal technique (J. W. Li et al. 2007, 2008, 2009). In their synthesis of red-emitting $CaAlSiN_3{:}Eu^{2+}$, the CaAlSi alloy powder precursor was first prepared by arc-melting Si, Ca, Al, and Eu metal shots in an Ar atmosphere. Then, the alloy was converted to sodium ammonometallates (referring to metal amides, imidoamides, and their partially polymerized compounds) in 100 Mpa supercritical ammonia at 400°C, followed by decomposing into the nitride by releasing NH_3 at 800°C. The reaction is shown as below:

$$2Na + 2NH_3 \rightarrow 2NaNH_2 + H_2 \qquad (4.15)$$

$$Ca_xEu_{1-x}AlSi + NaNH_2 + NH_3$$

$$\rightarrow \text{amides and imidoamides} + H_2$$

$$\rightarrow \text{more polymerized compounds} + NH_3$$

$$\rightarrow Ca_xEu_{1-x}AlSiN_3 + NH_3 \qquad (4.16)$$

Zeuner et al. (2009) have applied the ammonothermal technique to synthesize $M_2Si_5N_8{:}Eu^{2+}$ (M = Ca, Sr, Ba) phosphors. In their synthesis, metal amides $M(NH_2)_2$ (M = Eu, Ca, Sr, Ba) were first prepared by dissolution of the

TABLE 4.36

Comparison of the Water System for Oxide Synthesis and the
Ammonia System for Nitride Synthesis (M = Metal)

Water System	Ammonia System
Acids: e.g., HF, HCl	Acids: e.g., NH_4F, NH_4Cl
Bases: e.g., NaOH, KOH	Bases: e.g., $NaNH_2$, KNH_2
Hydrolysis forming hydroxide	Ammonolysis forming amide
e.g., $M\text{-}Cl + H_2O \rightarrow M\text{-}OH + HCl$	e.g., $M\text{-}Cl + 2NH_3 \rightarrow M\text{-}NH_2 + NH_4Cl$
$M + H_2O \rightarrow M\text{-}OH + 0.5H_2$	$M + NH_3 \rightarrow M\text{-}NH_2 + 0.5H_2$
Hydroxide converting to oxide	Amide converting to nitride
$2M\text{-}OH \rightarrow M\text{-}O\text{-}H + H_2O$	$3M\text{-}NH_2 \rightarrow M\text{-}N + 2NH_3$

Source: Data from Li, J. W., Watanabe, T., Sakamoto, N., Wada, H.,
Setoyama, T., and Yoshimura, M., *Chem. Mater.*, 20, 2095–2105, 2008.

respective metals in supercritical ammonia at 150°C and 300 bar. Then, the
target $M_2Si_5N_8$:Eu^{2+} phosphors were synthesized by reacting metal amides
with silicon diimide at 1,150 to 1,400°C. The ammonothermal reactions are
given by

$$M + 2NH_3 \rightarrow M(NH_2)_2 + H_2 \tag{4.17}$$

$$2M(NH_2)_2 + 5Si(NH)_2 \rightarrow M_2Si_5N_8 + 6NH_3 \tag{4.18}$$

The nitridosilicate phosphors prepared by the ammonothermal synthesis
method exhibit high phase purity and fine particle size. The morphology
of phosphor powders varies in a variety of forms, depending on the Na/Ca
ratio (J. W. Li et al. 2009) or the thermal treatment (Zeuner et al. 2009).

4.4.5 Direct Nitridation

The direct nitridation method has been used to synthesize silicon nitride and
aluminum nitride, and is now an industrial approach to produce commercial aluminum nitride powders. This method is involved with the reaction
between the silicon or aluminum powder and nitrogen gas. To enhance the
reaction, the powder is also mixed with a small amount of catalysts, such
as fluorides.

Piao et al. (2007a, 2007b) investigated the synthesis of red-emitting
$CaAlSiN_3$:Eu^{2+} and $Ba_2Si_5N_8$:Eu^{2+} phosphors by using the direct nitridation
process. The (Ca,Eu)AlSi or (Ba,Eu)$_2$Si$_5$ alloy powder was used as the starting
powder, and reacted with flowing nitrogen gas at ~1,050°C. The resultant
powder was then calcinated at 1,350–1,550°C to form final nitride phosphors.
The direct nitridation for $CaAlSiN_3$:Eu^{2+} is given by

$$2Ca_{1-x}Eu_xAlSi + 3N_2 \rightarrow 2Ca_{1-x}Eu_xAlSiN_3 \ (x = 0\text{--}0.2) \tag{4.19}$$

TABLE 4.37

Summary of Synthetic Methods for Nitridosilicate Phosphors

Method	Starting Power	Gas	Temperature	Comments
Solid-state reaction	Binary nitrides	N_2	1,400–2,050°	Industrial scale, expensive, high quality, glove box required, pulverization required
Gas reduction and nitridation	Oxide	NH_3-CH_4	1,350–1,550°C	Industrial scale, cheap, fine particle, postannealing required
Carbothermal reduction	Oxide, carbon	N_2	1,400–1,600°C	Industrial scale, cheap, carbon removal necessary
Ammonothermal synthesis	Metal, amide, imide,	NH_3	600–800°C	Laboratory scale, very fine particle, postannealing required
Direct nitridation	Metal silicates	N_2	1,350–1,550°C	Industrial scale, cheap, explosion hazard, pulverization required

The synthesized $CaAlSiN_3$:Eu^{2+} powder consisted of irregular particles with a particle size of 6–9 μm, which is larger than that prepared by the solid-state reaction. The luminescence intensity of the phosphor prepared by the direct nitridation was comparable to that prepared by the solid-state reaction.

Table 4.37 summarizes the often used synthetic methods for nitridosilicates luminescent materials. Among these methods, the solid-state reaction is now accepted by industries to synthesize commercial phosphor powders. Besides the synthetic methods mentioned above, there are some other approaches that are attempted to prepare nitride and oxynitride luminescent materials, such as combustion synthesis (Zhou et al. 2008), spark plasma sintering (Choi and Hong 2010), etc. The powder characteristics (particle size, particle size distribution, particle morphology, crystallinity, impurity levels), photoluminescence properties (quantum efficiency, absorption), and cost of phosphors are closely tied with synthetic approaches. Cost-effective and simple synthetic methods for mass production of high-efficiency nitride phosphors are therefore continuously pursued by phosphor researchers and engineers.

References

Bachmann, V., Ronda, C., Oeckler, O., Schnick, W., and Meijerink, A. 2009. Color point tuning for (Sr,Ca,Ba)SiON:Eu for white light LEDs. *Chem. Mater.* 21:316–325.

Bemporad, E., Pecchio, C., De Rossi, S., and Carassiti, F. 2004. Characterisation and wear properties of industrially produced nanoscaled CrN/NbN multilayer coating. *Surf. Coating Technol.* 188:319–330.

Braun, C., Seibald, M., Börger, S. L., Oeckler, O., Boyko, T. D., Moewes, A., Miehe, G., Tuckes, A., and Schnick, W. 2010. Material properties and structural characterization of $M_3Si_6O_{12}N_2:Eu^{2+}$ (M = Ba, Sr)—A comprehensive study on a promising green phosphor for pc-LEDs. *Chem. Euro. J.* 16:9646–9657.

Caicedo, J. C., Amaya, C., Yate, L., Nos, O., Gomez, M. E., and Prieto, P. 2010. Hard coating performance enhancement by using [Ti/Tin]n, [Zr/ZrN]n and [TiN/ZrN]n multilayer system. *Mater. Sci. Eng. B* 171:56–61.

Cao, G. Z., and Metselaar, R. 1991. α'-sialon ceramics, a review. *Chem Mater.* 3:242–252.

Cao, G. Z., Metselaar, R., and Haije, W. G. 1993. Neutron diffraction study of yttrium a'-sialon. *J. Mater. Sci. Lett.* 12:459–460.

Chen, P., Xiong, Z., Luo, J., Lin, J., and Tan, K. L. 2002. Interaction of hydrogen with metal nitrides and imides. *Nature* 420:302–304.

Choi, S. W., and Hong, S. H. 2010. Luminescence properties of Eu^{2+}-doped Ca-alpha-SiAlON synthesized by spark plasma sintering. *J. Electrochem. Soc.* 157:J297–J300.

Cole, M., O'Reilly, K. P. J., Redington, M., and Hampshire, S. 1991. EXAFS study of a hot-pressed α'-sialon ceramic containing erbium as the modifying cation. *J. Mater. Sci.* 26:5143–5148.

Dierre, B., Xie, R.-J., Hirosaki, N., and Sekiguchi, T. 2007. Blue emission of Ce^{3+} in lanthanide silicon oxynitride phosphors. *J. Mater. Res.* 22:1933–1941.

Dierre, B., Yuan, X. L., Inoue, K., Hirosaki, N., Xie, R.-J., and Sekiguchi, T. 2009. Role of Si in the luminescence of AlN:Eu,Si phosphors. *J. Am. Ceram. Soc.* 92:1272–1275.

Duan, C. J., Wang, X. J., Otten, W. M., Delsing, A. C. A., Zhao, J. T., and Hintzen, H. T. 2008. Preparation, electronic structure, and photoluminescence properties of Eu^{2+}- and Ce^{3+}/Li^+-activated alkaline earth silicon nitride $MSiN_2$ (M = Sr, Ba). *Chem. Mater.* 20:1597–1605.

Ekstrom, T., and Nygen, M. 1992. SiAlON ceramics. *J. Am. Ceram. Soc.* 75:259–276.

Feng, W. R., Yan, D. R., He, J. N., Zhang, G. L., Chen, G. L., Gu, W. C., and Yang, S. 2005. Microhardness and toughness of the TiN coating prepared by reactive plasma spraying. *Appl. Surf. Sci.* 243:204–213.

Fukuda, Y. 2007. Luminescence of Eu-doped Sr-SiAlON phosphor. Paper presented at Phosphor Global Summit 2007, Seoul, Korea.

Gal, Z. A., Mallinson, P. M., Orchard, H. J., and Clarke, S. J. 2004. Synthesis and structure of alkaline earth silicon nitrides: $BaSiN_2$, $SrSiN_2$, and $BaSiN_2$. *Inorg. Chem.* 43:3998–4006.

Gregory, D. H. 1999. Structural families in nitride chemistry. *J. Chem. Soc. Dalton Trans.* 3:259–270.

Grins, J., Esmaeilzadeh, S., Svensson, G., and Shen, Z. J. 1999. High-resolution electron microscopy of a Sr-containing sialon polytypoid phase. *J. Euro. Ceram. Soc.* 19:2723–2730.

Grins, J., Shen, Z., Esmaeilzadeh, S., and Berastegui, P. 2001. The structure of the Ce and La N-phases $RE_3Si_{8-x}Al_xN_{11-x}O_{4+x}$ (x ≈ 1.75 for RE = Ce, x ≈ 1.5 for RE = La), determined by single-crystal x-ray and time-of-flight neutron powder diffraction, respectively. *J. Mater. Chem.* 11:2358–2362.

Grins, J., Shen, Z., Nygren, M., and Ekstrom, T. 1995. Preparation and crystal structure of $LaAl(Si_{6-z}Al_z)N_{10-z}O_z$. *J. Mater. Chem.* 5:2001–2006.

Grun, R. 1979. The crystal structure of β-Si_3N_4. *Acta Crystallogr. B* 35:800–804.

Hampshire, S. Park, H. K., Thompson, D. P., and Jack, K. H. 1978. α-sialon ceramics. *Nature (London)* 274:880–883.

Harris, R. K., Leach, M. J., and Thompson, D. P. 1992. Nitrogen-15 and oxygen-17 NMR spectroscopy of silicates and nitrogen ceramics. *Chem. Mater.* 4:260.

Hatfield, G. R., Li, B., Hammond, W. B., Reidinger, F., and Yamanis, J. 1990. Preparation and characterization of lanthanum silicon nitride. *J. Mater. Sci.* 25:4032–4035.

Hecht, C., Stadler, F., Schmidt, P. J., auf der Gunne, S. J., Baumann, V., and Schnick, W. 2009. SrAlSi$_4$N$_7$:Eu^{2+}—A nitridoalumosilicate phosphor for warm white-light (pc)LEDs with edge-sharing tetrahedra. *Chem. Mater.* 21:1595–1601.

Higashi, M., Abe, R., Takata, T., and Domen, K. 2009. Photocatalytic overall water splitting under visible light using ATaO$_2$N (A = Ca, Sr, Ba) and WO$_3$ in a IO$_3^-$/I$^-$ shuttle redox mediated system. *Chem. Mater.* 21:1543–1549.

Hirosaki, N., Xie, R. -J., Inoue, K., Sekiguchi, T., Dierre, B., and Tamura, K. 2007. Blue-emitting AlN:Eu^{2+} nitride phosphor for field emission displays. *Appl. Phys. Lett.* 91:061101.

Hirosaki, H., Xie, R-J., Kimoto, K., Sekiguchi, T., Yamamoto, Y., Suehiro, T., and Mitomo, M. 2005. Characterization and properties of green-emitting βirosaki, H^{2+} powder phosphors for white light-emitting diodes. *Appl. Phys. Lett.* 86:211905-1–211905-3.

Hirosaki, N., Xie, R.-J., and Sakuma, K. 2006. Development of new nitride phosphors for white LEDs. *Bull. Ceram. Soc. Jpn.* 41:602–606 (in Japanese).

Hoppe, H. A., Lutz, H., Mory, P., Schnick, W., and Seilmeier, A. 2000. Luminescence in Eu^{2+}-doped Ba$_2$Si$_5$N$_8$: Fluorescence, thermoluminescence, and upconversion. *J. Phys. Chem. Solids.* 61:2001–2006.

Hoppe, H. A., Stadler, F., Oeckler, O., and Schnick, W. 2004. Ca[Si$_2$O$_2$N$_2$]—A novel layer silicate. *Angew. Chem. Int. Ed.* 43:5540–5542.

Huppertz, H., and Schnick, W. 1997a. Edge sharing SiN$_4$ tetrahedra in the highly condensed nitridosilicate BaSi$_7$N$_{10}$. *Chem. Eur. J.* 3:249–252.

Huppertz, H., and Schnick, W. 1997b. Synthesis, crystal structure, and properties of the nitridosilicates SrYbSi$_4$N$_7$ and BaYbSi$_4$N$_7$. *Z. Anorg. Allg. Chem.* 623:212–217.

Inoue, K., Hirosaki, N., Xie, R.-J., and Takeda, T. 2009. Highly efficient and thermally stable blue-emitting AlN:Eu^{2+} phosphor for ultraviolet white light-emitting diodes. *J. Phys. Chem. C* 133:9392–9397.

Inoue, Z., Mitomo, M., and Nobuo, I. 1980. A crystallographic study of a new compound of lanthanum silicon nitride, LaSi$_3$N$_5$. *J. Mater. Sci.* 15:2915–2920.

Izumi, F., Mitomo, M., and Suzuki, J. 1982. Structure refinement of yttrium α-sialon from x-ray powder profile data. *J. Mater. Sci. Lett.* 1:533–535.

Jack, K. H. 1976. Review: Sialons and related nitrogen ceramics. *J. Mater. Sci.* 11:1135–1158.

Jansen, M., and Letschert, H. P. 2000. Inorganic yellow-red pigments without toxic metals. *Nature* 404:980–982.

Juza, R. and Jacobs, H. 1966. Ammonothermal synthesis of magnesium and beryllium amides. *Ang. Chem. Int. Ed.* 5:247–248.

Kechele, J. A., Hecht, C., Oeckler, O., auf der Günne, J. S., Schmidt, P. J., and Schnick, W. 2009a. Ba$_2$AlSi$_5$N$_9$—A new host lattice for Eu^{2+}-doped luminescent materials comprising a nitridoalumosilicate framework with corner- and edge-sharing tetrahedra. *Chem. Mater.* 21:1288–1295.

Kechele, J. A., Oeckler, O., Stadler, F., and Schnick, W. 2009b. Structure elucidation of $BaSi_2O_2N_2$—A host lattice for rare earth doped luminescent materials in phosphor-converted (pc)-LEDs. *Solid State Sci.* 11:537–543.

Kim, Y. I., Woodward, P. M., Baba-Kishi, K. Z., and Tai, C. W. 2004. Characterization of the structural, optical and dielectrical properties of oxunitride perovskite AMO_2N (A = Ba, Sr, Ca; M = Ta, Nb). *Chem. Mater.* 16:1267–1276.

Kimoto, K., Xie, R.-J., Matsui, Y., Ishizuka, K., and Hirosaki, N. 2009. Direct observation of single dopant atom in light-emitting phosphor of β-sialon Eu^{2+}. *Appl. Phys. Lett.* 94:041908.

Kohatsu, I., and McCauley, J. W. 1974. Re-examination of the crystal structure of α-Si_3N_4. *Mater. Res. Bull.* 9:917–920.

Kojima, Y., and Kawai, Y. 2004. Hydrogen storage of metal nitride by a mechanochemical reaction. *Chem. Comm.* 2210–2211.

Lauterbach, R., Irran, E., Henry, P. F., Weller, M. T., and Schnick, W. 2000. High-temperature synthesis, single-crystal x-ray and neutron powder diffraction, and materials properties of $Sr_3Ln_{10}Si_{18}Al_{12}O_{18}N_{36}$ (Ln = Ce, Pr, Nd)—Novel sialons with an ordered distribution of Si, Al, O, and N. *J. Mater. Chem.* 10:1357–1364.

Lauterbach, R., and Schnick, W. 1998. Synthesis, crystal structure, and properties of a new sialon—$SrSiAl_2O_3N_2$. *Z. Anorg. Allg. Chem.* 624:1154–1158.

Le Toquin, R., and Cheetham, A. K. 2006. Red-emitting cerium-based phosphor materials for solid state lighting applications. *Chem. Phys. Lett.* 423:352–356.

Li, H. L., Xie, R.-J., Hirosaki, N., Suehiro, T., and Yajima, Y. 2008a. Phase purity and luminescence properties of fine Ca-α-sialon:Eu phosphors synthesized by gas reduction and nitridation method. *J. Electrochem. Soc.* 155:J175–J179.

Li, H. L., Xie, R.-J., Hirosaki, N., and Yajima, Y. 2008b. Synthesis and photoluminescence properties of $Sr_2Si_5N_8$:Eu^{2+} red phosphor by a gas-reduction and nitridation method. *J. Electrochem. Soc.* 155:J378–J381.

Li, H. L., Xie, R.-J., Zhou, G. H., Hirosaki, N., and Sun, Z. 2010. A cyan-emitting $BaSi_7N_{10}$:Eu^{2+} phosphor prepared by gas reduction and nitridation for UV-pumping white LEDs. *J. Electrochem. Soc.* 157:J251–J255.

Li, J. W., Watanabe, T., Sakamoto, N., Wada, H., Setoyama, T., and Yoshimura, M. 2008. Synthesis of a multinary nitride, Eu-doped $CaAlSiN_3$, from alloy at low temperatures. *Chem. Mater.* 20:2095–2105.

Li, J. W., Watanabe, T., Wada, H., Setoyama, T., and Yoshimura, M. 2007. Low-temperature crystallization of Eu-doped red-emitting $CaAlSiN_3$ from alloy-derived ammonometallates. *Chem. Mater.* 19:3592–3594.

Li, J. W., Watanabe, T., Wada, H., Setoyama, T., and Yoshimura, M. 2009. Synthesis of Eu-doped $CaAlSiN_3$ from ammonometallates: Effects of sodium content and pressure. *J. Am. Ceram. Soc.* 92:344–349.

Li, Y. Q., de With, G., and Hintzen, H. 2008c. The effect of replacement of Sr by Ca on the structural and luminescence properties of the red-emitting $Sr_2Si_5N_8$:Eu^{2+} LED conversion phosphor. *J. Solid State Chem.* 181:515–524.

Li, Y. Q., Fang, C. M., de With, G., and Hintzen, H. T. 2004. Preparation, structure and photoluminescence properties of Eu^{2+} and Ce^{3+}-doped $SrYSi_4N_7$. *J. Solid State Chem.* 177:4687–4694.

Li, Y. Q., Hirosaki, N., and Xie, X.-J. 2007. Structural and photoluminescence properties of $CaSiN_2$:Ce^{3+}, Li^+. Paper presented at the Japan Society of Applied Physics, the 67th autumn meeting, Sapporo, Japan.

Li, Y. Q., Hirosaki, N., Xie, R.-J., Takeda. T., and Mitomo, M. 2008a. Crystal and electronic structures, luminescence properties of Eu^{2+}-doped $Si_{6-z}Al_zO_zN_{8-z}$ and $M_ySi_{6-z}Al_{z-y}O_{z+y}N_{8-z-y}$ (M = 2Li, Mg, Ca, Sr, Ba). *J. Solid State Chem.* 181:3200–3210.

Li, Y. Q., Hirosaki, N., Xie, R.-J., Takeda, T., and Mitomo, M. 2008b. Yellow-orange-emitting $CaAlSiN_3$:Ce^{3+} phosphor: Structure, photoluminescence, and application in white LEDs. *Chem. Mater.* 20:6704–6714.

Li, Y. Q., Hirosaki, N., Xie, R.-J., Takeda, T., and Mitomo, M. 2009. Synthesis, crystal and local electronic structures, and photoluminescence properties of red-emitting $CaAl_zSiN_{2+z}$:Eu^{2+} with orthorhombic structure. *Int. J. Appl. Ceram. Technol.* DOI:10.1111/j.1744-7402.2009.02393.x.

Li, Y. Q., van Steen, J. E. J., van Krevel, J. W. H., Botty, G., Delsing, A. C. A., DiSalvo, F. J., de With, G., and Hintezen, H. T. 2006. Luminescence properties of red-emitting $M_2Si_5N_8$:Eu^{2+} (M = Ca, Sr, Ba) LED conversion phosphors. *J. Solid State Compd.* 417:273–279.

Liu, M. Y., You, W. S., Lei, Z. B., Takata, T., Domen, K., and Li, C. 2006. Photocatalytic water splitting to hydrogen over a visible light-driven $LaTaON_2$ catalyst. *Chin. J. Catalysis* 27:556–558.

Liu, Y., Horikawa, K., Fujiyosi, M., Imanishi, N., Hirona, A., and Takeda, Y. 2004. The effect of doped elements on the electrochemical behavior of hexagonal $Li_{2.6}Co_{0.4}N$. *J. Electrochem. Soc.* 151:A1450–A1455.

Mandal, H. 1999. New developments in α-sialon ceramics. *J. Euro. Ceram. Soc.* 19:2349–2357.

Marchand, R. 1998. Ternary and higher order nitride materials. In *Handbook on physics and chemistry of rare earths*, ed. K. A. Gschneidner and L. Eyring, 51–99. Vol 25. New York: Elsevier Science B.V.

Marchand, R., Laurent, Y., Guyader, J., L'Haridon, P., and Verdier, P. 1991. Nitrides and oxynitrides: Preparation, crystal chemistry and properties. *J. Euro. Ceram. Soc.* 8:197–213.

Metselaar, R. 1994. Progress in the chemistry of ternary and quaternary nitrides. *Pure Appl. Chem.* 66:1815–1822.

Michiue, Y., Shioi, K., Hirosaki, N., Takeda, T., Xie, R.-J., Sato, A., Onoda, M., and Mtsushita, Y. 2009. *Acta Cryst.* B65:567–575.

Morkoc, H., Strite, S., Gao, G. B., Lin, M. E., Sverdlov, B., and Burns, M. 1994. Large-band-gap SiC, III-V nitride, and II-VI ZnSe-based semiconductor device technologies. *J. Appl. Phys.* 76:1363–1398.

Nakamura, S. 1993. InGaN blue-light-emitting diodes, *J. Inst. Electron. Comm. Eng. Jpn.* 76:3911–3915.

Nakatsuka, A., Yahiasa, A., and Yamanaka, T. 1999. Cation distribution and crystal chemistry of $Y_3Al_{5-x}Ga_xO_{12}$ (0 ≤ x ≤ 5) garnet solid solutions. *Acta Crysta.* B55:266–272.

Niewa, R., and DiSalvo, F. J. 1998. Recent developments in nitride chemistry. *Chem. Mater.* 10:2733–2752.

Oeckler, O., Kechele, J. A., Koss, H., Schmidt, P. J., and Schnick, W. 2009. $Sr_5Al_{5+x}Si_{21-x}N_{35-x}O_{2+x}$:$Eu^{2+}$ (x ≈ 0)—A novel green phosphor for white light pcLEDs with disordered intergrowth structure. *Chem. Eur. J.* 15:5311–5319.

Oeckler, O., Stadler, F., Rosenthal, T., and Schnick, W. 2007. Real structure of $SrSi_2O_2N_2$. *Solid State Sci.* 9:205–212.

Orth, M., and Schnick, W. 1999. On $LiSi_2N_3$—Synthesis and crystal structure refinement. *Z. Anorg. Allg. Chem.* 625:1426–1428.

Oyama, Y. and Kamigaito, O. 1971. Solid solubility of some oxides in silicon nitride, *Jpn. J. Appl. Phys.* 10:1637–1642.

Piao, X. Q., Horikawa, T., Hanzawa, H., and Machida, K. 2006. Characterization and luminescence properties of $Sr_2Si_5N_8$:Eu^{2+} phosphor for white light-emitting-diode illumination. *Appl. Phys. Lett.* 88:161908.

Piao, X. Q., Machida, K., Horikawa, T., and Hanzawa, H. 2007a. Self-propagating high temperature synthesis of yellow-emitting $Ba_2Si_5N_8$:Eu^{2+} phosphors for white light-emitting diodes. *Appl. Phys. Lett.* 91:041908.

Piao, X. Q., Machida, K., Horikawa, T., Hanzawa, H., Shimonura, Y., and Kijima, N. 2007b. Preparation of $CaAlSiN_3$:Eu^{2+} phosphors by SHS. *Chem. Mater.* 194:4592–4599.

Ponce, F. A., and Bour, D. P. 1997. Nitride-based semiconductors for blue and green light-emittinig devices. *Nature* 386:351–359.

Prosini, P. P. 2003. A composite electrode based on graphite and β-Li_3N for Li-ion batteries. *J. Electrochem. Soc.* 150:A1390–A1393.

Ruan, J., Xie, R.-J., Hirosaki, N., and Takeda, T. 2010. Nitrogen gas pressure sintering and photoluminescence properties of orange-red $SrAlSi_4N_7$:Eu^{2+} phosphors for white light-emitting diode. *J. Am. Ceram. Soc.* DOI: 10.1111/j.1551–2916.2010.04104.x.

Sakuma, K., Hirosaki, N., and Xie, R.-J. 2007. Red-shift of emission wavelength caused by reabsorption mechanism of europium activated Ca-α-sialon ceramic phosphors. *J. Lumin.* 126:843–852.

Schlieper, T., Milius, W., and Schnick, W. 1995. Nitrido silicates. II. High temperature synthesis and crystal structure of $Sr_2Si_5N_8$ and $Ba_2Si_5N_8$. *Z. Anorg. Allg. Chem.* 621:1380–1384.

Schlieper, T., and Schnick, W. 1995a. Nitrido silicates. I. High temperature synthesis and crystal structure of $Ca_2Si_5N_8$. *Z. Anorg. Allg. Chem.* 621:1037–1041.

Schlieper, T., and Schnick, W. 1995b. Nitrido-silicates. III[1]. High-temperature synthesis, crystal structure, and magnetic properties of $Ce_3[Si_6N_{11}]$. *Z. Anorg. Allg. Chem.* 621:1535–1538.

Schmolke, C., Bichler, D., Johrendt, D., and Schnick, W. 2009. Synthesis and crystal structure of the first chain-type nitridosilicates $RE_3Si_5N_9$ (RE = La, Ce). *Solid State Sci.* 11:389–394.

Schnick, W. 2001. Nitridosilicates, oxonitridosilicates (sions), and oxonitridoalumino-silicates (sialons)—New materials with promising properties. *Inter. J. Inorg. Mater.* 3:1267–1272.

Schnick, W., and Huppertz, H. 1997. Nitridosilicates—A significant extension of silicate chemistry. *Chem. Eur. J.* 3:679–683.

Schnick, W., Huppertz, H., and Lauterbach, R. 1999. High temperature synthesis of novel nitride- and oxonitrido-silicates and sialons using rf furnaces. *J. Mater. Chem.* 9:289–296.

Selvaduray, G., and Sheet, L. 1993. Aluminium nitride: Review of synthesis methods. *Mater. Sci. Technol.* 9:463–473.

Setlur, A. A., Heward, W. J., Hannah, M. E., and Happek, U. 2008. Incorporation of Si^{4+}-N^{3-N} into Ce^{3+}-doped garnets for warm white LED phosphors. *Chem. Mater.* 20:6277–6283.

Seto, T., Kijima, N., and Hirosaki, N. 2009. A new yellow phosphor $La_3Si_6N_{11}$:Ce^{3+} for white LEDs. *ECS Trans.* 25:247–252.

Shioi, K., Michiue, Y., Hirosaki, N., Xie, R.-J., Takeda, T., Matsushita, Y., Tanaka, M., and Li, Y. Q. 2011. Synthesis and photoluminescence of a novel Sr-SiAlON:Eu^{2+} blue-green phosphor (Sr$_{14}$Si$_{68-s}$Al$_{6+s}$O$_s$N$_{106-s}$:Eu^{2+} (s ≈ 7)). *J. Alloy Compds.* 509:332–337.

Starosvetsky, D., and Gotman, I. 2001. TiN coating improves the corrosion behavior of superelastic NiTi surgical alloy. *Surf. Coating Technol.* 148:268–276.

Stoeva, Z., Jager, B., Gomez, R., Messaoudi, S., Yahia, M. B., Rocquefelte, X., Hix, G. B., Wolf, W., Titman, J. J., Gautier, R., Herzig, P., and Gregory, D. 2007. Crystal chemistry and electronic structure of the metallic lithium ion conductor, LiNiN. *J. Am. Chem. Soc.* 129:1912–1920.

Strite, S., and Morkoc, H. 1992. GaN, AlN, and InN: A review. *J. Vac. Sci. Technol. B* 10:1237–1266.

Su, Y. L., and Yao, S. H. 1997. On the performance and application of CrN coating. *Wear* 205:112–119.

Suehiro, T., Hirosaki, N., and Xie, R.-J. 2011. Yellow-emitting (La,Ca)$_3$Si$_6$N$_{11}$:Ce^{3+} fine powder phosphors for warm-white light-emitting diodes. *ACS Appl Mater Intel.* In print.

Suehiro, T., Hirosaki, N., Xie, R.-J., and Mitomo, M. 2005. Powder synthesis of Ca-α-sialon as a host material for phosphors. *Chem. Mater.* 17:308–314.

Suehiro, T., Hirosaki, N., Xie, R.-J., and Sato, T. 2009. Blue-emitting LaSi$_3$N$_5$:Ce^{3+} fine powder phosphor for UV-converting white light diodes. *Appl. Phys. Lett.* 95:051903.

Suehiro, T., Tatami, J., Meguro, T., Matsuo, S., and Komeya, K. 2002. Synthesis of spherical AlN particles by gas-reduction-nitridation method. *J. Euro. Ceram. Soc.* 22:521–526.

Suzuki, S., and Shodai, T. 1999. Electronic structure and electrochemical properties of electrode material Li$_{7-x}$MnN$_4$. *Solid State Ionics* 116:1–9.

Takahashi, K., Hirosaki, N., Xie, R.-J., Harada, M., Yoshimura, K., and Tomomura, Y. 2007. Luminescence properties of blue La$_{1-x}$Ce$_x$Al(Si$_{6-z}$Al$_z$)(N$_{10-z}$O$_z$)(z ~ 1) oxynitrides phosphors and their application in white light-emitting diode. *Appl. Phys. Lett.* 91:091923.

Takahashi, J., Yamane, H., Hirosaki, N., Yamamoto, Y., Suehiro, T., Kamiyama, T., and Shimada, M. 2003. Crystal structure of La$_4$Si$_2$O$_7$N$_2$ analyzed by the Rietveld method using the time-of flight neutron powder diffraction data. *Chem. Mater.* 15:1099.

Takeda, Y., Nishijima, M., Yamahata, M., Takeda, K., Imanishi, N., and Yamamoto, O. 2000. Lithium secondary batteries using a lithium cobalt nitride, Li$_{2.6}$Co$_{0.4}$N, as the anode. *Solid State Ionics* 130:61069.

Tessier, F., and Marchand, R. 2003. Ternary and higher order rare-earth nitride materials: Synthesis and characterization of ionic-covalent oxynitride powders. *J. Solid State Chem.* 171:143–151.

Titeux, S., Gervais, M., Verdier, P., and Laurent, Y. 2000. Synthesis and x-ray diffraction study of substituted La$_{10}$(Si$_6$O$_{22}$N$_2$)O$_2$ silicon apatites. *Mater. Sci. Forum* 17:325–326.

Uheda, K., Hirosaki, N., and Yamamoto, H. 2006a. Host lattice materials in the system Ca$_3$N$_2$-AlN-Si$_3$N$_4$ for white light emitting diode. *Phys. Stat. Sol. A* 11:2712–2717.

Uheda, K., Hirosaki, N., Yamamoto, Y., Naoto, A., Nakajima, T., and Yamamoto, H. 2006b. Luminescence properties of a red phosphor, CaAlSiN$_3$:Eu^{2+}, for white light-emitting diodes. *Electrochem. Solid State Lett.* 9:H22–H25.

Uheda, K., Shimooka, S., Mikami, M., Imura, H., and Kijima, N. 2008. Synthesis and characterization of new green oxonitridosilicate phosphor, $(Ba,Eu)_3Si_6O_{12}N_2$, for white LED. Paper presented at the 214th ECS Meeting, Honolulu, Hawaii.

Uheda, K., Takizawa, H., Endo, T., Yamane, H., Shimada, M., Wang, C. M., and Mitomo, M. 2000. Synthesis and luminescent property of Eu^{3+}-doped $LaSi_3N_5$ phosphor. *J. Lumin.* 87–89:967–969.

van Krevel, J. W. H., Hintzen, H. T., Metselaar, R., and Meijerink, A. 1998. Long wavelength Ce^{3+} emission in Y-Si-O-N materials. *J. Alloys Compd.* 268:272–277.

van Krevel, J. W. H., van Rutten, J. W. T., Mandal, H., Hintzen, H. T., and Metselaar, R. 2002. Luminescence properties of terbium-, cerium-, or europium-doped α-sialon materials. *J. Solid State Chem.* 165:19–24.

Wang, B. N. and Callahan, M. J. 2006. Ammonothermal synthesis of III-nitride crystals. *Crystal Growth & Design.* 6:1227–1246.

Watanabe, H., Yamane, H., and Kijima, N. 2008a. Crystal structure and luminescence of $Sr_{0.99}Eu_{0.01}AlSiN_3$. *J. Solid State Chem.* 181:1848–1852.

Watanabe, H., Wada, H., Seki, K., Itou, M., and Kijima, N. 2008b. Synthetic method and luminescence properties of $Sr_xCa_{1-x}AlSiN_3:Eu^{2+}$ mixed nitride phosphors. *J. Electrochem. Soc.* 155:F31–F36.

Weyl, W. 1864. Ueber metallammonium-verbindungen. *Annalen der Physik.* 197:601–612.

Woike, M., and Jeitschko, W. 1995. Preparation and crystal structure of the nitridosilicates $Ln_3Si_6N_{11}$ (Ln = La, Ce, Pr, Nd, Sm) and $LnSi_3N_5$ (Ln = Ce, Pr, Nd). *Inorg. Chem.* 34:5105–5108.

Xie, R.-J., and Hirosaki, N. 2007a. Silicon-based oxynitride and nitride phosphors for white LEDs—A review. *Sci. Technol. Adv. Mater.* 8:588–600.

Xie, R.-J., Hirosaki, N., Li, H. L., Li, Y. Q., and Mitomo, M. 2007. Synthesis and photoluminescence properties of β-sialon:Eu^{2+} $(Si_{6-s}Al_zO_zN_{8-z}:Eu^{2+})$—A promising green oxynitride phosphor for white light-emitting diodes. *J. Electrochem. Soc.* 154:J314–J319.

Xie, R.-J., Hirosaki, N., Liu, X.-J., Takeda, T., and Li, H.-L. 2008. Crystal structure and photoluminescence of Mn^{2+}-Mg^{2+} codoped gamma aluminum oxynitride (γ-alon): A promising green phosphor for white light-emitting diodes. *Appl. Phys. Lett.* 92:201905.

Xie, R.-J., Hirosaki, N., Mitomo, M., Sakuma, K., and Kimura, N. 2006a. Wavelength-tunable and thermally stable Li-α-SiAlON: Eu^{2+} oxynitride phosphors for white light-emitting diodes. *Appl. Phys. Lett.* 89:241103.

Xie, R.-J., Hirosaki, N., Mitomo, M., Suehiro, T., Xu, X., and Tanaka, H. 2005a. Photoluminescence of rare-earth-doped Ca-α-sialon phosphors: Composition and concentration dependence. *J. Am. Ceram. Soc.* 88:2883–2888.

Xie, R.-J., Hirosaki, N., Mitomo, M., Takahashi, K., and Sakuma, K. 2006b. Highly efficient white-light-emitting diodes fabricated with short-wavelength yellow oxynitride phosphors. *Appl. Phys. Lett.* 88:101104.

Xie, R.-J., Hirosaki, N., Mitomo, M., Uheda, K., Suehiro, T., Xu, X., Yamamoto, Y., and Sekiguchi, T. 2005b. Strong green emission from α-sialon activated by divalent ytterbium under blue light irradiation. *J. Phys. Chem. B* 109:9490–9494.

Xie, R.-J., Hirosaki, N., Mitomo, M., Yamamoto, Y., and Suehiro, T. 2004a. Optical properties of Eu^{2+} in α-SiAlON. *J. Phys. Chem. B* 108:12027–12031.

Xie, R.-J., Hirosaki, N., Mitomo, M., Yamamoto, Y., Suehiro, T., and Ohashi, N. 2004b. Photoluminescence of cerium-doped α-sialon material. *J. Am. Ceram. Soc.* 87:1368–1370.

Xie, R.-J., Hirosaki, N., Sakuma, K., Yamamoto, Y., and Mitomo, M., 2004c. Eu^{2+}-doped Ca-alpha-SiAlON: A yellow phosphor for white light-emitting diodes. *Appl. Phys. Lett.* 84:5404–5406.

Xie, R.-J., Hirosaki, N., Suehiro, T., Xu, F.-F., and Mitomo, M. 2006c. A simple, efficient synthetic route to $Sr_2Si_5N_8$:Eu^{2+}-based red phosphors for white lighting-emitting diodes. *Chem. Mater.* 18:5578–5583.

Xie, R.-J., Mitomo, M., Uheda, K., Xu, F.-F., and Akimune, Y. 2002. Preparation and luminescence spectra of calcium- and rare-earth (R = Eu, Tb, and Pr)-codoped α-SiAlON ceramics. *J. Am. Ceram. Soc.* 85:1229–1234.

Xiong, Z., Chen, P., Wu, G., Lin, J., and Tan, K. L. 2003. Investigations into the interaction between hydrogen and calcium nitrides. *J. Mater. Chem.* 13:1676–1680.

Yamada, S. 2010. Properties of SiAlON powder phosphors for white LEDs. Paper presented at the 3rd International Symposium on SiAlONs and Non-Oxides, Cappadocia, Turkey.

Yamane, H., and DiSalvo, F. J. 1996. Preparation and crystal structure of a new barium silicon nitride, $Ba_5Si_2N_6$. *J. Alloys. Compds.* 240:33–36.

Yamane, H., Kikkawa, S., and Koizumi, M. 1985. Lithium aluminum nitride, Li_3AlN_2, as a lithium solid electrolyte. *Solid State Ionic* 15:51–54.

Yamane, H., Kikkawa, S., and Koizumi, M. 1987. Preparation of lithium silicon nitrides and their lithium ion conductivity. *Solid State Ionic* 25:183–191.

Yen, W. M., Shionoya, S., and Yamamoto, H. 2006. *Phosphor handbook.* 2nd ed. Boca Raton, FL: CRC Press.

Zhang, H. C., Horikawa, T., Hanzawa, H., Hamaguchi, A., and Machida, K. 2007. Photoluminescence properties of α-sialon:Eu^{2+} prepared by carbothermal reduction and nitridation method. *J. Electrochem. Soc.* 154:J59–J61.

Zeuner, M., Schmidt, P. J., and Schnick, W. 2009. One-pot synthesis of single-source precursors for nanocrystalline LED-phosphors $M_2Si_5N_8$:Eu^{2+} (M= Sr, Ba). *Chem. Mater.* 21:2467–2473.

Zhou, Y., Yoshizawa, Y. I., Hirao, K., Lences, Z., and Sajgalik, P. 2008. Preparation of Eu-doped β-SiAlON phosphors by combustion synthesis. *J. Am. Ceram. Soc.* 91:3082–3085.

5

Structural Analysis of Nitride Phosphors

Structural analysis is the essential procedure for a better understanding of the underlying nature of nitride phosphors. It is also important to explore the ways to optimize and further improve the luminescence performance of nitride phosphors. Similar to oxide-based phosphors (Blasse and Grabmaier 1994; Yen et al. 2006), the luminescence properties of nitride phosphors are strongly dependent on the composition and crystal structure, in particular the local structure of the dopant sites, i.e., the environment around the activator ions, like Eu^{2+} and Ce^{3+}, because their 5d electrons are very sensitive to the type of first- and second-nearest coordination atoms, point symmetry, and bond lengths. Numerous techniques have been used for structural analysis. In this chapter, we will focus on the introduction of the applications of x-ray diffraction (XRD), electron microscopy (e.g., scanning electron microscopy (SEM), transmission electron microscopy (TEM), and scanning transmission electron microscopy (STEM)), and x-ray absorption near-edge structure/near-edge x-ray absorption fine-structure (XANES/EXAFS) techniques for the typical nitride phosphors. We will not, however, cover these specific techniques in detail. Furthermore, as the emerging powerful tools in a combination of experiments, we will discuss the theoretical approaches of first-principles calculations based on the density functional theory (DFT) approximation and molecular orbitals (MOs) cluster calculations on the selected nitride phosphor systems.

5.1 X-ray Diffraction and Rietveld Refinement

X-ray diffraction (XRD) techniques are the most widely used methods to analyze nitride phosphors because the nitride phosphors are generally powdered polycrystalline materials. A beam of x-rays is directed on a crystalline material that may experience diffraction as a result of the interaction with a series of parallel atomic planes. For a given set of lattice planes with an interplanar spacing of d, the condition for a diffraction peak to occur can be represented by Bragg's law:

$$2d\sin\theta = n\lambda \qquad (5.1)$$

where λ is the wavelength of x-ray, θ is the scattering angle, and *n* is an integer order of the diffraction peak that is directly related to the atomic distances in the lattices. Additionally, the interplanar spacing is a function of the Miller indices and the lattice parameters, as well as the crystal structure (Clearfield et al. 2008; Pecharsky and Zavalij 2008). Because the wavelength of x-rays (i.e., CuKα = 1.542 Å) is comparable to the size of atoms, x-ray diffraction is significantly suitable for determining the atomic arrangement in the nitride phosphors. The profile of x-ray diffraction patterns, such as the peak position, intensity, width, and shape, provides important information about the chemical components of the phosphors as well as the basic crystal structure of a target phase (Pecharsky and Zavalij 2008). Typical x-ray powder diffraction applications for the nitride phosphors are related to qualitative and quantitative phase analysis, precise lattice parameters, and crystal structure determination for a doping material.

5.1.1 Qualitative X-ray Diffraction Analysis

Relatively, qualitative x-ray powder diffraction analysis is the most important approach to identify the obtained phases in phosphors, by comparison with the reference patterns from the International Centre for Diffraction Data (ICCD) database (i.e., powder diffraction file (PDF) card), using simply a search-match process. The positions and intensities of the diffraction peaks have been used for identifying the underlying polycrystalline phase of the nitride phosphors. For example, the x-ray diffraction lines of calcium silicon nitride of cubic $CaSiN_2$ are quite different from orthorhombic $CaSiN_2$, even though they are made of Ca, Si, and N with a small amount of impurity oxygen in composition that results in two different structures (Lee et al. 1997; Zoltán et al. 2004; Le Toquin and Cheetham 2006). Cubic $CaSiN_2$ is crystallized in the space group F23 with the lattice parameter a = 14.8806(1) Å. Orthorhombic $CaSiN_2$ has a space group Pbca with the lattice parameters a = 5.1295(1) Å, b = 10.3004(2) Å, and c = 14.5320(2) Å. The x-ray powder diffraction patterns of $CaSiN_2$ for the two phases are shown in Figure 5.1. The phase identification is important because the optical properties of luminescent materials are highly dependent on the crystal structure. As an example, due to the different structures of cubic and orthorhombic phases, the luminescence properties of $CaSiN_2:Eu^{2+}$ and $CaSiN_2:Ce^{3+}$ of the two modifications show significant differences, as listed in Table 5.1.

As seen in Table 5.1, the luminescence properties and the thermal stability of Eu^{2+} and Ce^{3+} significantly change from the cubic to the orthorhombic structure of $CaSiN_2$ due to the large differences of the environment around the activator ions (Li et al. 2007, 2009). $CaSiN_2:Eu^{2+}$ is a red-emitting phosphor for both cubic and orthorhombic phases, while $CaSiN_2:Ce^{3+}$ is a red- and a yellowish-orange-emitting phosphor for the cubic and orthorhombic structures, respectively. In addition, the absorption bands of the cubic phase for both Eu^{2+} and Ce^{3+} are at longer wavelengths than they are for the

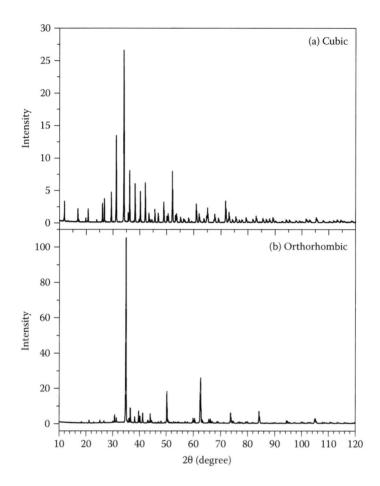

FIGURE 5.1
X-ray powder diffraction patterns of $CaSiN_2$: (a) cubic phase and (b) orthorhombic phase.

TABLE 5.1

Photoluminescence Properties of Eu^{2+} and Ce^{3+}-Doped $CaSiN_2$ for Two Structures

	Cubic		Orthorhombic	
Structure Composition	**$CaSiN_2$:Eu^{2+}**	**$CaSiN_2$:Ce^{3+}**	**$CaSiN_2$:Eu^{2+}**	**$CaSiN_2$:Ce^{3+}**
Host optical band gap	3.44 eV		4.16 eV	
Absorption band	440–530 nm	405, 550 nm	400 nm	430 nm
Excitation, λ_{exc}	400 nm	555 nm	410 nm	425 nm
Emission, λ_{em}	640 nm	645 nm	630 nm	593 nm
Quantum efficiency	15%	18%	23%	28%
Thermal quenching, $T_{1/2}$	~200°C	~250°C	<107°C	<155°C

orthorhombic phase, and the band of the former phosphor can even extend to the green absorption range (~540 nm). This typical example demonstrates that phase identification is an essential procedure to determine the existing phases in nitride phosphors. It is worth noting that effective phase identification should be supported by high-quality x-ray diffraction data that consist of careful specimen preparation and good setting conditions of x-ray powder diffractometers.

5.1.2 Quantitative X-ray Diffraction Analysis

Quantitative analysis of x-ray diffraction data is mainly used to determine the amounts of particular phases in the multiphase systems, the crystallite size, and the degree of crystallinity, as well as the lattice parameters of a specific phase, in the synthesized phosphor products. Quantitative x-ray powder diffraction analysis requires precise determination of the peak position and the intensity of the x-ray diffraction patterns. Usually, the peak position is related to the crystal structure and symmetry of the phase, and the peak intensity is associated with the total scattering from the plane and the atom distribution in a structure. Like other phosphors, the nitride phosphors are usually required to be a single-phase material in order to avoid all possible scattering loss arising from impurity phases. It is especially important for pure nitride phosphors, e.g., $Sr_2Si_5N_8$:Eu^{2+} and $CaAlSiN_3$:Eu^{2+}. Oxygen contamination in the lattices would remarkably decrease the quantum efficiency of those red-emitting phosphors and degrade their thermal stability at high temperature, as found in $Sr_2Si_5N_8$:Eu^{2+} with Sr_2SiO_4 as the second phase (Xie et al. 2006b). Therefore, quantitative estimation of the impurity phases, particularly for oxide impurities, is necessary to check the phase formation of the nitride phosphors.

There are many quantitative analysis methods that have been widely used for the estimation of a given phase: the absorption-diffraction method, the internal standard method, and the reference intensity ratio (RIR) method, in which those methods are based on the peak intensities. For detailed analysis of the methods, refer to related literature (Clearfield et al. 2008; Pecharsky and Zavalij 2008). Moreover, the full pattern analysis, i.e., the Rietveld method, is also extensively used for multi-phase-content calculations based on known structural models, apart from the structural refinement (Clearfield et al. 2008; Pecharsky and Zavalij 2008).

Accurate determination of the lattice parameters is a useful routine for understanding the formation of solid solutions and further modification of the luminescence behaviors in the design of the nitride phosphors. As a typical example, $M_2Si_5N_8$:Eu^{2+} (M = Ca, Sr, Ba) is a family of red-emitting phosphors with unusual long-wavelength emission of Eu^{2+} (620–680 nm), having strong absorption bands in the blue region (Hoppe et al. 2000; Li et al. 2006). Due to the structural variations with the M type, as expected, the solid solubility of Eu^{2+} is different in those hosts with a change of M (Li et al. 2006). Eu^{2+}-doped $Ca_2Si_5N_8$ forms a very limited solid solution with a maximum

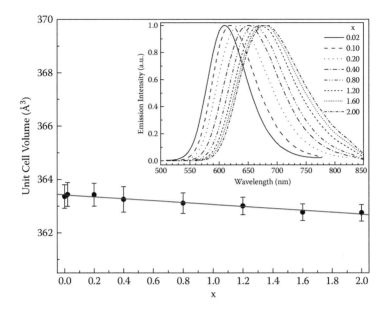

FIGURE 5.2

Unit cell volume of $Sr_{2-x}Eu_xSi_5N_8$ as a function of the Eu concentration (x). Inset shows the emission band of $Sr_2Si_5N_8:Eu^{2+}$ varying with the Eu concentration (x).

solubility of about 7 mol%. The Eu^{2+} ion, however, can be totally incorporated into $Sr_2Si_5N_8$ and $Ba_2Si_5N_8$, forming complete solid solutions, because the $M_2Si_5N_8$ (M = Sr, Ba) compounds are isostructural with $Eu_2Si_5N_8$ (Schlieper and Schnick 1995; Huppertz and Schnick 1997). Figure 5.2 shows the relationship between the Eu^{2+} concentration and the lattice parameters (i.e., unit cell volume) of $Sr_2Si_5N_8:Eu^{2+}$.

The lattice parameters were calculated by the least squares method using Si as an internal standard reference mixed with nitride phosphors of $Sr_2Si_5N_8:Eu^{2+}$ (Li et al. 2006). Consistent with Vegard's law, it is clearly seen that the unit cell volume linearly decreases with an increase of the Eu^{2+} concentration in $Sr_2Si_5N_8:Eu^{2+}$, suggesting that Eu^{2+} can dissolve into the host lattice and form a complete substitution solid solution. This can be explained by the fact that the ionic size of Eu^{2+} is very similar to that of Sr^{2+} (Shannon 1976). Accordingly, owing to the structural relaxation, the emission band of Eu^{2+} shows a red-shift and the position of the emission bands of $Sr_2Si_5N_8:Eu^{2+}$ can be tuned from 610 to 680 nm by varying the Eu^{2+} concentration, which is very flexible for the design of pc-LED devices, i.e., brightness, Commission International del'Eclairge (CIE) color point, and color rendering index (Li et al. 2006). Apart from the adjustment of Eu^{2+} content in $Sr_2Si_5N_8:Eu^{2+}$, the emission peaks can also be varied by partial replacement of Sr by Ca or Ba, and structural changes (the lattice parameters) can be measured by quantitative analysis of x-ray powder diffraction data (Li et al. 2008a).

5.1.3 Rietveld Refinement of X-ray Powder Diffraction

The Rietveld refinement method (also named full-pattern analysis) of x-ray powder diffraction is a more powerful technique than other quantitative analyses based on the relative intensity of a few identified diffraction peaks, since in the Rietveld refinement analysis the diffraction pattern combines both instrumental and specimen effects (Clearfield et al. 2008; Pecharsky and Zavalij 2008; Rietveld 1969). In addition, the Rietveld refinement of x-ray diffraction can get more detailed crystal structure information of studied phosphors, which enables us to confirm a hypothetical crystal structure and refine the lattice parameters, atomic positions, and fractional occupancy, as well as thermal or displacement parameters of constitutional atoms. Furthermore, it can determine the local structure around the dopant sites in the lattices, e.g., interatomic distances, angles, and point symmetry.

The Rietveld method, based on the x-ray powder diffraction analysis, refines the selected parameters to minimize the difference between experimental XRD data and a structural model on the hypothesized crystal structure and the calculated data. The assumed model is refined by minimizing the residual by a least square process:

$$R = \sum_i w_i \left| y_{io} - y_{ic} \right|^2 \tag{5.2}$$

where y_{io} and y_{ic} are the observed and calculated intensities, respectively, at the ith step in the data point, and w_i is the weight associated with the standard deviation of the peak and the background intensities (Rietveld 1969; Young 1995). For a successful refinement, it is important to obtain high-quality experimental XRD data, an initial structure model, and suitable peak and background functions. Since the Rietveld method tries to fit a multivariable structure, background, and profile model to the experimental pattern, in general, it needs to refine the most important variables, and then other remaining variables step by step, in order to make the residual R-factors as small as possible in the course of the refinement. As a general refinement process, the refinement parameters normally include the scale factor, zero shift, linear background, lattice parameters, peak width, atomic positions, preferred orientation, isotropic displacement parameters, and profile parameters. Many free Rietveld refinement programs for x-ray powder diffraction are widely used, such as Generalized Structure and Analysis Software (GSAS) (Larson and Von Dreele 2000), FullProf (Rodriguez-Carvajal 1993), and RIETveld Analysis (RIETAN)-FP (Izumi and Momma 2007) packages. In the following section, we will demonstrate the Rietveld refinement of powder XRD applications for the analysis of the crystal structures of some typical nitride phosphors using the GSAS package combined with Experiment Graphical User Interface (EXPUI) (Larson and Von Dreele 2000; Toby 2001).

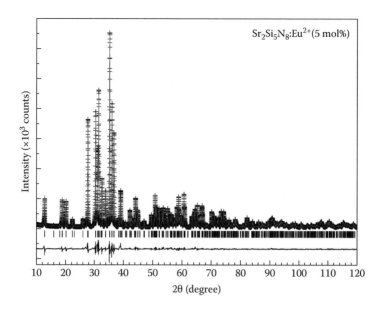

FIGURE 5.3

Rietveld refinement of x-ray powder diffraction patterns of $Sr_2Si_5N_8$:Eu^{2+} (5 mol%). (Reprinted from Li, Y. Q., de With, G., and Hintzen, H. T., *J. Solid State Chem.*, 181, 515–524, 2008a. With permission.)

5.1.3.1 $Sr_2Si_5N_8$:Eu^{2+}

$Sr_2Si_5N_8$:Eu^{2+} is a well-known red phosphor with a broad emission band of Eu^{2+} ranging from 610 to 680 nm, depending on the Eu^{2+} concentration when excited by blue light at 450 nm (Li et al. 2006). Moreover, $Sr_2Si_5N_8$:Eu^{2+} possesses high quantum efficiency (~80%) and excellent thermal stability, and has already been proved to be an excellent wavelength-conversion phosphor for pc-LED applications. The unusual long-wavelength emission of Eu^{2+} in a nitride host lattice is strongly dependent on its novel crystal structure and high electronegative N^{3-}, with a high formal charge in the host directly connected with the activator ion of Eu^{2+}. In order to design or tune the luminescence properties of $Sr_2Si_5N_8$:Eu^{2+}, it is important to fully understand the detailed crystallography and crystal chemistry of Eu^{2+}-doped $Sr_2Si_5N_8$. Figure 5.3 shows the Rietveld refinement of the x-ray powder diffraction pattern of $Sr_2Si_5N_8$:Eu^{2+} (5 mol%). The crystal structure data of $Sr_2Si_5N_8$:Eu^{2+} (5 mol%) are given in Table 5.2.

$Sr_2Si_5N_8$:Eu^{2+} is crystallized in an orthorhombic crystal system with space group $Pmn2_1$ (2006; Schlieper and Schnick 1995), in which the activator ions of Eu^{2+} occupy the Sr sites in the lattice due to their similar ionic radii (e.g., Sr^{2+}: 1.21 Å, Eu^{2+}: 1.20 Å for coordination number (CN) = 7 (Shannon 1976)). As shown in Figure 5.4, the SiN_4 tetrahedra make up the three-dimensional

TABLE 5.2

Crystallographic Data for $Sr_2Si_5N_8:Eu^{2+}$ (5 mol%)

Formula	$Sr_{1.9}Eu_{0.1}Si_5N_8$
Space group	Orthorhombic $Pmn2_1$ (No. 31)
Lattice parameters	$a = 5.7069(1)$ Å, $b = 6.8142(1)$ Å, $c = 9.3269(1)$ Å
	$V = 362.71(1)$ Å^3
Z	2
R_{wp}	0.082
R_p	0.057
χ^2	3.8

Atom	Wyck.	SOF	x	y	z	U_{iso} [Å^2]
(Sr/Eu)1	2a	0.95/0.05	0	0.8708(4)	−0.0014(1)	0.01392
(Sr/Eu)2	2a	0.95/0.05	0	0.8790(4)	0.3665	0.01415
Si1	4b	1.0	0.2525(3)	0.6646(3)	0.6816(12)	0.01311
Si2	2a	1.0	0	0.0552(4)	0.6716(8)	0.01293
Si3	2a	1.0	0	0.4165(13)	0.4562(10)	0.0129
Si4	2a	1.0	0	0.4103(13)	0.8986(9)	0.01145
N1	2a	1.0	0	0.1975(25)	0.5234(28)	0.01275
N2	4b	1.0	0.2466(7)	0.9063(7)	0.6856(19)	0.01416
N3	4b	1.0	0.2496(10)	0.4523(9)	0.0083(8)	0.01196
N4	2a	1.0	0	0.5820(12)	0.7693(8)	0.01309
N5	2a	1.0	0	0.1773(24)	0.8444(27)	0.00426
N6	2a	1.0	0	0.4226(11)	0.2619(9)	0.00647

Source: Data from Li, Y. Q., de With, G., and Hintzen, H. T., *J. Solid State Chem.*, 181, 515–524, 2008a.

network by corner-sharing N atoms in the form of $N^{[2]}$ and $N^{[3]}$. There are two individual Sr/Eu crystallographic sites in this structure, and the Sr/Eu ions are located in the silicon-nitrogen channels along (100) surrounded by the six-membered rings of the SiN_4 tetrahedra, and they are directly connected by nitrogen atoms with the coordination numbers of 6 and 7, respectively, for (Sr/Eu)1 and (Sr/Eu)2 within the range of 3.0 Å (Figure 5.4). The average bond lengths of (Sr/Eu)1 and (Sr/Eu)2 are about 2.783 and 2.811 Å, respectively. Besides the framework of a rigid lattice, i.e., a high degree of cross-linking of the SiN_4 units, a pure nitrogen environment with appropriate coordination numbers, as well as a suitable ratio of Sr metal/silicon (Sr/Si = 0.4), results in the lowest of the excited levels of Eu^{2+} (the 4f → 5d transition), downward to lower energy, due to a high *nephelauxetic* effect and large crystal field splitting. Moreover, $Sr_2Si_5N_8:Eu^{2+}$ phosphor has high chemical and thermal stability owing to its peculiar structure characterizations. On the other hand, the nearest interatomic distance of (Sr/Eu)1-(Sr/Eu)2 is relatively short, about 3.4322 Å, which also implies that the interaction between two Eu^{2+} doping sites is also probably an important factor of the luminescence properties of $Sr_2Si_5N_8:Eu^{2+}$, especially the energy transfer between Eu^{2+}-Eu^{2+} ions, which needs to be further studied for the two individual Sr/Eu sites.

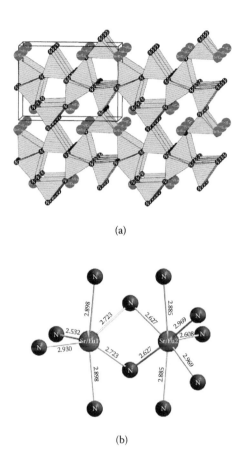

(a)

(b)

FIGURE 5.4

(a) Projection of the crystal structure of $Sr_2Si_5N_8$:Eu^{2+}, viewed along (100). (b) The local coordination of the Sr/Eu atom's coordination with N, viewed along (010).

5.1.3.2 $CaAlSiN_3$:Eu^{2+}

$CaAlSiN_3$:Eu^{2+} is another well-developed red-emitting nitride phosphor having a broad emission band of about 640–670 nm with the variation of the Eu^{2+} concentration under excitation in the UV and blue spectral range (Uheda et al. 2006a, 2006b). Figure 5.5 depicts the Rietveld refinement of x-ray diffraction patterns of $CaAlSiN_3$:Eu^{2+} (1 mol%). The crystal structure data, atomic positions, and selected interatomic distances and bond angles are summarized in Tables 5.4 and 5.5. The host lattice of $CaAlSiN_3$ is isostructural, with Si_2N_2O crystallized in an orthorhombic unit cell having a $Cmc2_1$ space group (Uheda et al. 2006b; Ottinger et al. 2004), in which the metal ions of Ca have only one crystallographic site with a fivefold coordination with the nitrogen atoms (within the range of 3 Å) in the lattice (Figure 5.6). The Si and Al atoms randomly occupy the equivalent site at the 8b Wyckoff sites. The

TABLE 5.3

Selected Atomic Distances of $Sr_2Si_5N_8$:Eu^{2+} (5 mol%)

Sr/Eu1-N5	2.536(18)	Si1-N2	1.648(5)
Sr/Eu1-N2	2.728(12) × 2	Si1-N6	1.705(6)
Sr/Eu1-N1	2.900(2) × 2	Si1-N3	1.750(6)
Sr/Eu1-N4	2.906(8)	Si1-N4	1.802(9)
Mean	2.783 ± 0.148	Mean Si1-N	1.726(7)
Sr/Eu2-N1	2.617(19)	Si2-N1	1.689(25)
Sr/Eu2-N2	2.661(12) × 2	Si2-N2	1.740(5) × 2
Sr/Eu2-N5	2.887(3) × 2	Si2-N5	1.814(25)
Sr/Eu2-N3	2.981(7) × 2	Mean Si2-N	1.745(15)
Mean	2.811 ± 0.159		
Si3-N1	1.618(19)	Si4-N5	1.666(18)
Si3-N3	1.755(7) × 2	Si4-N4	1.681(11)
Si3-N6	1.812(10)	Si4-N3	1.777(7) × 2

Source: Data from Li, Y. Q., de With, G., and Hintzen, H. T.,
         J. Solid State Chem., 181, 515–524, 2008a.

Eu^{2+} ions are partially substituted for Ca^{2+} at the 4a site in $CaAlSiN_3$:Eu^{2+}, and the solid solubility of Eu^{2+} is relatively lower in $CaAlSiN_3$ due to the large difference in ionic radius between Ca^{2+} (1.0 Å for C.N. = 6 (Shannon 1976)) and Eu^{2+} (1.17 Å for C.N. = 6). The three-dimensional network is composed of $(Si,Al)N_4$ tetrahedra, and the (Si,Al)-N channels along (001) are built up by the corner-shared $(Si,Al)N_4$ tetrahedral rings (like the conformation of a six-membered ring). Because of the presence of (Si,Al)-N bonds in $CaAlSiN_3$:Eu^{2+}, it is expected that the mixed $(Si,Al)N_4$ tetrahedra is larger than SiN_4 of $M_2Si_5N_8$:Eu^{2+} (see Tables 5.3 and 5.5). The activator ions of Eu^{2+} situate at the channels and connect by five nitrogen atoms in the range of 3 Å (Figure 5.6), with an average bond length of Ca/Eu-N of about 2.4949 Å, which is much shorter than those of Ca/Eu-N in $Ca_2Si_5N_8$:Eu^{2+} (~2.571 Å) (Li et al. 2006).

Accordingly, it is expected that the Eu^{2+} ions experience much stronger crystal field strength around the dopant Ca sites in the $CaAlSiN_3$ lattice, which, in contrast to $M_2Si_5N_8$:Eu^{2+} (M = Ca, Sr, Ba), leads to the emission band of $CaAlSiN_3$:Eu^{2+} extending to the deeper red region, even at a low Eu^{2+} concentration, because of those structural characterizations. Besides the structural factors, the electronic structures also can shed light on this issue (see Sections 5.3 and 5.4).

5.1.3.3 Ca-α-SiAlON:Eu²⁺

As described in the previous chapters, Ca-α-SiAlON:Eu^{2+} is an alternative yellow emission phosphor of YAG:Ce^{3+} (van Krevel et al. 2002; Xie et al. 2002, 2004a, 2004b, 2005, 2006a). According to the nitrogen and oxygen ratio, it is

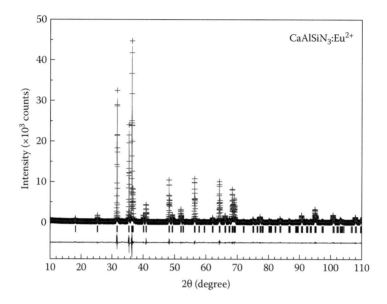

FIGURE 5.5
Rietveld refinement of x-ray powder diffraction patterns of CaAlSiN$_3$:Eu^{2+} (1 mol%).

TABLE 5.4

Crystal Structure Data of CaAlSiN$_3$:Eu^{2+} (1 mol%)

Formula	Ca$_{0.99}$Eu$_{0.01}$AlSiN$_3$
Crystal system	Orthorhombic
Space group	Cmc2$_1$ (36)
Z	4
Lattice parameters	$a = 9.7818(2)$, $b = 5.6510(1)$, $c = 5.0590(1)$
	$V = 279.65(1)$
R_{wp}	0.098
R_p	0.070
χ^2	5.0

Atom	Wyck.	x	y	z	SOF	U (100 Å^2)
Ca1	4a	0	0.3160(4)	0.4955	0.99	1.48(1)
Eu1	4a	0	0.3160(4)	0.4955	0.01	1.48(1)
Al1	8b	0.1717(1)	0.1541(4)	0.0221(5)	0.5	0.35(2)
Si1	8b	0.1717(1)	0.1541(4)	0.0221(5)	0.5	0.35(2)
N1	8b	0	0.1259(11)	0.3767(6)	1.0	0.13(2)
N2	4a	0	0.2402(8)	−0.0233(13)	1.0	0.17(1)

termed both a yellow (high oxygen content) and an orange (high nitrogen content) phosphor. The crystal structure of α-SiAlON can be derived from the α-Si$_3$N$_4$ lattice by partial substitution of Si^{4+} for Al^{3+}, and the charge balance can be achieved either through the substitution of N^{3-} by O^{2-} or by the introduction of metal ions, forming a solid solution of M$_{m/v}$Si$_{12-m-n}$Al$_{m+n}$O$_n$N$_{16-n}$

TABLE 5.5

Selected Interatomic Distances (Å) and Angles (°)
of $CaAlSiN_3:Eu^{2+}$ (1 mol%)

(Ca/Eu)1-N1	2.4168(28) × 2
(Ca/Eu)1-N2	2.4720(7)
(Ca/Eu)1-N2	2.5100(5)
(Ca/Eu)1-N2	2.6590(6)
Mean	2.4949
(Si/Al)1-N2	1.7631(22)
(Si/Al)1-N1	1.790(6)
(Si/Al)1-N1	1.835(7)
(Si/Al)1-N1	1.845(4)
Mean	1.8083
N2-(Ca/Eu)1-N2	160.75(22)
N1-(Ca/Eu)1-N1	118.76(32)
N1-(Ca/Eu)1-N2	115.78(18) × 2
N2-(Ca/Eu)1-N2	102.15(19)
N1-(Ca/Eu)1-N2	71.53(9) × 2
N1-(Ca/Eu)1-N2	99.65(10) × 2
N2-(Ca/Eu)1-N2	97.10(22)

(M = Ca, Li, Mg, Y, and v is the valency of M) (Jack and Wilson 1972; Jack 1976; Ekstrom and Nygren 1992; Petzow 2002; Cao and Metselaar 1991). Having the same crystal structure of α-Si_3N_4 (Jack and Wilson 1972; Petzow and Herrmann 2002), Ca-α-SiAlON:Eu^{2+} crystallizes in a trigonal system with the space group P31c, while the unit cell is much larger than that of Si_3N_4 due to the incorporation of Al-O or Al-N and metal cations of M in the lattice. As a typical example, Figure 5.7 shows the Rietveld refinement of x-ray powder diffraction patterns of a nitrogen-rich Ca-α-SiAlON:Eu^{2+} with a formula of $Ca_{1.47}Eu_{0.03}Si_9Al_{2.91}B_{0.09}N_{16}$, in which small amounts of B partially replace Al using BN as a source material. As usual, it is assumed that Al and B are statistically occupied on the Si sites. In case of a high-oxygen-containing Ca-α-SiAlON, similarly, the O atoms are supposed to be randomly distributed on the N sites. The detailed crystal structure data and selected bond lengths and angles are summarized in Tables 5.6 and 5.7. Like the common Ca-α-SiAlON, the crystal structure of $Ca_{1.47}Eu_{0.03}Si_9Al_{2.91}B_{0.09}N_{16}$ mainly consists of the (Si,Al)-N network by bridging the corners of the (Si,Al) N_4 tetrahedra (in the present example, actually (Si,Al,B)N_4 tetrahedra), and the modified metal cations of Ca, together with the activator ions of Eu^{2+} occupied on the Ca sites, are located at the interstices of the tetrahedral framework, as shown in Figure 5.8. However, the two large voids cannot be fully filled by Ca/Eu because a high metal content would destabilize the α-structure (in general, the m value is about 1 to 4 for a single-phase Ca-α-SiAlON formation range). The Ca/Eu atoms are coordinated by seven

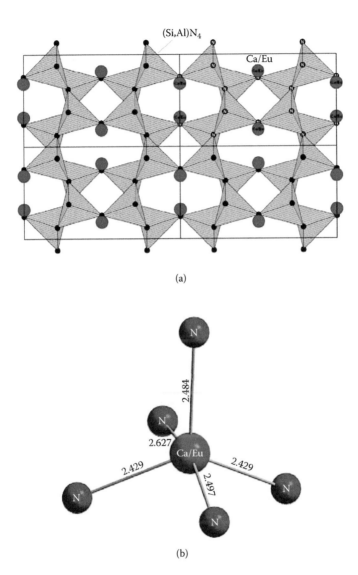

(a)

(b)

FIGURE 5.6
(a) Crystal structure of CaAlSiN$_3$:Eu^{2+} (1 mol%), viewed along (001). (b) The local coordination of the Ca/Eu atom with the N atoms.

nitrogen atoms in the range of 3 Å, and the average Ca/Eu-N distance is about 2.612 Å (see Figure 5.9). In addition, it is unusual in other nitride compounds that three (Si,Al,B) (mainly (Si,Al)) atoms are also present in this range, suggesting that nonmetal cations of (Si,Al) probably have somewhat of an influence, e.g., polarization effects on the activator ion of Eu^{2+}, on the luminescence properteis of M-α-SiAlON:Eu^{2+}. This is in agreement with the experimental results that the quantum efficiency is nearly linearly

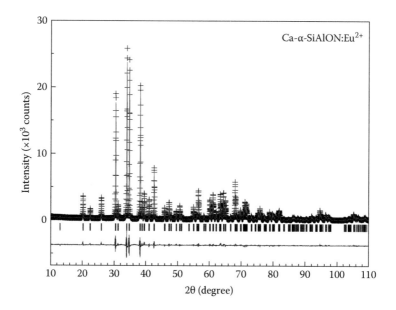

FIGURE 5.7
Rietveld refinement of x-ray powder diffraction patterns of the Ca-a-SiAlON:Eu^{2+} phosphor with the composition of Ca$_{1.5}$Si$_9$Al$_{2.91}$B$_{0.09}$N$_{16}$:Eu^{2+} (2 mol%).

increased with an increase in the m value in M$_{m/2}$Si$_{12-m-n}$Al$_{m+n}$O$_n$N$_{16-n}$, as illustrated in Figure 5.8. On the one hand, a high Ca/Eu content can enhance the absorption of Eu^{2+}, but on the other hand, a high m value can introduce more Al, which can significantly expand the unit cell as a result of increasing Ca/Eu-Si/Al distances, due to the replacement of Si-N with the longer Al-(N,O) bonds in the lattice. For the above-mentioned reasons, a high Ca content (i.e., high m value) is always desired for achieving highly efficient yellow-emitting M-α-SiAlON:Eu^{2+} (M = Ca, Mg, Li) phosphors, especially for orange-emitting Ca-α-SiAlON:Eu^{2+} phosphors.

5.1.3.4 β-SiAlON:Eu^{2+}

β-SiAlON:Eu^{2+} is a green phosphor with an emission peak of about 530 nm when excited by near-UV or blue light. Unlike Ca-α-SiAlON:Eu^{2+}, highly efficient green-emitting β-SiAlON:Eu^{2+} can only be realized in a nitrogen-rich composition. As a solid solution of β-Si$_3$N$_4$, Eu^{2+}-activated β-SiAlON, for example, with a formulation of Eu$_x$Si$_{6-z}$Al$_{z-x}$O$_{z+x}$N$_{8-z-x}$, also crystallizes in a centerosymmetric hexagonal system with the space group P6$_3$/m (or P6$_3$) having an enlarged unit cell owing to the replacement of Si-N by longer Al-O (Jack and Wilson 1972; Jack 1976; Ekstrom and Nygren 1992; Petzow 2002). For space group P6$_3$/m, (Si/Al) occupies a 6h and two different N occupy the 6h and 2c sites, respectively, in which the corner-connected (Si/Al)N$_4$ tetrahedra build up

TABLE 5.6

Crystal Structure Data for Ca-α-SiAlON:Eu^{2+}

Formula	$Ca_{1.47}Eu_{0.03}Si_9Al_{2.91}B_{0.09}N_{16}$
Crystal system	Trigonal
Space group	P31c (No. 159)
Z	1
Lattice parameters	$a = 7.89523(7)$ Å, $c = 5.72900(6)$ Å
	$V = 309.271(5)$ Å^3
R_{wp}	0.09
R_p	0.068
χ^2	4.9

Atom	x	y	z	U_{iso}*100	Wyck.	SOF
Ca1	0.3333	0.6667	0.2287(5)	2.12	2b	0.7350
Eu1	0.3333	0.6667	0.2287(5)	2.12	2b	0.0150
Si1	0.5117(2)	0.0835(2)	0.2094(3)	0.43	6c	0.7500
Al1	0.5117(2)	0.0835(2)	0.2094(3)	0.43	6c	0.2425
B1	0.5117(2)	0.0835(2)	0.2094(3)	0.43	6c	0.0075
Si2	0.1704(1)	0.2502(1)	−0.0041	0.27	6c	0.7500
Al2	0.1704(1)	0.2502(1)	−0.0041	0.27	6c	0.2425
B2	0.1704(1)	0.2502(1)	−0.0041	0.27	6c	0.0075
N1	0.0000	0.0000	0.0004(16)	0.06	2a	1.0000
N2	0.3333	0.6667	0.6468(11)	0.06	2b	1.0000
N3	0.3399(4)	−0.0541(4)	−0.0152(10)	0.09	6c	1.0000
N4	0.3223(5)	0.3187(5)	0.2476(11)	0.17	6c	1.0000

TABLE 5.7

Selected Atomic Distances (Å) and Angles (°) of Ca-α-SiAlON:Eu^{2+}

(Ca/Eu)1-N2	2.395(6)	N2-(Ca/Eu)1-N3	122.67(13) × 3
(Ca/Eu)1-N3	2.589(5) × 3	N2-(Ca/Eu)1-N4	87.70(15) × 3
(Ca/Eu)1-N4	2.707(4) × 3	N3-(Ca/Eu)1-N3	93.62(18)
Mean	2.612	N3-(Ca/Eu)1-N3	93.61(18)
		N3-(Ca/Eu)1-N4	149.62(25)
(Si/Al/B)1-N2	1.7614(17)	N3-(Ca/Eu)1-N4	66.82(12)
(Si/Al/B)1-N3	1.789(5)	N3-(Ca/Eu)1-N4	66.17(11)
(Si/Al/B)1-N3	1.786(6)	N3-(Ca/Eu)1-N3	93.61(18)
(Si/Al/B)1-N4	1.7561(32)	N3-(Ca/Eu)1-N4	66.15(11)
Mean	1.7731	N3-(Ca/Eu)1-N4	149.63(25)
		N3-(Ca/Eu)1-N4	66.82(12)
(Si/Al/B)2-N1	1.7477(12)	N3-(Ca/Eu)1-N4	66.81(12)
(Si/Al/B)2-N3	1.7829(28)	N3-(Ca/Eu)1-N4	66.15(11)
(Si/Al/B)2-N4	1.778(5)	N3-(Ca/Eu)1-N4	149.63(25)
(Si/Al/B)2-N4	1.747(5)	N4-(Ca/Eu)1-N4	119.826(21)
Mean	1.7639	N4-(Ca/Eu)1-N4	119.841(21)
		N4-(Ca/Eu)1-N4	119.855(21)

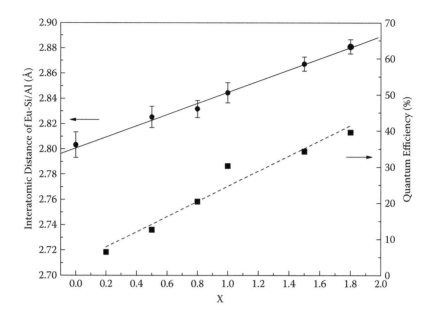

FIGURE 5.8
The relationship between the interatomic distance of Eu-Si/Al and the quantum efficiency of Ca-α-SiAlON:Eu^{2+} with the formulation of $Ca_xSi_{12-2x}Al_{2x}N_{16}$:$Eu^{2+}$ (2 mol% Eu^{2+}, x = 0.2–2.0).

the framework with a four-ring channel along (001) in the unit cell. Since the smallest radius of this channel is estimated to be about 1.586 Å, the Eu^{2+} ions cannot enter it due to the fact that the Eu-(N/O) bond length is normally above 2.5 Å (Huppertz and Schnick 1997; Marchand et al. 1978; Haferkorn and Meyer 1998) for Eu^{2+}. Therefore, the Eu^{2+} ion was presumably placed in the most promising site at 2b (0, 0, 0) in the β-Si_3N_4 lattice, where the six-membered rings of (Si/Al)N_4 tetrahedra construct a relative large channel within a 2 × 2 × 2 supercell model (Figure 5.10a). Considering a reasonable interatomic distance of Eu-Eu, if the Eu ions occupy the 2b site, Eu must be at either (0, 0, 0) or (0, 0, 0.5), but not both, in order to meet a reasonable Eu-Eu distance, which is found in many of the Eu compounds (the nearest Eu^{2+}-Eu^{2+} distance is greater than 3.4 Å at least) (Huppertz and Schnick 1997; Marchand et al. 1978; Haferkorn and Meyer 1998; Stadler et al. 2006; Li et al. 2004; Jacobsen et al. 1994) (Figure 5.10b). Subsequently, the crystal structure was refined by the Rietveld method, yielding a reasonable structural model for $Eu_xSi_{6-z}Al_{z-x}O_{z+x}N_{8-z-x}$ (Li et al. 2008b). This conclusion was further confirmed by the first-principles total energy calculations on different positions for Eu^{2+} along the (0, 0, Z) channel, with varying Z (i.e., the atomic position of Eu^{2+}) on the basis of a 2 × 2 × 2 supercell model with 113 atoms, as shown in Figure 5.10c. The total energy calculations clearly indicate that the Eu^{2+} ions at the (0, 0, 0) or (0, 0, 0.5) site have the lowest total energy (–918.72 eV) of β-SiAlON:Eu^{2+}, implying that the Eu^{2+} ions are favorable for occupying either the (0, 0, 0) or (0, 0, 0.5) site in the lattice.

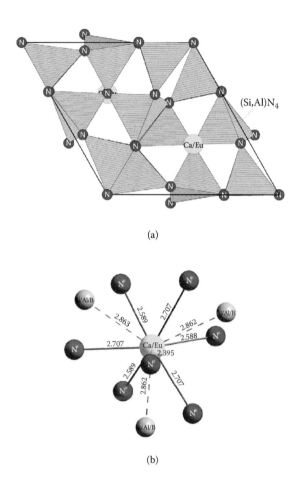

FIGURE 5.9
(a) Crystal structure of Ca-α-SiAlON:Eu^{2+}, viewed along (001). (b) The local coordination of the Ca/Eu atom with N and Si/Al/B.

Figure 5.11 depicts the Rietveldt refinement of x-ray diffraction patterns of β-SiAlON:Eu^{2+}. The detailed structure data and the atomic positions, as well as selected interatomic distances and bond angles, are given in Tables 5.8 and 5.9. In this model, Eu^{2+} is directly coordinated with six (N/O) atoms having the same bond length of 2.4932(18) Å (Table 5.9).

In addition, similar to M-α-SiAlON (M = Ca, Li), where the distances between M-(N/O) are relatively short (~2.833 Å), the nearest distances between Eu and (Si/Al) are as short as ~2.7767 (7) Å. In this way, the influence of the change of the Si/Al ratio on Eu^{2+} is expected to be effective. As a result, the luminescence properties are very sensitive to the ratio of Si/Al. It is worth noting that the nearest Eu-Eu distance is just about 2.9101(3) Å in $Eu_xSi_{6-z}Al_{z-x}O_{z+x}N_{8-z-x}$ (x = 0.013, z = 0.15), which is much shorter than that in regular Eu^{2+} compounds. This may explain why the maximum

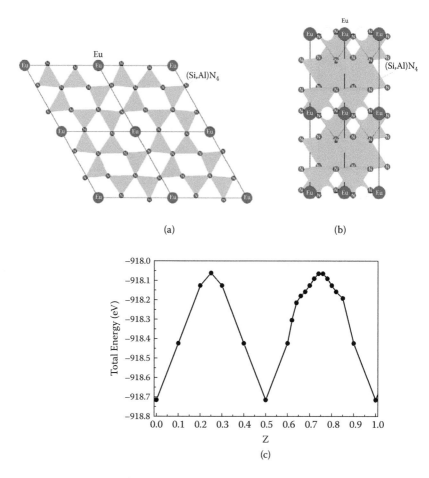

FIGURE 5.10

(See color insert.) Crystal structure of β-SiAlON:Eu^{2+}, viewed along (a) (001) and (b) (100). (Reprinted from Li, Y. Q., Hirosaki, N., Xie, R.-J., Takeda, T., and Mitomo, M., *J. Solid State Chem.*, 181, 3200–3210, 2008b. With permission.) (c) The total energy of β-SiAlON:Eu^{2+} as a function of the fractional position of the Eu atom along (0, 0, Z).

solid solubility of Eu^{2+} ($x \leq 0.03$) is very low and the single-phase range of Eu$_x$Si$_{6-z}$Al$_{z-x}$O$_{z+x}$N$_{8-z-x}$ is very narrow, from the structure point of view.

5.2 Electron Microscopy Observations

It is well known that the physical properties of phosphors are strongly related to the microstructure of a material, i.e., phases, defects, grain shape and size, and orientation. Electron microscopy can detect detailed

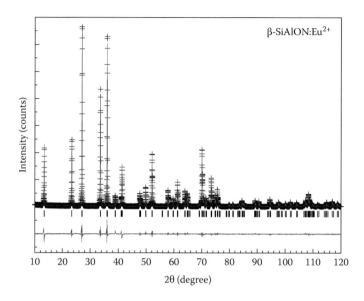

FIGURE 5.11

Rietveld refinement of x-ray powder diffraction patterns of β-SiAlON:Eu^{2+}. (Reprinted from Li, Y. Q., Hirosaki, N., Xie, R.-J., Takeda, T., and Mitomo, M., *J. Solid State Chem.*, 181, 3200–3210, 2008b. With permission.)

TABLE 5.8

Structure Data for β-SiAlON:Eu^{2+} (Eu$_x$Si$_{6-z}$Al$_{z-x}$O$_{z+x}$N$_{8-z-x}$ (z = 0.15, x = 0.013))

Formula	Eu$_{0.013}$Si$_{5.85}$Al$_{0.137}$O$_{0.163}$N$_{7.837}$
Crystal system	Hexagonal
Space group	P6$_3$/m (No. 176)
Z	2
a (Å)	7.6071(1)
c (Å)	2.9097(1)
V (Å^3)	145.82(1)
R$_{wp}$	12.8%
R$_p$	8.8%
χ^2	3.4

Atom	Wyck.	x	y	z	SOF	U (100 Å^2)
Si1	6h	0.1744(1)	−0.2311(1)	0.2500	0.9750	0.33
Al1	6h	0.1744(1)	−0.2311(1)	0.2500	0.0229	0.33
N1	6h	0.3279(1)	0.0311(1)	0.2500	0.9796	0.11
O1	6h	0.3279(1)	0.0311(1)	0.2500	0.0204	0.11
N2	2c	0.3333	0.6667	0.2500	0.9796	0.53
O2	2c	0.3333	0.6667	0.2500	0.0204	0.53
Eu1	2b	0.0000	0.0000	0.0000	0.0067	1.90

Source: Data from Li, Y. Q., Hirosaki, N., Xie, R.-J., Takeda, T., and Mitomo, M., *J. Solid State Chem.*, 181, 3200–3210, 2008b.

TABLE 5.9

Selected Interatomic Distances (Å) and Angles (°) for
$Eu_xSi_{6-z}Al_{z-x}O_{z+x}N_{8-z-x}$ (z = 0.15, x = 0.013)

(Si/Al)1-(N/O)1	1.7348(12) × 2
(Si/Al)1-(N/O)1	1.7356(21)
(Si/Al)1-(N/O)2	1.7340(8)
Eu1-(N/O)1	2.4932(18) × 6
(N/O)1-(Si/Al)1-(N/O)1	107.21(10) × 2
(N/O)1-(Si/Al)1-(N/O)1	113.99(12)
(N/O)1-(Si/Al)1-(N/O)2	107.21(10)
(N/O)1-(Si/Al)1-(N/O)2	110.44(8) × 2
(N/O)1-Eu1-(N/O)1	68.142(11) × 6
(N/O)1-Eu1-(N/O)1	111.858(11) × 6
(N/O)1-Eu1-(N/O)1	179.92 × 2
(N/O)1-Eu1-(N/O)1	179.98

Source: Data from Li, Y. Q., Hirosaki, N., Xie, R.-J., Takeda, T., and
Mitomo, M., *J. Solid State Chem.*, 181, 3200–3210, 2008b.

structure information over a wide range of magnification, from 10^6 to 10^{-1} nm resolution, enabling us to examine and investigate the microstructure even to atomic levels of materials, to subsequently correlate the properties between phosphor and the microstructure, and to further design new phosphors. Due to the easy preparation of specimens and operation of the instrument, scanning electron microscopy (SEM), covering the magnification range from 10 to 1 nm, is an extremely useful tool in the characterization of materials, for observation of the grain size and shape, morphology, phase or grain homogeneity, texture, and particle agglomeration of phosphors. Furthermore, to achieve ultra-microstructure characters (dislocation, grain boundary, unit cell dimensions, atom or ion arrangement, and size in the lattices) ranging from 10^3 to 10^{-1} nm, transmission electron microscopy (TEM), scanning transmission electron microscopy (STEM), and high-resolution transmission electron microscopy (HRTEM) are usually applied to phosphor materials because an angstrom-scale resolution is readily reached. With the accessory equipment, a combination of SEM or TEM with energy-dispersive x-ray spectroscopy (EDX) or electron probe microanalysis (EPMA) allows us to determine the chemical compositions of phosphors via qualitative and semiquantitative analysis, especially for the identification of an unknown phosphor. It is worth noting that STEM/HRTEM observations can provide additional insight information on the crystal structure of phosphors beyond the capability of the x-ray diffraction techniques, which, combined with the XRD method, is significantly useful for probing the accurate structure. In this section, we briefly demonstrate the microstructure characterizations of some typical nitride phosphors based on the electron microscopy observations.

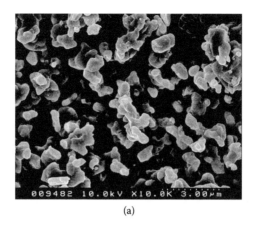

(a)

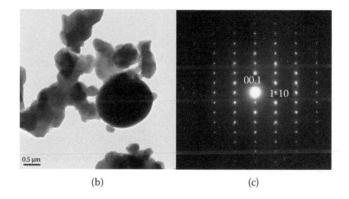

(b)   (c)

FIGURE 5.12
(a) Scanning electron microscopy, (b) transmission electron microscopy, and (c) SAED patterns indexed along (110) of prepared $Sr_2Si_5N_8:Eu^{2+}$ phosphors. (Reprinted from Xie, R.-J., Hirosaki, N., Suehiro, T., Xu, F.-F., and Mitomo, M., *Chem. Mater.*, 18, 5578–5583, 2006b. With permission.)

5.2.1 Red Nitride Phosphors

5.2.1.1 $M_2Si_5N_8:Eu^{2+}$ (M = Ca, Sr)

Figure 5.12a shows a SEM image of the $Sr_2Si_5N_8:Eu^{2+}$ phosphor. The synthesized red nitride phosphor powder consists of uniformly spheroidal particles with a mean diameter of approximately 1 to 2 μm. The transmission electron microscopy (TEM) observations (Figure 5.12b) further indicate that the primary particles of $Sr_2Si_5N_8:Eu^{2+}$ are fine crystallite grains with facets of about 0.5–0.8 μm in size. The sharp diffraction spots of the selected area electron diffraction (SAED) patterns, as shown in Figure 5.12c, demonstrate that $Sr_2Si_5N_8:Eu^{2+}$ is a well-crystallized phosphor.

Interestingly, it is observed that in different host lattices of alkaline earth silicon nitrides, the concentration of the activator ions of Eu^{2+} may affect

FIGURE 5.13
Scanning electron microscopy images of $Sr_2Si_5N_8$:Eu^{2+} prepared at 1,400°C with different Eu concentrations: (a) Eu = 1 mol% and (b) = 5 mol%.

(a) Eu = 1 mol%   (b) Eu = 5 mol%

FIGURE 5.14
Scanning electron microscopy images of $Ca_2Si_5N_8$:Eu^{2+} prepared at 1,400°C with different Eu concentrations: (a) Eu = 1 mol% and (b) = 5 mol%.

the morphology of the grains of $M_2Si_5N_8$:Eu^{2+} (M = Ca, Sr). As shown in Figure 5.13, the influence of the Eu content on the morphology and grain size is very limited for M = Sr, with an increase of Eu concentration.

However, at high Eu concentration, small amounts of abnormal grains with long plate-like morphology ($\sim$20 × 3 µm) were formed in the host of $Ca_2Si_5N_8$ apart from the round-shaped grains ($\sim$1–2 µm) (Figure 5.14), which may be ascribed to the different chemical composition due to the relatively low eutectic temperature point of $Ca_2Si_5N_8$:Eu^{2+}. In addition, the degree of agglomeration of $Ca_2Si_5N_8$:Eu^{2+} is relatively less than that of $Sr_2Si_5N_8$:Eu^{2+}.

5.2.1.2 CaAlSiN₃:Eu²⁺

With respect to the determination of the space group of $CaAlSiN_3$:Eu^{2+}, it is hard to choose the space group from only x-ray diffraction data, since many candidate space groups can meet systematic absences (Clearfield

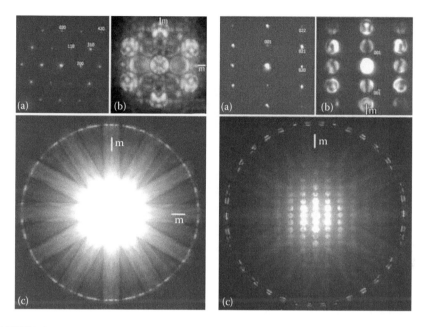

FIGURE 5.15
Electron diffraction (a) and convergence beam electron diffraction (b) patterns for $CaAlSiN_3$ (c) with the incident beam along the c-axis (left) and a-axis (right).

et al. 2008; Pecharsky and Zavalij 2008). With the aid of TEM observations, for example, the convergent beam electron diffraction (CBED) technique of TEM, we can determine the real space group. As displayed in Figure 5.15, the electron diffraction patterns along the c-axis (left images) indicate that $CaAlSiN_3:Eu^{2+}$ has a twofold rotational symmetry and two kinds of mirror symmetries. Moreover, the electron diffraction patterns along the a-axis (right images) indicate only one kind of mirror symmetry, and there is no twofold mirror symmetry in this mirror plane. Accordingly, the point group is mm2 of the orthorhombic lattice for $CaAlSiN_3:Eu^{2+}$. In addition, based on an orthorhombic system, the indexed diffraction patterns turn out the cell type of $CaAlSiN_3:Eu^{2+}$ to be the C type due to meeting the extinction rule of $h + k = 2n$. Furthermore, it is observed that the reflection of $(0, 0, 1)$ and $(0, 0, -1)$ follows the dynamic extinction law, implying the presence of 2_1 screw axes along the c-axis. Finally, the space group is determined to be $Cmc2_1$ (No. 36) by convergent beam electron diffraction.

Figure 5.16 shows the scanning electron microscope image of the $CaAlSiN_3:Eu^{2+}$ phosphor. The phosphor powder of $CaAlSiN_3:Eu^{2+}$ also processes spheroidal shape particles. Compared to $M_2Si_5N_8:Eu^{2+}$ red nitride phosphors, the degree of crystallite is high and the particle size of $CaAlSiN_3:Eu^{2+}$ is large, with the average size ranging from 4 to 7 μm in diameter. This can be explained by its high-temperature synthetic processes (1,600–1,800°C). As a result, from the SEM image it is clear that the particles are tightly bonded

FIGURE 5.16
Scanning electron microscopy image of $CaAlSiN_3:Eu^{2+}$.

together to form a large agglomeration with a size of about 10–20 μm. Therefore, the morphology of the particles of the $CaAlSiN_3:Eu^{2+}$ phosphor generally needs to be improved by adopting appropriate flux during the firing process (Blasse and Grabmaier 1994; Yen et al. 2006; Uheda et al. 2006a, 2006b).

5.2.2 Green Nitride Phosphors

5.2.2.1 $SrSi_2O_2N_2:Eu^{2+}$

Compared to nitride phosphors, the $SrSi_2O_2N_2:Eu^{2+}$ phosphor contains oxygen in the host; therefore in the Sr-Si-O-N system the eutectic point is much lower than in the nitride phosphors, which results in the grains in $SrSi_2O_2N_2:Eu^{2+}$ easily growing in the presence of a liquid phase at low temperature. Figure 5.17 is the SEM image of a green nitride phosphor of $SrSi_2O_2N_2:Eu^{2+}$. Two kinds of particles can be observed: spheroidal particles and plate-like particles. The particle sizes range from 5 to 8 μm, and they normally gather together to form a large agglomeration (>30 μm). A more detailed structure study with TEM-SAD (selected area diffraction) observations can also be found in Oeckler et al. (2007).

5.2.2.2 β-SiAlON:Eu^{2+}

Like β-Si_3N_4, in order to form a stable modification of β-SiAlON:Eu^{2+}, it is necessary to fire the phosphors at high temperature (1,900–2,000°C), compared with the modification of α-SiAlON:Eu^{2+}. As exhibited in Figure 5.18, β-SiAlON:Eu^{2+} powder phosphor is generally crystallized in long-plate-shaped grains with

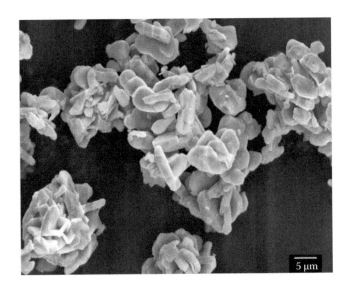

FIGURE 5.17
Scanning electron microscopy image of $SrSi_2O_2N_2:Eu^{2+}$ phosphor.

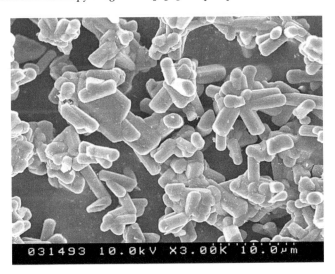

FIGURE 5.18
Scanning electron microscopy image of β-SiAlON:Eu^{2+} phosphor.

dimensions of about 3–6 μm in length and 0.3–0.6 μm in width (Hirosaki et al. 2005).

The fine structure of β-SiAlON:Eu^{2+} crystal was further analyzed by high-resolution transmission electron microscopy (HRTEM), as shown in Figure 5.19. The interplanar d spacing is about 0.3 nm, corresponding to the (100) plane of β-SiAlON:Eu^{2+}. On the surface of a crystal, a thin amorphous

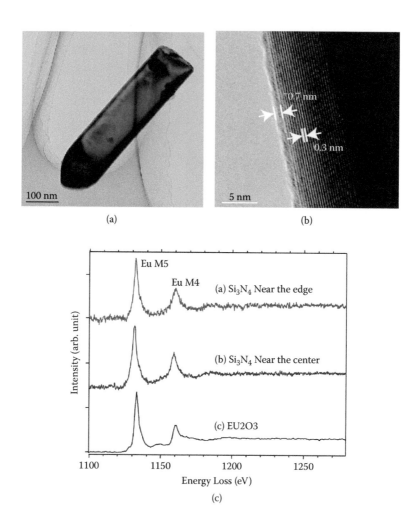

FIGURE 5.19
(a) Transmission electron microscopy and (b) high-resolution transmission electron microscopy images of β-SiAlON:Eu²⁺ crystals, as well as (c) electron energy loss spectra of β-SiAlON:Eu²⁺ at the different areas, along with that of Eu₂O₃. (Reprinted from Hirosaki, N., Xie, R. J., Kimoto, K., Sekiguchi, T., Yamamoto, Y., Suehiro, T., and Mitomo, M., *Appl. Phys. Lett.*, 86, 211905-1–211905-3, 2005. With permission.)

layer with a thickness of 0.7 nm is observed. Additionally, the particle of β-SiAlON:Eu²⁺ is a single crystal and has nearly no defects of dislocation (Figure 5.19a). The electron energy-loss spectra (TEM-EELS) show that Eu atoms are homogeneously distributed near the edge and at the center of the β-SiAlON:Eu²⁺ particles (Figure 5.19c). Those results suggest that the Eu ions are segregated neither on the amorphous surface layer nor at the defect sites, but in the regular β-SiAlON lattice. Indeed, a subsequent study further confirmed that the Eu²⁺ ions are really located at the channels along

(001) (see Figure 5.10). By using advanced scanning transmission electron microscopy (STEM) analysis, we can directly "see" a single Eu dopant atom in the phosphor of the β-SiAlON lattice (Kimoto et al. 2009). Figure 4.28 shows the STEM images of coherent bright-field (BF) and annular dark-field (ADF) β-SiAlON:Eu^{2+} phosphor. From the STEM micrographs it is found that the positions of the Si atom columns are schematically overlapped as black circles, and the Si atom columns are observed as black and white dots in the STEM-BF and STEM-ADF images, respectively. The ADF image (Figure 4.28b) clearly shows a single Eu atom at the origin, as indicated by a white arrow. The calculated results are in good agreement with the experimental results, as illustrated by the bottom insets in broken-line rectangles. By employing the STEM images, we can estimate the observed bond length between Eu^{2+} and N/O to be 0.26 nm, according to the radius of the atomic channel of Si/Al-N/O, which is well consistent with the data of the Rietveld refinement of x-ray powder diffraction (~0.25 nm; see Table 5.9) (Li et al. 2008b).

5.2.3 Yellow Phosphors

The morphology of Ca-α-SiAlON:Eu^{2+} particles is very crystalline and mostly uniform equiaxed shaped, along with some hexagonal-shaped crystals (Figure 5.20a). The mean diameter of the particle is approximately 1–2 μm; the particles grow together and from larger aggregates. However, those agglomerates can be easily broken into fine phosphor powder, and the median particle size is about $d_{50} = 1.6$ μm, with a narrow particle size distribution. It is worth noting that the morphology of $Ca_{m/2}Si_{12-m-n}Al_{m+n}O_nN_{16-n}$:Eu^{2+} phosphors can also be realized by changing m and n values in compositions (Figure 5.21b) and by long-time heat treatment at high temperatures (Yen et al. 2006; Xie et al. 2002, 2004a, 2004b, 2005, 2006a).

5.2.4 Blue Nitride Phosphor

As discussed in previous chapters, $LaAl(Si_{6-z}Al_z)N_{10-z}O_z$:Ce^{3+} (z ≈ 1), i.e., JEM:Ce^{3+}, emitting blue light at about 475 nm, originated from the 5d → 4f transition of Ce^{3+} under excitation at 405 nm. As shown in Figure 5.21, the JEM:Ce^{3+} phosphor has spheroid-shaped particles, and those particles are closely accumulated together. Among the nitride phosphors, the particle size of JEM:Ce^{3+} is much bigger, with the diameter in the range of 10–30 μm.

5.3 XANES/EXAFS Techniques

XANES/EXAFS is a kind of x-ray absorption spectroscopy (XAS), and the absorption of x-rays by a sample is measured across the edge of the

(a)

(b)

FIGURE 5.20
Scanning electron microscopy of (a) Ca-α-SiAlON:Eu^{2+} and (b) Ca-α-SiAlON:Ce^{3+} (m = 3.5) phosphors.

photoemission of a core K or L electron. The typical XANES/EXAFS spectrum consists of three regions in energy (Bianconi 1980; Bianconi et al. 1982; Newville 2010; Bunker 2010). The pre-edge region where the electron is excited without emitting can give the character of some elements and provide information on the oxidation state. X-ray absorption near-edge structure (XANES) within the energy region up to about 50 eV beyond the edge (i.e., hard X-rays) and edge x-ray can exhibit some information on the oxidation state of the absorbing atoms and the local structure around the atom. Near-edge x-ray absorption fine structure (EXAFS or NEXAFS) covers high-energy regions extending up to ~1 keV (i.e., soft x-rays), which provides information on the local environment of the atom, coordination number, and bond length. In short, XANES/EXAFS is a unique

FIGURE 5.21
Scanning electron microscopy of a JEM:Ce^{3+} phosphor synthesized at 1,850°C.

technique to probe structural information surrounding x-ray-absorbing atoms that can be used to determine the average nearest-neighbor bond length, bond angle, ligands, and valence state, as well as the site symmetry (http://www.u-picardie.fr/alistore/platforms/EXAFS.htm). It is especially useful when the structure is temporaily unavailable for a newly developed phosphor; the XANES/EXAFS techniques will be valuable tools to explore the local environment around the activator ions, like Eu^{2+} and Ce^{3+}, and the initial geometric states. The typical XANES/EXAFS experiments are carried out on a synchrotron source using small amounts of the sample (10–100 mg); furthermore, the concentrtions of the target atoms can be measured in a wide range from ppm to 100%, which can disclose detailed information around the luminescent centers of the nitride phosphors.

Figure 5.22 shows the Eu L_3-edge XANFS spectra of Eu-α-SiAlON with a nominal composition of $Eu_{0.325}Si_{11.35-n}Al_{0.65+n}O_nN_{16-n}$ (n = 0–0.325). The oxide state of Eu was determined by XANES spectra using $Eu_2Si_5N_8$ and Eu_2O_3 as the references of Eu^{2+} and Eu^{3+}, respectively. From the XANES spectra it can be clearly observed that the dominant peaks are at about 6,973 eV in spectra corresponding to the Eu^{2+} characterstic. In addition, there is a small shoulder in spectra at about 6,970 eV that is attributed to Eu^{3+}. The results show that Eu ions in Eu-α-SiAlON mainly exist in a divalent state with relative amounts of 87–92 mol%. Besides Eu^{2+}, there are small amounts of Eu^{3+} in Eu-α-SiAlON, even within this strong reductive host lattice, in comparison with oxide lattices. Depending on the composition, for example, with increasing oxygen (i.e., n value) in the host lattice, the Eu^{3+} content is markedly increased, as shown in Figure 5.23. As a result, the emission

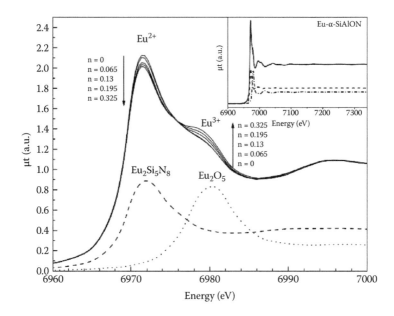

FIGURE 5.22
XANES spectra of $Eu_{0.325}Si_{11.35-n}Al_{0.65+n}O_nN_{16-n}$ with different n values. Inset shows the spectra in a wide energy range. (Reprinted from Shioi, K., Hirosaki, N., Xie, R. J., Takeda, T., and Li, Y. Q., *J. Alloys Compd.*, 504, 579–584, 2010. With permission.)

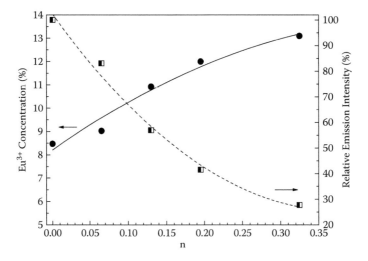

FIGURE 5.23
Variation of the oxygen content (n) with the Eu^{3+} concentration and relative emission intensity of Eu-α-SiAlON with a composition of $Eu_{0.325}Si_{11.35-n}Al_{0.65+n}O_nN_{16-n}$. (Reprinted from Shioi, K., Hirosaki, N., Xie, R. J., Takeda, T., and Li, Y. Q., *J. Alloys Compd.*, 504, 579–584, 2010. With permission.)

intensity of Eu^{2+} signficantly decreases due to a high Eu^{3+}/Eu^{2+} ratio, where Eu^{3+} plays a key role as a luminescent killer center, with the oxygen content increasing in composition (Shioi et al. 2010). Therefore, the Eu^{3+}/Eu^{2+} ratio or the Eu^{3+} concentration should be controlled to be as low as possible in order to obtain highly efficient Eu^{2+}-activated nitride phosphors.

5.4 First-Principles Calculations

It is well known that all physical and chemical properties of a given solid-state material can be obtained from the quantum mechanical wavefunction by solving the Schrödinger equation associated with the Hamiltonian, for example:

$$H\Psi = E\Psi \tag{5.3}$$

where H is the Hamiltonian, E is the energy eigenvalue, and ψ is the wavefunction for a time-independent system (Jensen 2006; Young 2001). Since the crystalline materials are many-particle systems, the approximate quantum mechanical approaches have to be introduced in the calculations. Therefore, many methods have been developed according to different approximations (Dronskowski 2005). Among them, a first-principles quantum mechanical simulation based on the density functional theory (DFT) has been widely used for the calculations (principally on the ground state) of the electronic, mechanical, and optical properties, as well as the predictions of the structure of solid-state materials (Jensen 2006; Young 2001; Dronskowski 2005; Milman et al. 2000). Regarding DFT, the total energy is expressed by the total one-electron density instead of the wavefunction. In the calculation, an approximate Hamiltonian and an approximate expression for the total electron density have been adopted, which make DFT methods possible to obtain very accurate results with relatively low computational costs, compared to traditional ways, which are based on the complicated many-electron wavefunction (Dronskowski 2005). In this section, we focus on recent developments in the application of first-principles calculations to study the crystal and electronic structures of nitride-based phosphors. Due to the limit of approximation modeling for dealing with the rare earth activator ions of Eu^{2+} or Ce^{3+}, these examples are only intended to represent the applications of calculations on undoped hosts of nitride phosphor in this field, enabling us to better understand the nature of the host lattice of promising nitride phosphors. Far more extensive examples of first-principles calculations on the structure and physical properties of binary and ternary nitride compounds can be found from recent literature elsewhere.

There are many excellent first-principles calculation packages based on the density functional theory (Dronskowski 2005). The Vienna Ab-initio

Simulation Program (VASP) is the most popular package of *ab initio* total energy and molecular dynamics programs for calculations of solid-state materials. In the following examples of nitride compounds, the electronic structures have been calculated by VASP code, which combines the total energy pseudopotential method with a plane-wave basis set (Milman et al. 2000; Kresse and Hafner 1993, 1994; Kresse and Furthmuller 1996a, 1996b). The projector augmented wave (PAW) method was employed, and the generalized gradient approximation (GGA) has been used for electron exchange-correlation energy (Perdew et al. 1992). The numerical integration of the Brillouin zone (BZ) was performed using a discrete form $4 \times 4 \times 4$ to $8 \times 8 \times 12$ Monkhorst–Pack k-point sampling, depending on the symmetry of the hosts, and the plane-wave cutoff energy was set to be 500 eV. The Fermi level is set at zero energy for all the calculations. First-principles calculations were first performed with relaxation of both atomic positions and lattice parameters for structure optimization. Based on the obtained equilibrium structure, the densities of states and band structures of the nitride compounds were calculated. The structure data, including lattice parameters and atomic coordination, were employed from the refinement of x-ray powder diffraction data using the Rietveld refinement method performed on the nitride hosts from our experiment data in most cases.

5.4.1 $M_2Si_5N_8$ (M = Sr, Ba)

Alkaline earth silicon nitride compounds have been investigated by using the localized spherical wave (LSW) method with a scalar-relativistic Hamiltonian for $Sr_2Si_5N_8$ (Fang et al. 2001a, 2001b, 2002), and the local density approximation (LDA) of the density functional theory (DFT) for $Ba_2Si_5N_8$ (Fang et al. 2003). In the present work, calculations based on the projector augmented wave (PAW) method as implemented in the VASP package turned out results similar to those in previously reported data (Fang et al. 2001a, 2001b, 2002, 2003). Figure 5.24 illustrates the partial and total densities of states (DOSs) of $Sr_2Si_5N_8$. As seen, the valence band is composed of the N-2p, Si-3s, Si-3p, and Sr-4p states and mainly distributes into two energy regions: –13.5 to –18 eV and 0 to –10 eV. The N-2s and Sr-4p states are at the lower energy part, and a sharp Si-3p state at about –14 eV is also presented in this region. The upper-valence band edge close to the Fermi energy level is dominated by the N-2p states, with a total bandwidth of about 10 eV, which is hybridized with Si-3p along with small amounts of Si-3s states, implying that Si-N can form covalent bonding of Si-N as expected. The N-2p states are almost fully occupied, while the Sr-4s states are nearly empty. The conduction band mainly consists of Sr-4p4d, Si-3s3p, and N-2s2p levels, ranging from 3.6 to 6 eV, with a small, sharp peak at the bottom of the conduction band at about 3.4 eV from the Sr-4d states.

Figure 5.25 shows the dispersion curve of the energy bands along the high-symmetry path through the first Brillouin zone of $Sr_2Si_5N_8$. The top of the

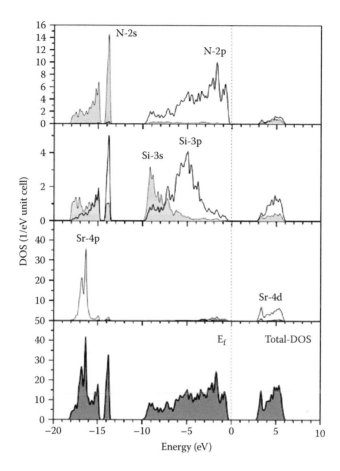

FIGURE 5.24
Total and atomic partial densities of states of $Sr_2Si_5N_8$.

valence band is around the Γ point, and the bottom of the conduction band is also at the Γ point. The band gap energy E_g is calculated to be 3.21 eV, which is in fair agreement with the experimental data ($E_g \sim 4.4$ eV). Therefore, $Sr_2Si_5N_8$ can be regarded as a direct wide-band-gap semiconductor. The different band gaps between the calculated and experimental data are attributed to the first-principles approaches based on the DFT approach, which always underestimates the band gap energy for semiconductors (Dronskowski 2005; Fang et al. 2001a). From the calculations, it is expected that with such a band gap, the $4f^65d \leftrightarrow 4f^7$ transitions of Eu^{2+} could lie in between the top of the valence bands and the bottom of the conduction bands.

Another alkaline earth silicon nitride, $Ba_2Si_5N_8$, is isostructural with $Sr_2Si_5N_8$ (Huppertz and Schnick 1997), with a relatively large unit cell. Owing to their structural similarity, their electronic structures also resemble each other. Figure 5.26 shows the partial and total densities of states of $Ba_2Si_5N_8$.

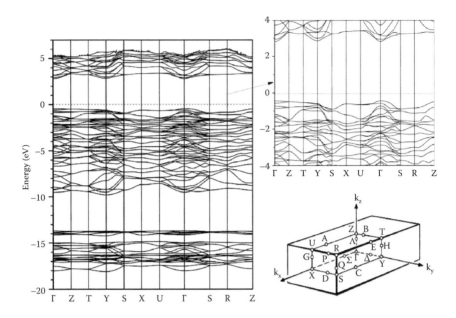

FIGURE 5.25
Band structure and high-symmetry points in the BZ of $Sr_2Si_5N_8$.

The valence bands mainly consist of N-2s2p, Si-3s3p, and Sr-4p states, in which the N-2s, Si-3s, and Ba-5p states are located at lower energy in the range of –11 to –18 eV. In contrast to $Sr_2Si_5N_8$, the Ba-5p states are positioned at relatively high energy (~12 eV). The upper portion of the valence bands is mainly determined by N-2p and Si-3p, from the Fermi energy level to –7 eV. The conduction band of $Ba_2Si_5N_8$ is composed of Ba-4d, Si-3s3p, and N-2s2p levels from 2.1 to 6 eV. The band structure calculation indicates that $Ba_2Si_5N_8$ is also a direct-band-gap material with an energy band gap E_g of about 2.87 eV at point Γ, as shown in Figure 5.27. For the same reasons as for the DFT calculations, which always underestimate the band gap, it is understandable that the calculated E_g is smaller than that of the experimental data: ~3.98 eV (Dronskowski 2005). On the other hand, this calculated band gap energy is smaller than that of $Sr_2Si_5N_8$ because the bottom of the conduction band shifts to lower energy from wide distribution of the Ba-4d, N-2s2p, and Si-3s3p states (Figure 5.26). Those results are also in agreement with the experimental data obtained from the reflection spectra, by which the optical band gap of $Ba_2Si_5N_8$ is estimated to be 4.0 eV and is smaller than that of $Sr_2Si_5N_8$ (~4.4 eV) (Li et al. 2006).

5.4.2 CaAlSiN₃

The atomic and electronic structures of the $CaAlSiN_3$ host have been studied by the first-principles pseudopotential method based on the density

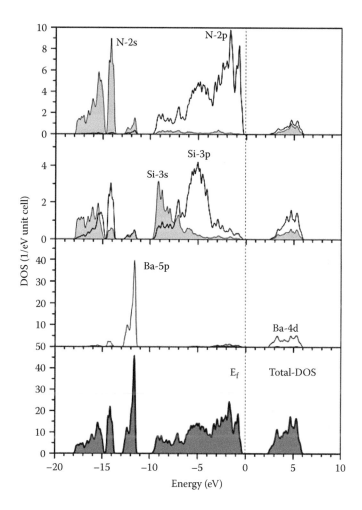

FIGURE 5.26
Total and atomic partial densities of states of the host of $Ba_2Si_5N_8$.

functional theory through the virtual crystal approximation and the assumption of the Al/Si-ordered distribution in the primitive unit cell to deal with the random distribution of Al and Si in the structure. The calculated electronic structures (density of states and band structure) appear similar for the two approaches (Mikami et al. 2006). In the present work, the crystal and electronic structures of $CaAlSiN_3$ were calculated by the approximation of a $2 \times 2 \times 2$ supercell modeling, in which Al and Si were separately arranged into different positions of 8b sites in the subcell of the orthorhombic lattice (space group $Cmc2_1$). After the structure optimization and lattice relaxation, the density of states and band structure of $CaAlSiN_3$ were calculated. It was found that those results are also consistent with the reported data (Mikami et al. 2006). Figure 5.28 depicts the partial and total densities of states of the

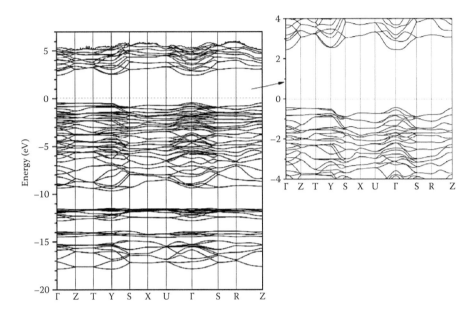

FIGURE 5.27
Band structure of the host of $Ba_2Si_5N_8$.

CaAlSiN$_3$ host. As shown in Figure 5.28, the density of states of CaAlSiN$_3$ can be divided into three energy regions. In the valence bands, the N-2p and N-2s states are the major components. At the lower energy part (–18 to –12 eV) of the valence bands are the N-2s, (Si,Al)-3s, and Ca-3s states. The valence bands in the range of –10 to 0 eV are mainly composed of the N-2p, (Si,Al)-3s3p, and Ca-3s3p states, in which the N-2p states dominate the upper portion of the valence bands close to the Fermi level and hybridize with (Si,Al)-3p and Ca-3s 3p near the zero energy, along with the (Si,Al)-3s states beyond the zero energy, from –10 to –3 eV. Additionally, the 2p states of N[2] are slightly higher than those of N[3] in energy. Accordingly, both the Ca and (Si,Al) atoms may form covalent bonds for both Ca-N and (Si,Al)-N in CaAlSiN$_3$. Similar to Ca$_2$Si$_5$N$_8$ (Fang et al. 2001a) and CaSiN$_2$ (Li et al. 2007), the Ca-3s3p states are located at higher energies in the valence bands, which is probably related to its shorter average bond of Ca-N (~2.5 Å), compared to Sr-N (~2.78 Å) and Ba-N (~2.86 Å) in M$_2$Si$_5$N$_8$ (M = Sr, Ba) (Hoppe 2000; Li et al. 2006; Schlieper and Schnick 1995). This feature is also one of the reasons that the emission band of pure CaAlSiN$_3$:Eu^{2+} is always in the deep-red range (640–670 nm), even at lower Eu^{2+} concentrations, due to its more covalent chemical bonding of Eu$_{Ca}$-N. The conduction bands are mainly composed of the Ca-3d states, along with the Ca-3s3p, (Si,Al)-3s3p, and N-2s2p states, ranging from 2.6 to 5.4 eV, hybridizing each other. Moreover, the Ca-3d levels are located at the bottom of the conduction bands. Because the largest component is Ca at the bottom of the conduction bands, the band gap in CaAlSiN$_3$ is mainly

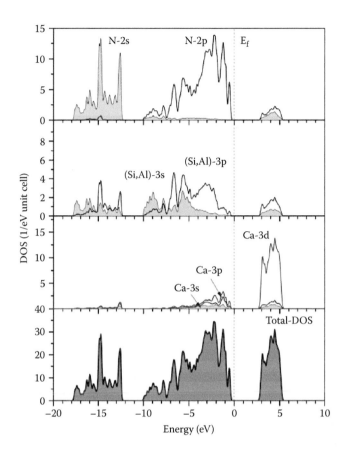

FIGURE 5.28
Total and atomic partial densities of states of the host of CaAlSiN$_3$.

determined by the Ca-N interaction, while the effect of (Si,Al)-N seems relatively small. Figure 5.29 shows the calculated band structure of CaAlSiN$_3$ based on the present model along the symmetry lines of the orthorhombic BZ. The band structure curves indicate that the CaAlSiN$_3$ host possesses a direct band gap at point Γ, and both the top of the valence bands and the bottom of the conduction bands are at the Γ points. Clearly, the present results are different from those of a previous report, in which CaAlSiN$_3$ is an indirect band gap, with the top of the valence bands located within the flat line of Γ-Z, not at Γ (Mikami et al. 2006). Nevertheless, the energy band gap at point Γ is about 3.4 eV, in agreement with the previous calculations (Mikami et al. 2006). Moreover, the calculated band gap is also very close to the experimental value (~3.82 eV) derived from the reflection spectrum of CaAlSiN$_3$. Another similarity to the previous study is that the top of the valence bands is very flat from the Y to Γ points, particularly in Γ-Z; however, the energy at Γ is indeed higher than that at Z, as shown in the inset of Figure 5.29.

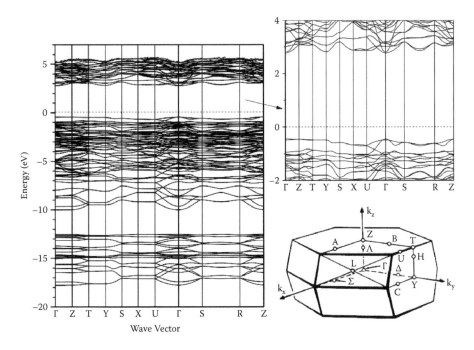

FIGURE 5.29
Band structure of CaAlSiN$_3$ and the high-symmetry points along the BZ of an orthorhombic lattice.

5.4.3 SrSi$_2$O$_2$N$_2$

An oxynitride compound of SrSi$_2$O$_2$N$_2$ is a layered structure built up by the corner-shared Si(N,O)$_4$ tetrahedra through connecting N (Figure 5.30) (Oeckler et al. 2007). The Sr atoms are partially located at the 1a sites in between two tetrahedral layers, and eight individual Sr atoms are coordinated with O and N with different coordination numbers (Oeckler et al. 2007). SrSi$_2$O$_2$N$_2$:Eu^{2+} is a high-efficiency green-emitting phosphor, and the emission band of Eu^{2+} peaks at about 540 nm under near-UV and blue excitation (Li et al. 2005; Mueller-Mach et al. 2005).

On the basis of a triclinic structure model (Oeckler et al. 2007), the density of states of the host of SrSi$_2$O$_2$N$_2$ has been calculated by first principles based on the DFT approximation, with the assumption that Sr is fully occupied at each crystallographic site of Sr in the structure. Figure 5.31 illustrates the calculated three-dimensional valence electron density distribution of SrSi$_2$O$_2$N$_2$. The total and atomic partial densities of states are plotted in Figure 5.32. Similar to strontium silicon nitride, the distribution of N, Si, and Sr states is nearly in the same energy range in the valence bands. O-2s, N-2s, Sr-4p, and partial Si-2s2p states are at a lower energy range, from –13 to –22 eV, and the O-2s states in the low energy range, compared to the N-2s states. A sharp peak at about –18.2 eV results from Si-3s and O-2s states overlapping

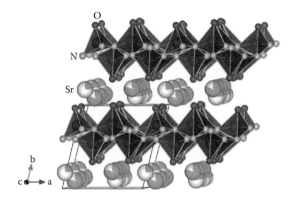

FIGURE 5.30
Projection of the crystal structure of $SrSi_2O_2N_2$, viewed along (001). The incomplete filled spheres of Sr represent partial occupancy on the sites.

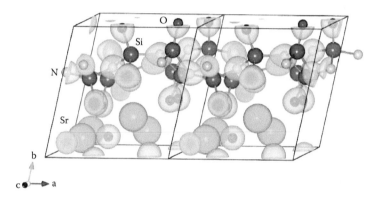

FIGURE 5.31
(See color insert.) Three-dimensional valence electronic density distribution image of $SrSi_2O_2N_2$, viewed along (001).

around this range, implying the formation of Si-O bonding. In addition, the chemical bonding of Sr-O has a more ionic character because the Sr-4p and O-2s2p states are well separated in the valence bands, while Sr-N appears to have a little covalent bonding due to the overlap of Sr-4p and N-2s in the range of –17.5 to –15 eV. Apart from the N-2p and Si-3p states, the O-2p states are also major components at the top of the valence bands, from –10 to 0 eV at the Fermi level, in this host of alkaline earth silicon oxynitride, similar to $Y_3Si_5N_9O$ (Xu et al. 2005), Si_2N_2O (Xu and Ching 1995), and β-SiAlON (Ching et al. 2000). Furthermore, in comparison with alkaline earth silicon nitrides (Fang et al. 2001a, 2001b, 2002, 2003; Mikami 2006), the covalent bonding of Si-N and Si-O can be formed because the Si-3p states are hybridized with both O-2p and N-2p states. Relatively, N-2p is mainly positioned at the upper part of the valence bands, close to the Fermi level, compared to

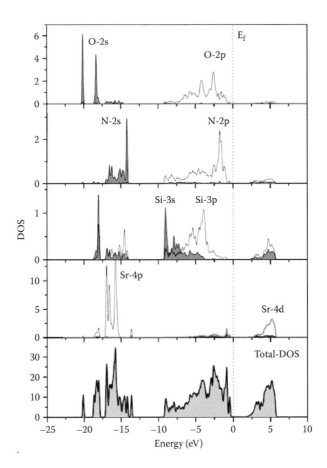

FIGURE 5.32
Total and atomic partial densities of states of the host of $SrSi_2O_2N_2$.

O-2p. The conduction bands primarily consist of the Sr-4d and Si-3s3p states, which also dominate the bottom of the conduction bands at about 1.95 eV. As a result, the energy band gap of $SrSi_2O_2N_2$ is essentially dependent on the Sr-N and Si-O interactions. The calculated energy band gap is about 2.3 eV at Γ, which is much smaller than the experimental data (~4.14 eV) resulting from the diffused reflection spectrum of $SrSi_2O_2N_2$ powder (Li et al. 2005). The errors can be ascribed to the limit of DFT approximation within the PAW method (Dronskowski 2005) and an inappropriate structural model, which actually is a defect structure rather than a perfect structure model.

5.4.4 $CaSiN_2$ and $Ca_2Si_5N_8$

There are two stable calcium nitride compounds in the Ca-Si-N system: $CaSiN_2$ and $Ca_2Si_5N_8$, which are also interesting hosts of the nitride phosphors for use in solid-state lighting (Li et al. 2006, 2009). Eu^{2+}-doped $CaSiN_2$ and $Ca_2Si_5N_8$ emit red light at about 630 and 610 nm, respectively, when excited by near-UV and blue light (Li et al. 2006, 2009). Figure 5.33 exhibits the total and partial densities of states as well as the band structure of the orthorhombic host of $CaSiN_2$. Like in $CaAlSiN_3$, the distribution of N-2s2p and Si-3s3p states is also mainly in the valence bands of an orthorhombic host of $CaSiN_2$, where the N-2s and Si-3s states are at lower energy and form DOS peaks in the lowest energy part, from –16.2 to –13.2 eV, with very small contributions from Ca-3d states. The top of the valence bands of $CaSiN_2$ are mainly composed of the N-2p and Si-3p states, as well as the Ca-3s3p3d states. Among them, N-2p and Si-3p are largely hybridized in the range of –5.5 to –4 eV, originating from the chemical bonding of Si-N. Similar to $CaAlSiN_3$, the metal cations of the Ca-3s3p3d states are also pronouncedly hybridized with the N-2p states near the Fermi energy level, from –3.9 to 0 eV, due to short Ca-N bonds, as discussed above. In the conduction bands, the Ca-3d states are the major components, together with small amounts of the Ca-3s3p and N-2p states, ranging from 3.95 to 6.45 eV. As a consequence of Ca-3d being the major character of the bottom of the conduction bands, the band gap of the orthorhombic form of $CaSiN_2$ is mainly determined by the Ca-N bonding. From the calculated results it turns out that $CaSiN_2$ is an indirect wide-band-gap semiconductor with an indirect band gap of about 3.51 eV, and the top of the valence bands is at point X and the bottom of the conduction bands is at point Γ. The energy band gap is about 3.58 eV at point Γ (Figure 5.33b). These calculated band gaps are in fair agreement with the experimental value of about 4.16 eV based on the diffused reflection spectrum of an orthorhombic $CaSiN_2$ (Li et al. 2009).

Figure 5.34 shows the density of states and the band structures of $Ca_2Si_5N_8$. With the same component elements in composition, $Ca_2Si_5N_8$, crystallized in a monoclinic system (Li et al. 2006; Huppertz and Schnick 1997; Fang et al. 2001a), has a different crystal structure than $CaSiN_2$. As discussed above, the N-2s2p, Si-3s3p, and Ca-3s3p3d states are quite similar in the Ca-containing silicon nitride hosts, where the Ca-3s3p3s states are sensitive to the coordination atoms of N. Therefore, the band gap of $Ca_2Si_5N_8$ is also determined by Ca-N bonding. The top of the valence bands is in between M and A, and the bottom of the conduction bands is located at the Γ points in the host of $Ca_2Si_5N_8$. Therefore, the $Ca_2Si_5N_8$ host is an indirect-band-gap material having a band gap of about 3.39 eV (the experiment optical band gap, E_g = 4.44 eV (Li et al. 2006); and a previous calculation using a different approach, E_g = 4.10 eV (Fang et al. 2001a)). The largely different band gap values are related to the first-principles method based on the DFT using the PAW approach, as mentioned above (Dronskowski 2005; Fang et al. 2001a).

As described in the above examples, we only dealt with undoped host lattices of nitride phosphors. When we consider the doping system, such as Eu^{2+}- and Ce^{3+}-activated nitride phosphors, as a general routine, the supercell approximation approaches have to be employed for modeling the

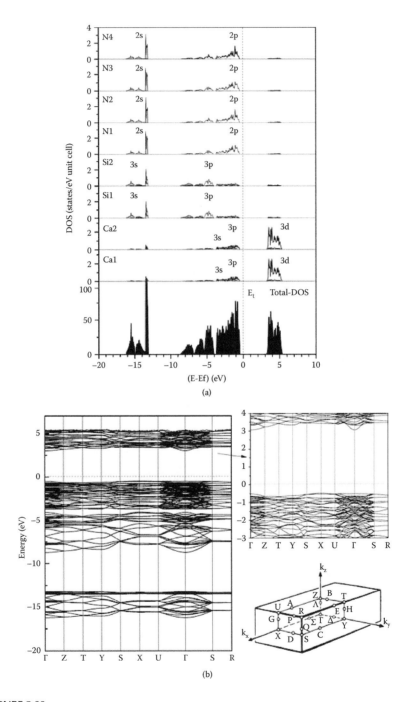

FIGURE 5.33
(a) Total and atomic partial densities of states of the orthorhombic CaSiN$_2$ host lattice. (b) Band structure and high-symmetry points of the first BZ of orthorhombic CaSiN$_2$.

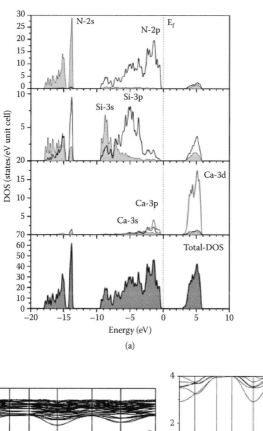

(a)

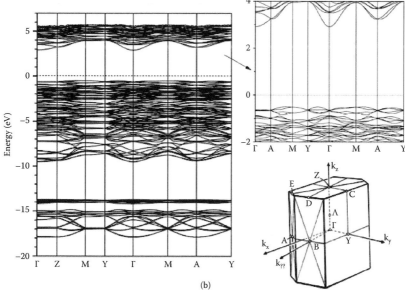

(b)

FIGURE 5.34

(a) Total and partial densities of states of the $Ca_2Si_5N_8$ host lattice. (b) Band structures and high-symmetry points of the first BZ of $Ca_2Si_5N_8$.

phosphor systems. This is a very *time-consuming calculation and needs a powerful supercomputer,* particularly for a phosphor system with very low concentrations of the activator ions having great numbers of atoms. Another problem for first-principles calculations based on plane waves and pseudopotential approximation is that the wavefunctions describe only the valence and the conduction electrons, while the core electrons are treated with pseudopotentials (Dronskowski 2005). As a result, they cannot yet well describe the rare earth ions. *For this reason, we need other economic and fast approaches to deal with the complex phosphor systems, by focusing on the local structures in the range of the first- or second-nearest-neighbor atoms just around the activator ions. That is the next topic we will discuss in detail regarding the calculations of the local electronic structures of some typical nitride phosphors.*

5.5 Molecular Orbital (MO) Cluster Calculations

The discrete variational $X\alpha$ (DV-$X\alpha$) molecular orbital (MO) method is an all-electron method, and it can find the wavefunctions of all of the electrons of the cluster models by the calculations of the molecular orbitals assumed by the Hartree-Fock-Slater approximation (Rósen et al. 1976; Adachi et al. 1978, 2005; Nakayasu et al. 1997; Ogasawara et al. 2001; Yoshida et al. 2006). Using Slater's potential, we can write the equation of the exchange-correlation term between the electrons as follows (Adachi et al. 1978, 2005):

$$V_{xc}(r) = -3\alpha \left[\frac{3}{8\pi} \rho(r) \right]^{1/3} \tag{5.4}$$

where $\rho(r)$ is the electron density at r, and the only parameter α is fixed at 0.7. The self-consistent charge approximation is employed for the MO calculations. The matrix elements of the Hamiltonian and the overlap integrals are calculated by a random sampling method. The molecular orbital wavefunction is treated with a linear combination of atomic orbitals (LCAO). Given a relatively small basis set, the DV-$X\alpha$ method can significantly reduce the computational cost to obtain accurate electronic states even for a large cluster model. Therefore, the DV-$X\alpha$ method has been successfully applied to the study of a variety of materials (Adachi et al. 2005). In this section, we will focus on the investigation of the local electronic structures and energy levels of Eu^{2+}-doped nitride phosphors by the relativistic DV-$X\alpha$ method based on the first-principles MO calculations for a designed cluster model around the activator ion. In the course of the relativistic DV-$X\alpha$ molecular orbital (MO) calculations, the typical computational procedures include selection of the cluster models around the activator ions of Eu^{2+}, determination of the symmetry of the cluster models, and generation of the Madelung potential by point charges outside the cluster to maintain the selected cluster in a charge-neutral state, as well as self-consistent and physical properties calculations. In the present work, all the structure models have been extracted from the experiment data of the Rietveld refinement of powder x-ray diffraction of our synthesized nitride phosphors. It should be noted that the structural relaxation of the cluster models was not incorporated in the calculations.

5.5.1 $Sr_2Si_5N_8:Eu^{2+}$ (5 mol%)

As discussed in Section 5.1, there are two individual Sr crystallographic sites in the $Sr_2Si_5N_8:Eu^{2+}$ structure. The coordination numbers and bond lengths are different for the two Sr sites (Li et al. 2006; Schlieper and Schnick 1995). For a better understanding of the luminescence properties of $Sr_2Si_5N_8:Eu^{2+}$, it is essential to correlate the local electronic structures of the dopant sites, i.e., the Eu_{Sr}-N environment. In general, small geometrical differences in the average distance of Eu_{Sr}-N and the coordination polyhedron volume of two Eu_{Sr}^{2+} crystallographic sites are hardly reflected in the photoluminescence spectra, showing nearly a single emission band of Eu^{2+}, as found in $M_2Si_5N_8:Eu^{2+}$ (M = Ca, Sr, Ba) (Li et al. 2006, 2008b). Therefore, the contributions of the individual luminescent centers to the overall luminescence properties are impossible to separate from the structural and optical evidence. However, the local electronic structures may provide some clues for the effects of the two different Eu^{2+} centers on the luminescence of $Sr_2Si_5N_8:Eu^{2+}$. Subsequently, the N and Si atoms have been included in two doping sites of Sr in the range of 6 Å, with the N atoms as the outermost layer. The two cluster models of $[Eu_{Sr}(I)Si_{10}N_{27}]^{-39}$ and $[Eu_{Sr}(II)Si_{10}N_{28}]^{-42}$ surrounding the two different Sr sites occupied by the activator ions of Eu^{2+} have been adopted from the Rietveld refinement structure of $Sr_2Si_5N_8:Eu^{2+}$ (5 mol%, i.e., $Sr_{1.9}Eu_{0.1}Si_5N_8$), as shown in Figure 5.35a and b. The point groups of both clusters are the C_i symmetry. The electron configurations of the atomic orbitals are Si:1s-2p, N:1s-3p, and Eu:1s-6p for the calculations. In addition, the Madelung potential was employed by introducing thousands of point charges outside of the clusters in the calculations, and the 4f energy of Eu^{2+} was set at zero. Figure 5.36 depicts the total and atomic densities of states of the clusters of $[Eu_{Sr}(I)Si_{10}N_{27}]^{-39}$ and $[Eu_{Sr}(II)Si_{10}N_{28}]^{-42}$ extracted from the $Sr_2Si_5N_8:Eu^{2+}$ (5 mol%) phosphor. Due to different coordination numbers and geometric sizes of the two clusters (the coordination number of $Eu_{Sr}(I)$ with the first-nearest N is six, in contrast to seven of $Eu_{Sr}(II)$), the profiles of their densities of states (DOSs) are quite different, while the positions of the energy levels for each atom are very similar. In the atomic energy level chart, the atomic orbital components are indicated by the length of the horizontal lines. The occupied and unoccupied levels are expressed by solid lines and dotted lines, respectively. With respect to the $[Eu_{Sr}(I)Si_{10}N_{27}]^{-39}$ and $[Eu_{Sr}(II)Si_{10}N_{28}]^{-42}$ clusters, as shown in Figure 5.36, the occupied MO mainly

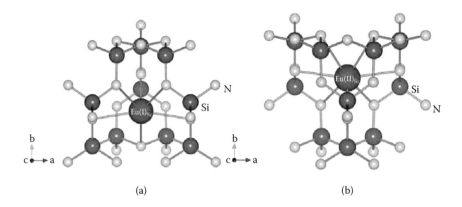

(a)   (b)

FIGURE 5.35
Cluster models of (a) $[Eu_{Sr}(I)Si_{10}N_{27}]^{-39}$ and (b) $[Eu_{Sr}(II)Si_{10}N_{28}]^{-42}$ for the two Eu_{Sr} sites of the $Sr_2Si_5N_8:Eu^{2+}$ (5 mol%) phosphor.

consists of N-2s and N-2p orbitals, where N-2s is at the lower energy part, below −20 eV. In addition, a small fraction of Si-3s3p orbitals are present in the range of −9 to −16 eV, hybridized with N-2p orbitals. The unoccupied MO is composed of Si-3s and Si-3p orbitals as well as Eu-5d orbitals, which are located in between Si-3p and Si-3s or Eu-4f orbitals. The energy gaps between the highest occupied molecular orbitals (HOMOs) and the lowest unoccupied molecular orbitals (LUMOs) are about 3.91 and 3.27 eV, respectively, for $Eu_{Sr}(I)$ and $Eu_{Sr}(II)$ clusters. The energy gaps between the lowest 5d levels of Eu^{2+} and the 4f levels at zero are about 1.74 and 1.27 eV, respectively, for $Eu_{Sr}(I)$ and $Eu_{Sr}(II)$ clusters. The local electronic structures indicate that the $Eu(II)_{Sr}$ center may mainly determine the overall luminescence properties of $Sr_2Si_5N_8:Eu^{2+}$ due to the lowest Eu-5d levels at the lower energy and the upper part of N-2p levels at higher energy. Additionally, the calculated bond overlap population (BOP) also reveals that the BOP of $Eu_{Sr}(II)$-N is about 0.014, which is higher than that of $Eu_{Sr}(I)$-N (BOP = 0.0127), suggesting that the Eu-5d orbitals have strong interactions with the N-2p orbitals for a large cluster of $[Eu_{Sr}(II)Si_{10}N_{28}]^{-42}$. However, considering the fact that the average $Eu(I)_{Sr}$-N bond length is smaller than that of $Eu(II)_{Sr}$-N (see Section 5.1), these trade-offs may explain why the observed luminescence properties over the two centers are actually quite similar. Here, it is worth noting that the local electronic structures only reflect the target domain in a limited range around Eu^{2+}, which may be markedly different from the average electronic structures from the whole crystal structure.

5.5.2 CaAlSiN₃:Eu²⁺ (1 mol%)

A cluster model of $[Eu_{Ca}(Si_4Al)_4N_{24}]^{-42}$ around the dopant site of Ca was chosen in the range of 5.6 Å using an initial structural model from the Rietveld

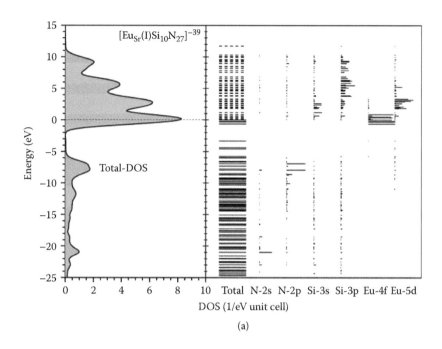

(a)

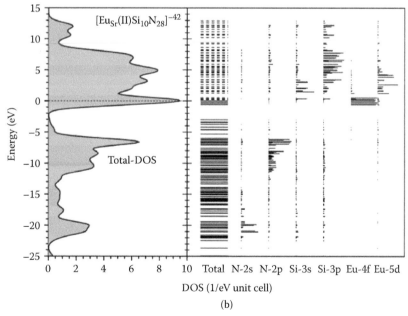

(b)

FIGURE 5.36
Densities of states and energy-level diagrams for the (a) $[Eu_{Sr}(I)Si_{10}N_{27}]^{-39}$ cluster and (b) $[Eu_{Sr}(II)Si_{10}N_{28}]^{-42}$ cluster of the $Sr_2Si_5N_8$:Eu^{2+} (5 mol%) phosphor.

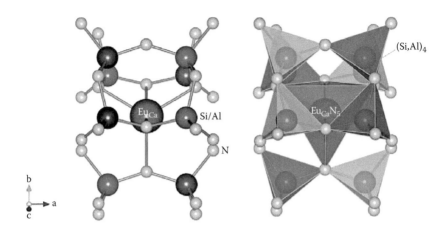

FIGURE 5.37
Cluster models of $[Eu_{Ca}(Si,Al)_4N_{24}]^{-42}$ for $CaAlSiN_3{:}Eu^{2+}$ (1 mol%) phosphor (left: a ball-and-stick model; right: a polyhedral model).

refinement of x-ray powder diffraction data. Owing to the disorder character of Al and Si in the structure, it is assumed that the arrangement of 4Al and 4Si is ordered at the (Si,Al) sites in a short range, as shown in Figure 5.37, in order to simplify the simulations. The point group of a selected cluster model is C_1 symmetry. The electron configurations of the atomic orbitals are 1s-4s for Ca, 1s-3p for Al and Si, 1s-2p for N, and 1s-6s for Eu^{2+}. The Eu-4f energy levels were fixed at zero energy in the calculations. To keep a charge-neutral cluster, the Madelung potential has been made for MO calculations.

Figure 5.38 shows the total and partial densities of states and atomic energy levels of the $[Eu_{Ca}(Si,Al)_4N_{24}]^{-42}$ cluster. The N-2s and N-2p orbitals are the major components of the occupied MO overlapped with small fractions of (Si,Al)-3s and (Si,Al)-3p orbitals. The overlap between the N-2p and (Si,Al)-3p orbital on the top of occupied MO indicates the formation of covalent banding of (Si,Al)-N, as expected. The unoccupied MO is dominated by wide distribution of (Si,Al)-3s and (Si,Al)-3p orbitals ranging from 0 to 15 eV. The lowest of the Eu-5d orbitals are located above Eu-4f of about 1.28 eV, and the top of Eu-5d is partially hybridized with the bottom of the (Si,Al)-3s2p levels. The calculated energy gap between the HOMOs and LUMOs of the $[Eu_{Ca}(Si,Al)_4N_{24}]^{-42}$ is estimated to be 2.1 eV, which is smaller than the calculated one for the host (~3.4 eV from first-principles calculations based on the DFT approach), as the dopant Eu^{2+} ions can reduce the band gap due to the introduction of impurity levels. The overall bond overlap population (BOP) is calculated to be about 0.007 for Eu_{Ca}-N, with 27 nitrogen atoms in the cluster model.

5.5.3 Ca-α-SiAlON:Eu²⁺

The local electronic structures, that is, the energy levels of the atomic orbitals and the density of states (DOS) of Eu^{2+}-doped Ca-α-SiAlON (nominal

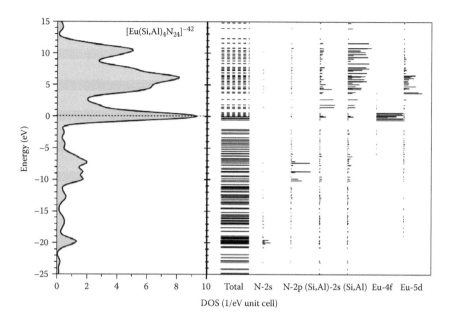

FIGURE 5.38
Density of states and atomic energy level diagram for the $[Eu_{Ca}(Si,Al)_8N_{24}]^{-42}$ cluster of $CaAlSiN_3$:Eu^{2+} (1 mol%).

composition $Ca_{1.764}Eu_{0.036}Al_{3.68}Si_{8.32}N_{16}$), were calculated by employing the relativistic DV-Xα method based on the first-principles molecular orbitals (MOs) approach. On the basis of the Rietveld refinement of x-ray powder diffraction data, the cluster models of $[Eu_{Ca}(Si,Al)_8N_{24}]^{-42}$ were designed for Ca-α-SiAlON:Eu^{2+} corresponding to the point groups of Ci symmetry, in which the Eu^{2+} ions are assumed to be substituted for the Ca ions (Figure 5.39). About 50,000 sampling points were used for all the clusters in the present calculations. The electron basis sets are 1s-4s for Ca, 1s-3p for Al and Si, 1s-2p for N, and 1s-6s for Eu^{2+}. The effective Madelung potential was taken into account with the point charges embedding outside of the clusters.

Figure 5.40 shows the total and partial densities of states, as well as the energy level of the $[Eu_{Ca}(Si,Al)_8N_{24}]^{-42}$ cluster calculated by the DV-Xα method. The bottom of the lowest unoccupied molecular orbitals (LUMOs) is mainly composed of the (Si,Al)-2s and the Eu-5d orbitals for the $[Eu_{Ca}(Si,Al)_8N_{24}]^{-42}$ cluster. (Si,Al)-3p orbitals are at relatively high energy and are widely dispersed in the unoccupied MOs. The localized 4f-electron levels are at about zero energy, while the 5d levels are positioned at about 2.7 eV higher energy. The occupied MO is mainly contributed from the N-2s and N-2p orbitals, in which the upper part of the occupied MO is determined by N-2p orbitals in the range of –10 to –2 eV, while the N-2s orbitals are at a lower energy part (–26 to –20 eV). In addition, small amounts of (Si,Al)-3p and (Si,Al)-2s overlap with N-2p (Figure 5.41) due to the chemical bonding of (Si,Al)-N. The energy

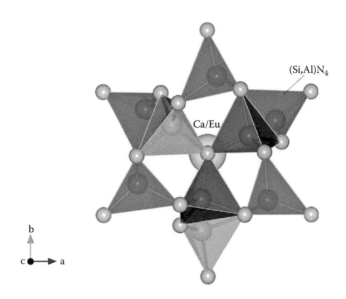

FIGURE 5.39
Cluster model of $[Eu_{Ca}(Si,Al)_{12}N_{29}]^{-40}$ for the Ca-α-SiAlON:Eu^{2+} phosphor.

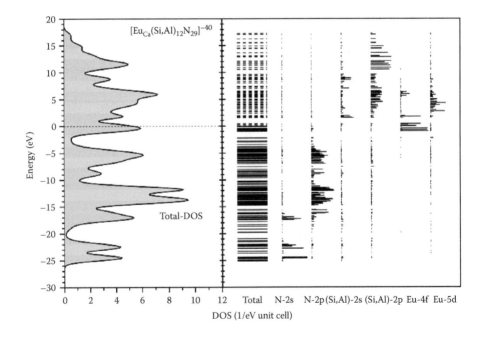

FIGURE 5.40
Density of states and energy levels of the cluster model of $[Eu_{Ca}(Si,Al)_{12}N_{29}]^{-40}$ for Ca-α-SiAlON:Eu^{2+} (2 mol%).

gap between the HOMOs and LUMOs is about 2.2 eV. It should be noted that the energy gap also strongly depends on the size of the selected clusters. As a result, the calculated energy gap is not good enough to represent the properties of the average structure of Ca-α-SiAlON:Eu^{2+} by the DV-Xα approach due to its limitations; nevertheless, it can enable us to deeply understand the differences in the local electronic structures due to different coordination numbers, bond lengths, and site symmetries of the target clusters.

5.5.4 β-SiAlON:Eu²⁺

The DV-Xα calculations have been performed on the electronic structure of $Eu_xSi_{6-z}Al_{z-x}O_{z+x}N_{8-z-x}$ ($x = 0, 0.013$ for $z = 0.15$) (Li et al. 2008b). The selected cluster model was extracted from the XRD refined structures, namely, $[EuSi_{12}N_{30}]^{-40}$ based on a supercell model, as shown in Figure 5.41. The point group of the cluster has C_i symmetry. For the sake of simplicity, a small amount of Al and O were not included in the selected clusters, and relaxation of the atoms was not performed in the calculations. The atomic orbitals are based on 1s-2p, 1s-3p, and 1s-6p orbitals for Si, N, and Eu, respectively. In addition, the Madelung potential generated by point charges outside the cluster was introduced into the calculations, and the Eu-4f energy level was set at zero.

Figure 5.42 shows the total and partial densities of states of the $[EuSi_{12}N_{30}]^{-40}$ cluster in $Eu_xSi_{6-z}Al_{z-x}O_{z+x}N_{8-z-x}$ ($x = 0.013$ for $z = 0.15$). In good agreement with the experimental data, the calculated band gap HOMOs and LUMOs decreases from the host of ca. 5.55 (Li et al. 2008b) to 5.45 eV as the Eu^{2+} is

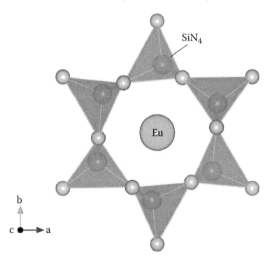

FIGURE 5.41
Cluster model of $[EuSi_{12}N_{30}]^{-40}$ for β-SiAlON:Eu^{2+} displayed in a tetrahedral model.

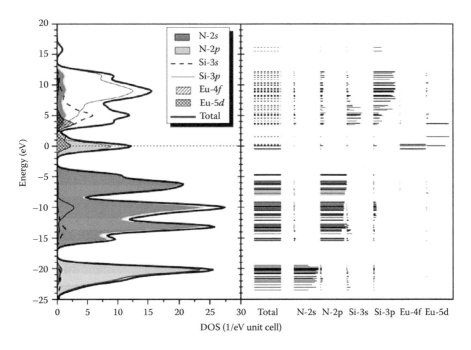

FIGURE 5.42

Density of states and atomic energy levels of the $[EuSi_{12}N_{30}]^{-40}$ cluster for β-SiAlON:Eu^{2+}. (Reprinted from Li, Y. Q., Hirosaki, N., Xie, R.-J., Takeda, T., and Mitomo, M., *J. Solid State Chem.*, 181, 3200–3210, 2008b. With permission.)

incorporated into $Si_{6-z}Al_zO_zN_{8-z}$ lattice. It should be noted that an accurate calculation of the band gap is not the main purpose of the present work, and actually, it strongly depends on the size of the selected cluster for the DV-Xα method. In $Eu_xSi_{6-z}Al_{z-x}O_{z+x}N_{8-z-x}$ (x = 0.013, z = 0.15), partial high levels of the Eu-5d orbitals are hybridized with the Si-3s, Si-3p, and N-2p orbitals at the bottom of the unoccupied MO, suggesting that the 5d electrons of Eu^{2+} take part in the formation of chemical bonding, which is influenced by its neighbor atoms, in particular the second-nearest Si (or Si/Al) atoms from the electronic structure viewpoints. The occupied MOs mainly consist of N-2s, and N-2p hybridized with small amounts of Si-3s and Si-3p distributed in lower energies for $Eu_xSi_{6-z}Al_{z-x}O_{z+x}N_{8-z-x}$ (x = 0, x = 0.013), with the N-2s orbital clearly separating from the other orbitals of the valence bands at much lower energies, whereas the N-2p orbitals are at the top of the valence band. The calculated 4f and 5d levels of Eu^{2+} are located in between the top of N-2p and the bottom of Si-3s3p. We should note that a precise calculation is not good enough without including Al and O, and the relative position of the 4f and 5d orbitals of Eu^{2+} is also far from satisfactory by only the DV-Xα method, which should be combined with other appropriate first-principles methods to produce accurate results. In short, the DV-Xα calculations strongly confirm our experimental results from the diffuse reflection spectra, that the band gap is slightly narrowed by Eu doping in $Si_{6-z}Al_zO_zN_{8-z}$ theoretically. Furthermore, the large hybridization of Eu-5d with Si-3s3p and N-2p orbitals indicates that the 5d states of Eu^{2+} are more sensitive to Si (or Si/Al), in agreement with the obtained structural data.

5.5.5 $CaSiN_2:Eu^{2+}$ and $Ca_2Si_5N_8:Eu^{2+}$

Similar to the luminescence properties, the local electronic structures are also very sensitive to the local structures, such as the point symmetry, coordination number, and bond length. Typical examples are red-emitting $CaSiN_2:Eu^{2+}$ (Li et al. 2010) and $Ca_2Si_5N_8:Eu^{2+}$ nitride phosphors. There are two individual Ca crystallographic sites in orthorhombic $CaSiN_2:Eu^{2+}$ and monoclinic $Ca_2Si_5N_8:Eu^{2+}$ having the same coordination numbers with the first-nearest N (five and six) and point symmetry (C_1) on the two luminescent centers, while the bond lengths are different for the $[CaN_n]^{-3n-2}$ ($n = 5, 6$) clusters. In the calculations, the selected cluster models, that is, $[Eu(I)N_6]^{-16}$ and $[Eu(II)N_5]^{-13}$ for $CaSiN_2:Eu^{2+}$ (0.1 mol%) and $[Eu(I)N_5]^{-13}$ and $[Eu(II)N_6]^{-16}$ for $CaSiN_2:Eu^{2+}$ (2 mol%), were obtained from the XRD refined crystal structures for the two individual sites in the range of 3.0 Å. The atomic orbitals are based on 1s-4p, 1s-2p, and 1s-6p orbitals for Ca, N, and Eu, respectively. Figure 5.43 shows the molecular orbital energy levels and the total and partial densities of states of $[EuN_5]^{-13}$ and $[EuN_6]^{-16}$ for the two individual Eu_{Ca} clusters of $CaSiN_2:Eu^{2+}$ and $Ca_2Si_5N_8:Eu^{2+}$. Obviously, due to different coordination numbers and symmetries over two Ca_{Eu} sites, both DOSs and energy levels are quite different for two cluster models of each nitride phosphor. The common features are that the N-2s orbitals are located at lower energies below −10 eV, while the N-2p orbitals are at high energy and dominate the occupied MO, particularly on top of the occupied MO. The Eu-4f and Eu-5d orbitals are positioned at high energy close to and above the Fermi level, hybridized with smaller fractions of N-2p orbitals. The energy gap between HOMOs and LUMOs seems to be related to the coordination number of Ca_{Eu} with N. In contrast to the $[Eu_{Ca}N_5]^{-13}$ cluster, the densities of states of the $[Eu_{Ca}N_6]^{-16}$ clusters in both $CaSiN_2:Eu^{2+}$ and $Ca_2Si_5N_8:Eu^{2+}$ give continuous bands for the N-2p and Eu-5d orbitals' crossover zero energy (Fermi level *Ef*), without obvious energy gaps between HOMOs and LUMOs, like metallic DOS dispersion covering a wide energy range. From this point of view, the Eu^{2+} ions on the Ca site having a sixfold coordination with N are probably much more easily quenched, even at low Eu^{2+} concentration, which may explain why the red emission of Eu^{2+} always shows a single emission band with high symmetry for $CaSiN_2:Eu^{2+}$ and $Ca_2Si_5N_8:Eu^{2+}$, which is probably attributed to one luminescent center being in effect.

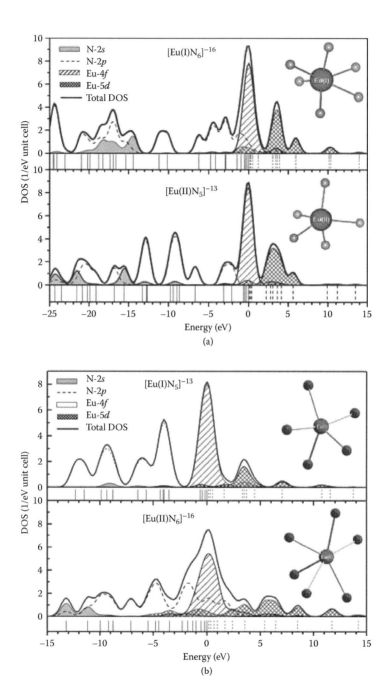

FIGURE 5.43
Density of states and total atomic energy levels of $[CaN_n]^{-3n-2}$ (n = 5, 6) for (a) $CaSiN_2$:Eu^{2+} and (b) $Ca_2Si_5N_8$:Eu^{2+}. (Reprinted from Li, Y. Q., Hirosaki, N., Xie, R. J., Takeda, T., and Mitomo, M., *Int. J. Appl. Ceram. Technol.*, 7:787–802, 2010. With permission.)

References

Adachi, H., Mukoyama, T., and Kawai, J. 2005. *Hartree-Fock-Slater method for materials science: The DV-Xα method for design and characterization of materials.* New York: Springer.

Adachi, H., Tsukada, M., and Satoko, C. 1978. Discrete variational Xα cluster calculations. I. Application to metal clusters. *Jpn. J. Phys. Soc.* 45:875–883.

Bianconi, A. 1980. Surface x-ray absorption spectroscopy: Surface EXAFS and surface XANES. *Appl. Surf. Sci.* 6:392–418.

Bianconi, A., Dell'Ariccia, M., Durham, P. J., and Pendry, J. B. 1982. Multiple-scattering resonances and structural effects in the x-ray-absorption near-edge spectra of Fe II and Fe III hexacyanide complexes. *Phys. Rev. B* 26:6502–6508.

Blasse, G., and Grabmaier, B. C. 1994. *Luminescent materials.* Berlin: Springer-Verlag.

Bunker, G. 2010. *Introduction to XAFS: A practical guide to x-ray absorption fine structure spectroscopy.* Cambridge: Cambridge University Press.

Cao, G. Z., and Metselaar, R. 1991. Alpha'-sialon ceramics—A review. *Chem. Mater.* 3:242–252.

Ching, W. Y., Huang, M.-Z., and Mo, S.-D. 2000. Electronic structure and bonding of β-SiAlON. *J. Am. Ceram. Soc.* 83:780–786.

Clearfield, A., Reibenspies, J. H., and Bhuvanesh, N. 2008. *Principles and applications of powder diffraction.* Chichester: Blackwell Publishing Ltd.

Dronskowski, R. 2005. *Computational chemistry of solid state materials.* Weinheim, Germany: Wiley-VCH Verlag.

Ekstrom, T., and Nygren, M. 1992. SiAlON ceramics. *J. Am. Ceram. Soc.* 75:259–276.

Fang, C. M., Hintzen, H. T., de Groot, R. A., and de With, G. 2001b. First-principles electronic structure calculations of $Ba_5Si_2N_6$ with anomalous Si_2N_6 dimeric units. *J. Alloys Compd.* 322:L1–L4.

Fang, C. M., Hintzen, H. T., and de With, G. 2002. First-principles electronic structure calculations of $BaSi_7N_{10}$ with both corner- and edge-sharing SiN_4 tetrahedra. *J. Alloys Compd.* 336:1–4.

Fang, C. M., Hintzen, H. T., and de With, G. 2003. Electronic structure of Ba-Si-N compounds with different Ba/Si ratios from first principles calculations. *Recent Res. Dev. Mater. Sci.* 4:283–291.

Fang, C. M., Hintzen, H. T., de With, G., and de Groot, R. A. 2001a. Electronic structure of the alkaline-earth silicon nitrides $M_2Si_5N_8$ (M = Ca and Sr) obtained from first-principles calculations and optical reflectance spectra. *J. Phys. Condens. Matter* 13:67–76.

Haferkorn, B., and Meyer, G. Z. 1998. Li_2EuSiO_4, in europium (II)-dithiosilicate: $Eu[(Li_2Si)O_4]$. *Anorg. Allg. Chem.* 624:1079–1081.

Hirosaki, N., Xie, R. J., Kimoto, K., Sekiguchi, T., Yamamoto, Y., Suehiro, T., and Mitomo, M. 2005. Characterization and properties of green-emitting β-SiAlON:Eu^{2+} powder phosphors for white light-emitting diodes. *Appl. Phys. Lett.* 86:211905-1–211905-3.

Hoppe, H. A., Lutz, H., Morys, P., Schnick, W., and Seilmeier, A. 2000. Luminescence in Eu^{2+}-doped $Ba_2Si_5N_8$: Fluorescence, thermoluminescence, and upconversion. *J. Phys. Chem. Solids* 61:2001–2006.

Huppertz, H., and Schnick, W. 1997. $Eu_2Si_5N_8$ and $EuYbSi_4N_7$. The first nitridosilicates with a divalent rare earth metal. *Acta Crystollgr.* C53:1751–1753.

Izumi, F., and Momma, K. 2007. Three-dimensional visualization in powder diffraction. *Solid State Phenom.* 130:15–20.

Jack, K. H. 1976. Review: Sialons and related nitrogen ceramics. *J. Mater. Sci.* 11:1135–1158.

Jack, K. H., and Wilson, W. I. 1972. Ceramics based on the Si-Al-O-N and related systems. *Nature Phys. Sci. (London)* 238:28–29.

Jacobsen, H., Meyer, G., Schipper, W., and Blasse, G. 1994. Synthesis, structures and luminescence of two new europium(II) silicate-chlorides, $Eu_2SiO_3Cl_2$ and $Eu_5SiO_4Cl_6$. *Z. Anorg. Allg. Chem.* 620:451–456.

Jensen, F. 2006. *Introduction to computational chemistry.* 2nd ed. New York: Wiley.

Kimoto, K., Xie, R. J., Matsui, Y., and Hirosaki, N. 2009. Direct observation of single dopant atom in light-emitting phosphor of β-sialon:Eu^{2+}. *Appl. Phys. Lett.* 94:041908.1–041908.3.

Kresse, G., and Furthmuller, J. 1996a. Efficiency of *ab initio* total energy calculations for metals and semiconductors using a plane-wave basis set. *Comput. Mater. Sci.* 6:15–50.

Kresse, G., and Furthmuller, J. 1996b. Efficient iterative schemes for *ab initio* total-energy calculations using a plane-wave basis set. *Phys. Rev. B* 54:11169–11186.

Kresse, G., and Hafner, J. 1993. *Ab initio* molecular dynamics for liquid metals. *Phys. Rev. B* 47:558–561.

Kresse, G., and Hafner, J. 1994. *Ab initio* molecular-dynamics simulation of the liquid-metal–amorphous-semiconductor transition in germanium. *Phys. Rev. B* 49:14251–14269.

Larson, A. C., and Von Dreele, R. B. 2000. *GSAS—General structure analysis system.* Report LAUR 86-748. Los Alamos, NM: Los Alamos National Laboratory.

Lee, S. S., Lim, S. S., Sum, S., and Wager, J. F. 1997. Photoluminescence and electroluminescence characteristics of CaSiN2:Eu phosphor. In *Proceedings of the Society of Photo-Optical Instrumentation (SPIE)*, ed. A. Hariz, V. K. Varadan, and O. Reinhold, 75–83. Vol. 3241. Bellingham, WA: ETATS-UNIS:SPIE.

Le Toquin, R., and Cheetham, A. K. 2006. Red-emitting cerium-based phosphor materials for solid-state lighting applications. *Chem. Phys. Lett.* 423:352–356.

Li, Y. Q., Delsing, A. C. A., de With, G., and Hintzen, H. T. 2005. Luminescence properties of Eu^{2+}-activated alkaline earth silicon oxynitride $MSi_2O_{2-\delta}N_{2+2/3\delta}$ (M = Ca, Sr, Ba): A promising class of novel LED conversion phosphors. *Chem. Mater.* 17:3242–3248.

Li, Y. Q., de With, G., and Hintzen, H. T. 2008a. The effect of replacement of Sr by Ca on the structural and luminescence properties of the red-emitting $Sr_2Si_5N_8$:Eu^{2+} LED conversion phosphor. *J. Solid State Chem.* 181:515–524.

Li, Y. Q., Fang, C. M., de With, G., and Hintzen, H. T. 2004. Preparation, structure and photoluminescence properties of Eu^{2+} and Ce^{3+}-doped $SrYSi_4N_7$. *J. Solid State Chem.* 177:4687–4694.

Li, Y. Q., Hirosaki, N., and Xie, X.-J. 2007. Structural and photoluminescence properties of CaSiN₂:Ce^{3+}, Li^+. Paper presented at the Japan Society of Applied Physics, the 67th autumn meeting, Sapporo, Japan.

Li, Y. Q., Hirosaki, N., Xie, R.-J., Takeda, T., and Mitomo, M. 2008b. Crystal and electronic structures, luminescence properties of Eu^{2+}-doped $Si_{6-z}Al_zO_zN_{8-z}$ and $M_ySi_{6-z}Al_{z-y}O_{z+y}N_{8-z-y}$. *J. Solid State Chem.* 181:3200–3210.

Li, Y. Q., Hirosaki, N., Xie, R.-J., Takeda, T., and Mitomo, M. 2010. Synthesis, crystal and local electronic structures, and photoluminescence properties of red-emitting $CaAl_zSiN_{2+z}$:Eu^{2+} with orthorhombic structure. *Int. J. Appl. Ceram. Technol.* 7:787–802.

Li, Y. Q., van Steen, J. E. J., van Krevel, J. W. H., Botty, G., Delsing, A. C. A., DiSalvo, F. J., de With, G., and Hintzen, H. T. 2006. Luminescence properties of red-emitting $M_2Si_5N_8$:Eu^{2+} (M = Ca, Sr, Ba) LED conversion phosphors. *J. Alloys Compd.* 417:273–279.

Marchand, R., l'Haridon, P., and Laurent, Y. 1978. Structure crystalline de $Eu_2(II)$ $SiO_4\beta$. *J. Solid State Chem.* 24:71–76

Mikami, M., Uheda, K., and Kijima, N. 2006. First-principles study of nitridoaluminosilicate $CaAlSiN_3$. *Phys. Stat. Sol. A* 203:2705–2711.

Milman, V., Winkler, B., White, J. A., Pickard, J., Payne, M. C., Akhmatskaya, E. V., and Nobes, R. H. 2000. Electronic structure, properties, and phase stability of inorganic crystals: A pseudopotential plane-wave study. *Int. J. Quantum Chem.* 77:895–910.

Mueller-Mach, R., Mueller, G., Krames, M. R., Hoppe, H. A., Stadler, F., Schnick, W., Juestel, T., and Schmidt, P. 2005. Highly efficient all-nitride phosphor-converted white light emitting diode. *Phys. Stat. Sol. A* 202:1727–1732.

Nakayasu, T., Yamada, T., Tanaka, I., Adachi, H., and Goto, S., 1997. Local chemical bonding around rare-earth ions in α- and β-Si_3N_4. *J. Am. Ceram. Soc.* 80:2525–2532.

Newville, M. The fundamentals of XAFS. http://xafs.org/Tutorials (accessed September 29, 2010).

Oeckler, O., Stadler, F., Rosenthal, T., and Schnick, W. 2007. Real structure of $SrSi_2O_2N_2$. *Solid State Sci.* 9:205–212.

Ogasawara, K., Iwata, T., Koyama, Y., Ishii, T., Tanaka, I., and Adachi, H. 2001. Relativistic cluster calculation of ligand-field multiplet effects on cation $L_{2,3}$ x-ray-absorption edges of $SrTiO_3$, NiO, and CaF_2. *Phys. Rev. B* 64:115413–115418.

Ottinger, F. 2004. Synthese, Struktur und analytische Detailstudien neuer stickstoffhaltiger Silicate und Aluminosilicate. PhD thesis, Universität Karlsruhe, Karlsruhe.

Pecharsky, V. K., and Zavalij, P. J. 2008. *Fundamentals of powder diffraction and structural characterization of materials.* 2nd ed. New York: Springer.

Perdew, J. P., Chevary, J. A., Vosko, S. H., Jackson, K. A., Pederson, M. R., Singh, D. J., and Fiolhais, C. 1992. Atoms, molecules, solids, and surfaces: Applications of the generalized gradient approximation for exchange and correlation. *Phys. Rev. B* 46:6671–6687.

Petzow, G., and Herrmann, M. 2002. Silicon nitride ceramics. *Struct. Bonding* 102:47–167.

Rietveld, H. M. 1969. A profile refinement method for nuclear and magnetic structures. *J. Appl. Cryst.* 2:65–71.

Rodriguez-Carvajal, J. 1993. Recent advances in magnetic structure determination by neutron powder diffraction. *Phys. B* 192:55–69.

Rosen, A., Ellis, D. E., Adachi, H., and Averill, F. W. 1976. Calculations of molecular ionization energies using a self-consistent-charge Hartree–Fock–Slater method. *J. Chem. Phys.* 65:3629–3634.

Schlieper, T., and Schnick, W. 1995. Nitrido-silicate. II. Hochtemperatur-synthesen und kristallstrukturen von $Sr_2Si_5N_8$ und $Ba_2Si_5N_8$. *Z. Anorg. Allg. Chem.* 621:1380–1384.

Shannon, R. D. 1976. Revised effective ionic radii and systematic studies of inter-atomic distances in halides and chalcogenides. *Acta Crystollgr.* A32:751–767.

Shioi, K., Hirosaki, N., Xie, R. J., Takeda, T., and Li, Y. Q. 2010. Synthesis, crystal structure and photoluminescence of Eu-α-SiAlON. *J. Alloys Compd.* 504:579–584.

Stadler, F., Oeckler, O., Hoeppe, H. A., Moeller, M. H., Poettgen, R., Mosel, B. D., Schmidt, P., Duppel, V., Simon, A., and Schnick, W. 2006. Crystal structure, physical properties and HRTEM investigation of the new oxonitridosilicate $EuSi_2O_2N_2$. *Chem. Eur. J.* 12:6984–6990.

Toby, B. H. 2001. EXPGUI, a graphical user interface for GSAS. *J. Appl. Cryst.* 34:210–213.

Uheda, K., Hirosaki, N., Yamamoto, Y., Naito, A., Nakajima, T., and Yamamoto, H. 2006a. Luminescence properties of a red phosphor, $CaAlSiN_3$:Eu^{2+}, for white light-emitting diodes. *Electrochem. Solid-State Lett.* 9:H22–H25.

Uheda, K., Hirosaki, N., Yamamoto, Y., and Yamamoto, H. 2006b. Host lattice materials in the system Ca_3N_2-AlN-Si_3N_4 for white light emitting diode. *Phys. Stat. Sol. A* 203:2712–2717.

van Krevel, J. W. H., van Rutten, J. W. T., Mandal, H., Hintzen, H. T., and Metselaar, R. 2002. Luminescence properties of terbium-, cerium-, or europium-doped α-sialon materials. *J. Solid State Chem.* 165:19–24.

Xie, R. J., Hirosaki, N., Mitomo, M., Takahashi, K., and Sakuma, K. 2006a. Highly efficient white-light emitting diodes fabricated with short-wavelength yellow oxynitride phosphors. *Appl. Phys. Lett.* 88:101104:1–101104:3.

Xie, R. J., Hirosaki, N., Mitomo, M., Uheda, K., Suehiro, T., Xu, X., Yamamoto, Y., and Sekiguchi, T. 2005. Strong green emission from α-sialon activated by divalent ytterbium under blue light irradiation. *J. Phys. Chem. B* 109:9490–9494.

Xie, R. J., Hirosaki, N., Mitomo, M., Yamamoto, Y., Suehiro, T., and Sakuma, K. 2004a. Optical properties of Eu^{2+} in α-sialon. *J. Phys. Chem. B* 108:12027–12031.

Xie, R. J., Hirosaki, N., Sakuma, K., Yamamoto, Y., and Mitomo, M. 2004b. Eu^{2+}-doped Ca-α-sialon: A yellow phosphor for white light-emitting diodes. *Appl. Phys. Lett.* 84:5404–5406.

Xie, R.-J., Hirosaki, N., Suehiro, T., Xu, F. F., and Mitomo, M. 2006b. A simple, efficient synthetic route to $Sr_2Si_5N_8$:Eu^{2+}-based red phosphors for white light-emitting diodes. *Chem. Mater.* 18:5578–5583.

Xie, R. J., Mitomo, M., Uheda, K., Xu, F. F., and Akimune, Y. 2002. Preparation and luminescence spectra of calcium- and rare-earth (R = Eu, Tb, and Pr)-codoped α-sialon ceramics. *J. Am. Ceram. Soc.* 85:1229–1234.

Xu, Y. N., and Ching, W. Y. 1995. Electronic structure and optical properties of α and β phases of silicon nitride, silicon oxynitride, and with comparison to silicon dioxide. *Phys. Rev. B* 51:17379–17389.

Xu, Y. N., Rulis, P., and Ching, W. Y. 2005. Electronic structure and bonding in quaternary crystal $Y_3Si_5N_9O$. *Phys. Rev. B* 72:113101-1–113101-4.

Yen, W. M., Shionoya, S., and Yamamoto, H. 2006. *Phosphor handbook.* 2nd ed. New York: CRC Press.

Yoshida, H., Yoshimatsu, R., Watanabe S., and Ogasawara, K. 2006. Optical transitions near the fundamental absorption edge and electronic structures of $YAl_3(BO_3)_4$:Gd^{3+}. *Jpn. J. Phys. Soc.* 45:146–151.

Young, D. C. 2001. *Computational chemistry: A practical guide for applying techniques to real-world problems.* New York: John Wiley & Sons.

Young, R. A. 1995. *The Rietveld method (International Union of Crystallography Monographs on Crystallography).* Oxford: Oxford University Press.

Zoltán, A. G., Phillip, M. M., Heston, J. O., and Simon, J. C. 2004. Synthesis and structure of alkaline earth silicon nitride: $BaSiN_2$, $SrSiN_2$, and $CaSiN_2$. *Inorg. Chem.* 43:3998–4006.

6

Some Key Issues in Understanding Nitride Phosphors

6.1 Introduction

As described in the preceding chapters, nitrides and oxynitrides activated with rare earth ions have, in many cases, luminescence properties suited well to white LED application. Particularly important are strong blue or near-UV absorption resulting in efficient visible luminescence and small thermal quenching of luminescence. It is still not clear enough, however, why these features are brought about, although research has been done to deepen understanding of such fundamental properties in parallel with the search for new materials. In this chapter, the features of nitride and oxynitride phosphors, in comparison with conventional phosphors, are revisited, with an expectation that such a comparison can make clear the physical properties of these new phosphors.

Among many activator ions, Eu^{2+} is discussed here because it has proved to be the most frequently used and the most important activator ion in nitride and oxynitride phosphors, as well as Ce^{3+}. Optical transitions of Eu^{2+} take place between the ground state of the $4f^7$ configuration and excited states of the $4f^65d$ configuration. The transition energies are decided by the energies of 5d levels in a host crystal.

6.2 Absorption and Luminescence Transitions at Long Wavelength

6.2.1 A Trend Observed in Nitride and Oxynitride Phosphors

In many halides and oxides, Eu^{2+} luminescence is found in the region of near UV to green, though in some oxides it is in orange or red. Typical examples are BaFCl:Eu^{2+}, with an emission peak at 380 nm for x-ray detection; blue-emitting $BaMgAl_{10}O_{17}$:Eu^{2+}, widely used in fluorescent lamps; and $SrAl_2O_4$:Eu^{2+},Dy^{3+},

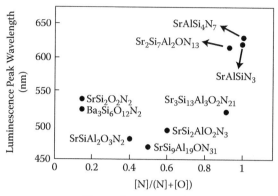

FIGURE 6.1
Correlation between luminescence peak wavelength and N atom fraction in the anions coordinated with Sr or Ba sites in some Eu^{2+} activated nitride or oxynitride compounds.

which is a long-persistent green phosphor. Meanwhile, in most sulfides or selenides, Eu^{2+} luminescence is shifted to green, orange, or red, as shown by $SrGa_2S_4$:Eu^{2+} emitting in green and CaS:Eu^{2+} with deep red luminescence. As indicated by these examples, there is a general trend that the absorption and emission transitions of a Eu^{2+} ion shift to longer wavelength, or are red-shifted, when a host has higher covalency in its chemical bonds.

Nitride hosts, which have covalent Eu-N bonds, seem to go along with this trend, because a Eu^{2+} ion shows orange or red luminescence as a consequence of visible absorption bands. In oxynitrides, Eu^{2+} shows the transition energies between oxides and nitrides. Figure 6.1 shows the luminescence wavelength of some Eu^{2+}-activated Sr nitrides and oxynitrides against a N atom fraction ratio in a coordinating sphere around the Sr^{2+} or Eu^{2+} ion. This plot indicates correlation between the luminescence wavelength and the N atomic fraction, though exceptional cases are found for oxynitrides. Also for oxynitrides, which are derived from MAl_2O_4:Eu^{2+} (M = Ca, Sr, and Ba) by cross-substitution of [Al-O]$^+$ pairs with [Si-N]$^+$ pairs, an increase in N/O atomic ratio was found to shift both excitation and luminescence spectra to longer wavelength (Li et al. 2006). This result also indicates the effect of covalency.

It is not straightforward, however, to understand or predict the optical transition energy for each compound, because there are more factors other than covalency. For example, it is not clear why $CaAlSiN_3$:Eu^{2+} shows deep red luminescence and why $CaSiN_2$:Eu^{2+} or $Ca_2Si_5N_8$:Eu^{2+} shows only orange-red luminescence. It is necessary to separate the covalency effect from other factors in order to know more about the optical transition energy. Here we briefly review factors deciding the energies of 5d levels in a Eu^{2+} ion in general, and then take examples of two red phosphors, $CaAlSiN_3$:Eu^{2+} and CaS:Eu^{2+}, to discuss possible reasons of their red luminescence.

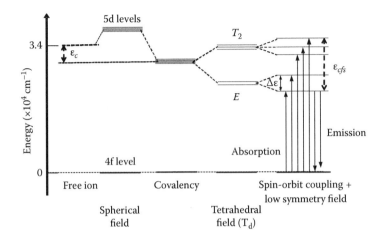

FIGURE 6.2
Schematic diagram of 5d energy levels Eu^{2+} in a crystal. From left to right, the levels in a free ion for reference, in a virtual spherical field, with the effect of covalency, in a tetrahedral field, and perturbed by spin-orbit interaction and a low-symmetry field. ε_c, ε_{cfs}, and ΔE denote energy differences indicated in the figure.

6.2.2 General Discussion on the Transition Energy of Eu^{2+} in a Crystal

A schematic diagram of 5d energy levels is shown in Figure 6.2. This Figure shows static energy levels and not relaxed states of excited levels. Excitation by a blue or violet LED and its reverse process, i.e., luminescence, corresponds to transitions between the lowest excited level and the ground state.

If an ion having a 5d electron or electrons such as Ce^{3+} or Eu^{2+} is introduced from the gas state into a solid, the energy of degenerate 5d levels is lowered by a magnitude denoted as ε_c, as denoted by Dorenbos (2002). Figure 6.2 schematically shows the effects of crystal-field potential and covalency on the 5d energy levels in a crystal. In an actual solid, degeneracy of 5d levels is lifted by nonspherical electrostatic interaction with surrounding ions, giving rise to two or more levels. In this situation, the barycenter, or "the centroid," of 5d levels is compared with the 5d levels of a free ion. It was reported that the barycenter of 5d levels goes down or the absolute value of ε_c is increased, when polarization of anions coordinated with Eu^{2+} is large. In other words, covalency of the Eu-anion bonds is large. This is because spatially spread electron clouds reduce electrostatic repulsion of a 5d electron with outermost electrons of the coordinated anions. This phenomenon is often called the *nephelauxetic effect*, or *cloud-expanding effect* (Section 2.2.1.4). Quite recently Mikami and Kijima (2010) discussed quantitatively the effect of polarization and the dielectric constant of a host crystal.

Under tetrahedral, cubic, or octahedral site symmetry, i.e., O_h or T_d symmetry, the degeneracy of 5d levels is lifted by crystal field potential, forming two levels denoted as t_{2g} and e_g or t_2 and e (Section 2.3.2). In such a case, the strength of the crystal field potential can be represented by the energy separation between these two energy levels, often called 10Dq, where parameters D and q are defined as follows in a framework of the point-charge model (see Equation 2.25):

$$D = \frac{35Ze}{4d_{OL}^5},$$

$$q = \frac{2e}{105} \int |d_{OL}(r)|^2 r^4 dr$$

Here Ze is the charge of each ligand ion, r is the radial position of the electron, and d_{OL} is the distance between the activator and the ligand ions (see Figure 2.7). The parameter q is proportional to the average of r^4 over a d orbital. Among the factors deciding 10Dq, the most important one is apparently the bond lengths, d_{OL}.

When the ion occupies the site in a distorted field of lower symmetry, the triply degenerated t_{2g} and doubly degenerated e_g states are split into three and two at the largest. As a result, the lowest excited state is further lowered by the value denoted as ΔE, as shown in Figure 6.2. A strong and asymmetric crystal field lowers the energy of the lowest excited state more, leading to photon absorption and emission at a longer wavelength.

This is the case with the most widely used phosphor, yellow-emitting $Y_3Al_5O_{12}$:Ce^{3+}, and its modifications. The dodecahedral site of Y^{3+}, which is occupied by Ce^{3+}, is distorted by different distances between neighboring O^{2-} ions. It was found that luminescence is red-shifted when this distortion is large (Wu et al. 2007). This unique ionic arrangement characteristic of the garnet type structure leads to yellow to orange luminescence, depending on the chemical composition, which is in striking contrast with any other Ce^{3+}-activated oxides, showing luminescence in the blue, violet, or UV region. The effect of distorted coordination geometry on the luminescence wavelength is also expected for Eu^{2+}-activated phosphors, though it is not quantitatively separated from the effect of the Stokes shift, discussed in Section 6.2.3.

As described above, a crystal field potential can be depicted as a sum of the high-symmetry component and a low-symmetry component, the latter being treated as a perturbation. The strength of each component can be measured by the energy splitting 10Dq and ΔE. In many cases, however, it is difficult to obtain the value of 10Dq, because overlap of fundamental absorption of a host crystal with 4f-5d absorption of an activator prevents us from identification of high crystal field components of 5d levels by a simple spectroscopic work. (Two-photon absorption spectroscopy may solve this problem.) Accordingly, we here notice the energy range of an excitation or absorption spectrum due to 4f-5d transitions, denoted as ε_{cfs} by Dorenbos (2002) (see Figure 6.2).

We can summarize possible reasons of the red-shift expected from the *static* energy levels of d electrons presented in Figure 6.2 as follows:

1. A decrease in the barycenter energy of the 5d levels by polarization of anions or bond covalency, denoted as ε_c.

2. A larger crystal field splitting of 5d levels in tetrahedral, cubic, or octahedral site symmetry, often denoted as $10Dq$.

3. A larger splitting of a degenerate 5d level, ΔE, by a distorted coordination of Eu^{2+}.

6.2.3 Effect of the Stokes Shift

The diagram in Figure 6.2 shows only static energy levels. Actually, a bond between an activator and a coordinated anion vibrates around the equilibrium length, resulting in *dynamical* energy change. In the one-dimensional configurational coordinate model, at very low temperature, luminescence originates from the minimum point of an excited state, where an activator-anion distance is elongated compared with the ground state, as indicated by the offset ΔR (see Figure 2.10). To reach the minimum point, an excited electron must use a part of its excess energy in pushing anions away. As a result, the energy of an excited electron becomes lower than the initial energy supplied by absorption of a photon. The difference in the electron energy, called the *Stokes shift*, is one of the factors determining the energy of an emitted photon or the luminescence wavelength. When static energy of an excited state is given, the luminescence wavelength is longer when the Stokes shift is larger. The large Stokes shift also broadens the luminescence band and lowers the thermal quenching temperature.

One can think of a simple picture where the Stokes shift is large when an anion coordinated to an activator can move farther away from the activator at an excited state. Keszler and his group reported a rule of thumb to find a large Stokes shift for Eu^{2+} luminescence in silicates or borates, including alkali earth elements (Akella and Keszler 1995; Diaz and Keszler 1996). In short, the Stokes shift is large if Eu^{2+} is doped in a soft or polarizable crystal. Among compounds with the same crystal structure, which includes a series of alkaline earth ions, Eu^{2+} luminescence shows the largest Stokes shift in Ba compounds. The reason is that Eu^{2+} finds the largest space at the Ba^{2+} site because of its smaller ionic radius than Ba^{2+}. If an anion coordinated to an activator has space to move in its coordination sphere, Eu^{2+} luminescence shows a large Stokes shift also.

Among many Eu^{2+}-doped oxides, $Ba_2Mg(BO_3)_2:Eu^{2+}$ shows that luminescence peaked probably at the longest wavelength: 617 nm. (An exceptional case is $MO:Eu^{2+}$, where M = Ca, Sr, or Ba, which is discussed in Section 6.2.4.) This compound also shows a Stokes shift of 14,000 cm^{-1}, which is much larger than the 5,600 cm^{-1} of $Sr_2Mg(BO_3)_2:Eu^{2+}$ or 4,300 cm^{-1} of $Ba_2Ca(BO_3)_2:Eu^{2+}$.

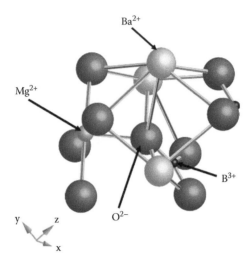

FIGURE 6.3

Ionic arrangement around O^{2-} ion in the $Ba_2Mg(BO_3)_2$ lattice. (Reprinted from Akella, A., and Keszler, D. A., *Mater. Res. Bull.*, 30, 105–111, 1995. With permission.)

A local atomic arrangement around an oxygen ion in $Ba_2Mg(BO_3)_2:Eu^{2+}$ is shown in Figure 6.3. One oxygen ion is connected to three Ba^{2+} ions, one Mg^{2+} ion, and one B^{3+} ion, and they form a distorted bipyramid. Behind the three Ba-O bonds, a large vacant space is created. According to the idea by Keszler's group, this local structure gives rise to a large Stokes shift (Diaz and Keszler 1996). Another red-emitting oxide is $Ba_3SiO_5:Eu^{2+}$, which has a luminescence peak at 590 nm and large Stokes shift of 5.8×10^3 cm^{-1} (see Figure 6.4; Yamaga et al. 2005). As shown in Figure 6.5a, the Ba^{2+} or Eu^{2+} site is coordinated with eight O^{2-} ions with two different kinds of bond lengths. The site symmetry is low, and splitting of the degenerate d levels is expected. The coordination around an O^{2-} ion is quite asymmetric, making vacant space in a way similar to that for $Ba_2Mg(BO_3)_2$ (see Figure 6.5b).

$Ba_2SiO_4:Eu^{2+}$ is also one of the oxides that shows luminescence at long wavelength, i.e., in green, though the luminescence wavelength varies with Eu concentration. A sample shown in Figure 6.4 has a luminescence band at 510 nm with a Stokes shift of 3.1×10^3 cm^{-1}. The structure of Ba_2SiO_4 has two unequivalent sites of the Ba^{2+} ion, and both are asymmetric. Accordingly, this case also seems to follow the rule by Keszler's group. A solid solution, $(Sr,Ba)_2SiO_4:Eu^{2+}$, has a high quantum efficiency and is commercially used as a green or yellow phosphor for white LEDs.

6.2.4 Dependence of Luminescence Wavelength on Eu Concentration

As mentioned above (Section 6.2.3), the luminescence of $Ba_2SiO_4:Eu^{2+}$ is shifted to a longer wavelength with an increase in Eu concentration

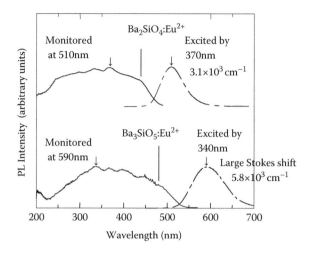

FIGURE 6.4
Luminescence and excitation spectra of $Ba_2SiO_4:Eu^{2+}$ and $Ba_3SiO_5:Eu^{2+}$. The Stokes shift is estimated by an energy difference between the excitation peak at the lowest energy and the luminescence peak. (Reprinted and revised from Yamaga, M., Masui, Y., Sakuta, S., Kodama, N., and Kaminaga, K., *Phys. Rev. B*, 71, 205102–205108, 2005. With permission.)

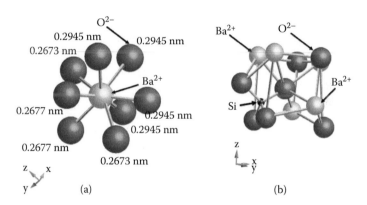

FIGURE 6.5
Coordination geometry in Ba_3SiO_5 around (a) Ba^{2+} and (b) O^{2-} ions.

(Figure 6.6). A similar shift is observed for many conventional phosphors. A result of $BaMgAl_{10}O_{17}:Eu^{2+}$ is shown in Figure 6.7. In this way, the Eu^{2+} concentration is one of the factors that determines the luminescence wavelength in most phosphors.

One reason for this shift may be reabsorption of luminescence. It will reduce luminescence intensity only in the short wavelength and make a shape of a luminescence band asymmetric. However, such a spectral change is not observed for the above phosphors. Reabsorption of luminescence is therefore not the main reason for the red-shift. Another possible reason is an increase

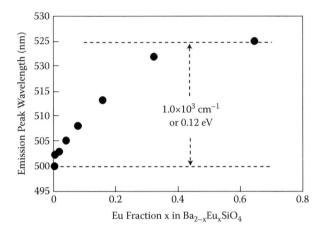

FIGURE 6.6
Luminescence peak wavelength as a function of a nominal Eu fraction in $Ba_{2-x}Eu_xSiO_4$.

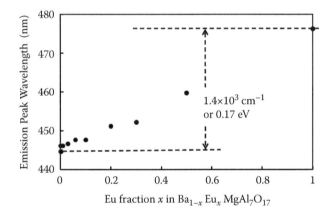

FIGURE 6.7
Luminescence peak wavelength as a function of a nominal Eu fraction in $Ba_{1-x}Eu_xMgAl_{10}O_{17}$.

in the crystal field splitting by a decrease in the lattice constants by substitution of Ba^{2+} with Eu^{2+}. This may be the case with $BaMgAl_{10}O_{17}:Eu^{2+}$, because $EuMgAl_{10}O_{17}$ forms a solid solution at all the Eu fractions, and also because the observed shift in the peak wavelength is nearly linear to the Eu fraction. More quantitative analysis is necessary to prove this assumption, however.

It is known that Eu^{2+} in Ca or Sr chalcogenides, including oxides, shows absorption in the visible region and luminescence in orange or red. Luminescence of $BaS:Eu^{2+}$ is quite different from the one of $CaS:Eu^{2+}$ or $SrS:Eu^{2+}$. Its luminescence band has a peak at 878 nm with a very broad band and large Stokes shift. This anomalous luminescence is originated in a bound exciton state related to Eu^{2+} (Smet et al. 2006). It is also known that the

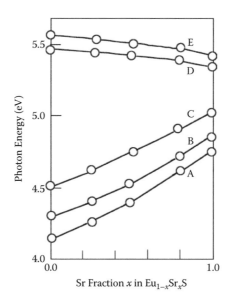

FIGURE 6.8
Exciton absorption energy as a function of a nominal Eu fraction in $Sr_xEu_{1-x}S$ at 2 K. The excitons A, B, and C originate in point X, and D and E in point Γ in k space. (Reprinted from Kaneko, Y., Morimoto, K., and Koda, T., *Oyobutsuri (Appl. Phys.)*, 50, 289–294, 1981. With permission.)

body color and emission color of these compounds largely shift to a longer wavelength with an increase in the Eu fraction, finally reaching a reddish black body color in pure EuS. The large dependence on the Eu fraction is shown in Figure 6.8, which depicts the photon energy of exciton absorption for $Sr_xEu_{1-x}S$ as a function of the Sr fraction ($0 \leq x \leq 1.0$) (Kaneko et al. 1981). The lowest-energy exciton, or "A exciton" in Figure 6.8, shows an increase in photon energy with the Eu fraction nearly linearly. The change in the energy is as large as 0.6 eV in the whole Eu fraction. This is much larger than the change observed for $BaMgAl_{10}O_{17}:Eu^{2+}$ or $Ba_2SiO_4:Eu^{2+}$, i.e., 0.17 eV or 0.12 eV. (Data for the luminescence wavelength as a function of the Eu concentration for Eu^{2+} in CaS or SrS are not available to the best of the author's knowledge.)

These sulfides have a NaCl type structure and EuS form solid solution with CaS and SrS in the whole fraction range. It is therefore unlikely that distortion arises around the Eu^{2+} ion at the Ca or Sr site. Most probably, crystal field strength, $10Dq$, is of normal value, because the bond length, 0.2845 nm for CaS and 0.3010 nm for SrS, is not particularly short. The covalency effect may partially contribute to the long-wavelength transitions, but it seems difficult to assume increased covalency with an increased Eu concentration.

It was found by both theoretical calculation and experiments that chalcogenides, including oxides, of Ca, Sr, or Ba with a NaCl type structure have an indirect band gap with the lowest conduction band at point X (Kaneko and Koda 1988). It was also found that 3d orbitals of Ca or 4d orbitals of Sr contribute most to the X point wavefunction, indicating that they can hybridize with Eu^{2+} 5d orbitals. If this is the case, two kinds of d orbitals provide bonding and antibonding orbitals with lowering of the bonding orbitals in energy. This hypothesis can explain the observed decrease in transition energies with Eu concentration.

As shown in Chapters 4 and 5, some Eu^{2+}-activated nitride phosphors show a considerable decrease in luminescence photon energy with an increase in Eu concentration. It is therefore meaningful to also consider the above reason for nitride phosphors.

6.2.5 Excitation and Luminescence Spectra of $CaAlSiN_3:Eu^{2+}$

With the above background, this section discusses red-shift of luminescence in $CaAlSiN_3:Eu^{2+}$. Among the many nitride and oxynitride phosphors so far developed, not many materials show such deep red luminescence as $CaAlSiN_3:Eu^{2+}$. Accordingly, it is interesting to consider the reason for the deep red luminescence of this phosphor. A similar luminescence color is shown by a conventional phosphor, $CaS:Eu^{2+}$. Figure 6.9 shows that they have a luminescence peak at nearly the same wavelength. On these phosphors, the factors deciding the luminescence wavelength are discussed below, with an aim to make clear the reasons for the spectral red-shift.

6.2.5.1 Consideration of the Crystal Field Strength

The crystal field strength at a site of high symmetry, $10Dq$, depends most on the nearest bond distance between an activator and a ligand ion, d_{OL},

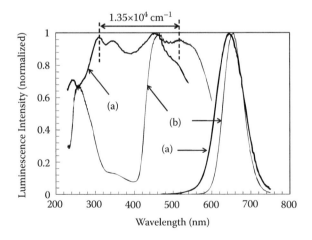

FIGURE 6.9
Excitation and luminescence spectra of (a) $CaAlSiN_3:Eu^{2+}$ (0.8 atom%) and (b) $CaS:Eu^{2+}$ at room temperature. The photon energy range of an excitation spectrum is indicated for $CaAlSiN_3:Eu^{2+}$.

as shown by Equation 2.25. The coordination around the Ca^{2+} site, which Eu^{2+} substitutes for, in $CaAlSiN_3$ is illustrated in Figure 5.6b. The Eu-N bond lengths are not short enough to assume strong crystal field potential. The shortest bond length is 0.2429 nm, which is comparable to the length of the Ca-F bond in CaF_2, 0.2365 nm, and longer than the shortest bond in $Ca_2Si_5N_8$, 0.2314 nm. This fact does not support a possibility of strong crystal field potential. The electric charge of a ligand ion, Ze, is larger for N^{3-} than for F^- or O^{2-}, if the formal charge is considered. For a covalent bond, however, the effective charge of the N ion is reduced, leading to a smaller value of Ze. Accordingly, we cannot find a good reason to assume large $10Dq$, which otherwise can decrease the energy of optical transitions.

On the other hand, the coordination geometry around a Eu^{2+} ion in $CaAlSiN_3$ is distorted from tetrahedral symmetry. Such geometry of low symmetry will split degenerate energy levels of irreducible representation E and lower the lowest excited state, though a quantitative value of the splitting, ΔE, is not clear. Asymmetric coordination geometry is also found in many nitride or oxynitride phosphors, e.g., $Ca_2Si_5N_8$ or $CaSiN_2$. These facts suggest that asymmetric geometry is one of the reasons for the low-energy optical transitions of Eu^{2+} in nitride hosts.

6.2.5.2 Consideration of the Center of 5d Levels

In order to determine the parameters, ε_{cfs}, ΔE, and the center of the 5d levels, it is necessary to identify all the split components of the 5d levels. This is, however, difficult for nitrides or sulfides having the fundamental absorption edge at a relatively long wavelength. Here we describe this feature by inspecting excitation spectra of $CaS:Eu^{2+}$ and $CaAlSiN_3:Eu^{2+}$.

In CaS, the Ca site, which Eu^{2+} replaces, has octahedral point symmetry, which splits 5d levels into two: one represented by E_g at the upper energy, and the other, T_{2g}, at lower energy. We can therefore expect that an excitation spectrum is composed of two bands, though each band is further split or broadened by coupling with the $4f^6$ configuration, as well as by vibronic interaction. In fact, the observed excitation spectrum in Figure 6.9 shows basically two strong bands. However, a distinct band peaking at about 260 nm probably originates in the direct absorption edge, which was previously reported to be 253 nm, and a shoulder at around 280 nm is due to indirect absorption at 288 nm (Kaneko and Koda 1988).

In $CaAlSiN_3$, Eu^{2+} is coordinated by five N atoms with four different bond lengths, as shown in Figure 5.6b. This geometry can be approximated as nearly tetrahedral coordination perturbed by an effect of the fifth N atom with the longest distance, 0.2627 nm. This approximation predicts that 5d levels are mainly split into two: T_2 at upper energy and E at lower energy, each being split into sublevels. The excitation spectrum in Figure 6.9 shows the main peaks at about 310, 340, and 440 nm. A shoulder at 510 nm is possibly ascribed to phonon coupling. Meanwhile, the absorption edge of

CaAlSiN$_3$ is found at about 310 nm by a reflection spectrum of an undoped host material. This value agrees with the excitation peak at 310 nm. Higher energy levels are therefore hidden in the fundamental absorption band, just like the spectrum of CaS:Eu^{2+}.

As a result of overlap with the fundamental absorption, it is difficult to find the splitting energy, ε_{cfs}, and the center of 5d levels by simple experiments for both CaAlSiN$_3$:Eu^{2+} and CaS:Eu^{2+}. However, the center of 5d levels is presumably located at low energy, because $10Dq$ is not particularly strong. To help evaluate the center of 5d levels, here we use the splitting energy, ε_{cfs}, in a host crystal with a wide band gap. The splitting energy, ε_{cfs}, is about 1.8×10^4 cm^{-1} in SrCl$_2$ (Reid et al. 2000; Pan et al. 2006) and 1.7×10^4 cm^{-1} in CaF$_2$ (Bayer and Schaack 1970; Downer et al. 1983). If we postulate a ε_{cfs} of 1.7×10^4 cm^{-1} and assume that the excitation peak at 450 nm corresponds to the lower energy level E, then we can assume the center of 5d levels to be about 325 nm, or 3.1×10^4 cm^{-1}.

6.2.5.3 Luminescence Peak Wavelength Dependent on Eu Fraction

A luminescence spectrum of CaAlSiN$_3$:Eu^{2+} shifts to a longer wavelength with an increase in Eu fraction, as shown in Figure 6.10. In the range below 0.20, or 20 atom%, a shift in the luminescence peak energy is about 1.4×10^3 cm^{-1}, or 0.17 eV, which is comparable to the shift of the exciton energy observed for SrS:Eu^{2+} in the same Eu fraction range (Figure 6.8). There is, however, a marked difference in the Eu fraction dependence between the two compounds. In CaAlSiN$_3$:Eu^{2+}, the luminescence peak energy decreases drastically in a range of the low Eu fraction, but it saturates at a high fraction near 0.20. This result contrasts with the dependence of the exciton energy on the Eu fraction in CaS, which is close to a linear relation (Figure 6.8). Figure 6.10

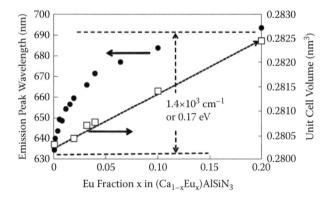

FIGURE 6.10
Luminescence peak wavelength and the unit cell volume as a function of nominal Eu fraction in (Ca$_{1-x}$Eu$_x$) AlSiN$_3$.

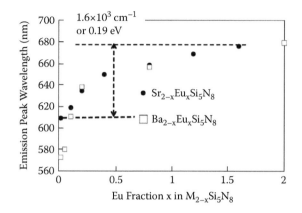

FIGURE 6.11
Luminescence peak wavelength as a function of nominal Eu fraction in $(M_{2-x}Eu_x) Si_5N_8$ (M = Sr, Ba).

also shows that the unit cell volume of $CaAlSiN_3:Eu^{2+}$ varies linearly with the Eu fraction, which indicates formation of a solid solution by incorporation of Eu^{2+}. The saturation of the peak energy in $CaAlSiN_3:Eu^{2+}$ may be caused by local deformation around Eu^{2+}, or by formation of an impurity level by interaction between 5d electrons.

A similar red-shift with the Eu fraction was reported for $M_2Si_5N_8:Eu^{2+}$ (M = Ca, Sr, Ba) (Li et al. 2006). The shift of the luminescence peak is plotted as a function of the Eu fraction for M = Sr and Ba in Figure 6.11. The magnitude of the shift is 0.19 eV for M = Sr and 0.34 eV for M = Ba.

6.2.5.4 Speculation on the Reasons for the Red-Shift in CaAlSiN₃:Eu²⁺

We can assume some possible reasons for the deep red luminescence of $CaAlSiN_3:Eu^{2+}$. The center of the 5d levels is lowered more than for other nitrides, such as $M_2Si_5N_8:Eu^{2+}$ (M = Ca, Sr, Ba). The crystal field splitting, $10Dq$, is most probably not strong, though distorted coordination around Eu^{2+} causes large splitting. Accordingly, the main reason for the red-shift is considered to be high covalency or local polarization around Eu^{2+}. The framework of $CaAlSiN_3$ can be regarded as wurtzite-type lattice made of AlN and GaN (Uheda et al. 2006a, 2006b). High covalency is indicated in this structure and the nearly tetrahedral coordination around Eu^{2+}. The emitting state is also lowered by some interaction dependent on the Eu concentration.

It is interesting that another nitridoalumosilicate, $SrAlSi_4N_7:Eu^{2+}$, shows red luminescence peaking at 632 nm. This compound has a condensed network of SiN_4 and AlN_4 tetrahedra which form infinite chains by edge sharing as well as corner sharing (Hecht et al. 2009). The covalency effect may be related to such a network of SiN_4 and AlN_4.

6.3 Quantum Efficiency and Thermal Quenching

6.3.1 Factors Reducing Quantum Efficiency

When an activator ion is excited by its localized absorption transition, a luminescence process is not sensitive to defects or impurities in a host lattice. As a result, basically high quantum efficiency of luminescence is expected. In actual cases, however, not all the materials have high quantum efficiency even under localized excitation. The main reasons are concentration quenching and thermal quenching.

Concentration quenching means a decrease in luminescence efficiency at a high activation concentration. This process occurs by migration of energy among activator ions, resulting in nonradiative recombination at defects or impurities. It begins to occur at an activator concentration below the optimum concentration, which is indicated by observations described in Section 6.3.2.2. It is often experienced that a phosphor with poor crystallinity has low quantum efficiency under localized excitation, e.g., blue light excitation. The low efficiency is most probably caused by concentration quenching involving nonradiative processes at lattice imperfections.

The phenomenon called *thermal quenching* is a decrease in efficiency by thermal energy. It is more or less observed for most luminescent materials. It is induced by different factors and, in some cases, related to concentration quenching. Thermal quenching is a serious issue to any phosphor, because it degrades efficiency at the operating temperature of a device, which uses a phosphor. It is particularly important to phosphors for high-power white LEDs, since the temperature of a phosphor often rises to 150°C or higher. Eu^{2+}-activated nitride or oxynitride phosphors are appreciated for the reason that many of them show small thermal quenching, which has not been found in conventional phosphors. Temperature dependence of luminescence intensity under 450 nm light excitation is shown for typical nitride phosphors in Figure 6.12. There is a remarkable difference even in imidosilicates, and $CaAlSiN_3:Eu^{2+}$ shows the least thermal quenching. Naturally, a question is raised as to the reason for this feature. In this section, the origin of this favorable and unusual feature is discussed based on information of conventional phosphors.

6.3.2 General Discussion on the Mechanism of Thermal Quenching

Energy dissipation by phonon emission can occur by different modes. In the following, these modes are briefly described first for a localized and then a delocalized luminescence mechanism.

6.3.2.1 Thermal Quenching Mechanism of a Localized Ion

For a rare earth ion such as Eu^{2+} and a transition metal ion, typically Mn^{2+} or Mn^{4+}, both absorption and emission of photons proceed basically in a

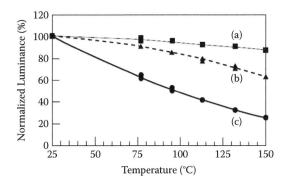

FIGURE 6.12
Temperature dependence of luminance under 450 nm light excitation for (a) CaAlSiN$_3$:Eu^{2+}, (b) Ca$_2$Si$_5$N$_8$:Eu^{2+}, and (c) CaSiN$_2$:Eu^{2+}, each with 1 atom% Eu.

localized ion without transfer of energy. This process can be pictured by the one-dimensional configurational coordinate diagram, which is shown in Figure 2.17. This diagram presents thermal quenching as a process where an electron passes over the crossing X with activation energy ΔE_{tq} to relax to the ground state without emitting a photon. The probability of this process exponentially depends on the inverse of the absolute temperature, as shown by Equation 2.23. This thermally activated process can be explained in more detail as follows.

The electron at an excited state has a certain probability to return to the ground state. If the energy separation between the ground and excited states, B and C in Figure 2.17, is wide enough, excess energy of the excited state can be emitted as a photon at low temperature. At high temperature, however, every chemical bond vibrates with its fundamental or multiple modes of frequency. Even at the ground state, the electron can have high kinetic energy acquired from a vibration mode of high multiple frequency. On the other hand, at an excited state, the total energy of the electron can be relatively low if its vibration has low frequency, and in such a situation, it can happen that the energy of an excited state is equal to the total energy of the ground state with high frequency. Accordingly, the electron at the excited state can transit to the ground state by changing a vibrational mode from low to high frequency. High vibrational energy is necessary for this transition, but it is compensated by lowering of the electronic energy. This transition does not emit a photon, but creates multiple phonons. If this transition occurs, luminescence efficiency is apparently decreased.

According to this simple picture, thermal quenching begins to occur at low temperature, when the equilibrium position at the excited state, point B in Figure 2.17, is shifted far away from the equilibrium position at the ground state, point A. The offset, the separation on the abscissa between B (or C) and A, is the distance between the two equilibrium positions. In other words, thermal quenching occurs more when the Stokes shift is larger. The correlation

TABLE 6.1

Mechanical Properties of Some Oxide and SiAlON Materials

Compound	Vickers Hardness Hv (GPa)	Young's Modulus (GPa)	Reference
Al_2O_3	17.5	380	Kyocera Catalog Library 2009
$Y_3Al_5O_{12}$	12	280	Kyocera Catalog Library 2009
α-SiAlON	20	305	Cother and Hodgson 1982
β-SiAlON	18	300	Mitomo et al. 1985

between thermal quenching and the Stokes shift was actually demonstrated for some oxide phosphors doped with Eu^{3+} as the emitting ion (Blasse and de Vries 1967). The excited state of Eu^{3+} responsible for thermal quenching is a charge-transfer state. It was reported that the thermal quenching occurred at a lower temperature, as the energy of the charge-transfer state was lower.

Another condition of low quenching temperature is that an activator and a ligand anion are bonded weakly by a small force constant. This is represented by a small curvature of a parabola for an excited state, which naturally lowers the position of the crossing X in Figure 2.17. A material containing weak chemical bonds is expected to have a large elastic constant, small hardness, and low melting point.

Nitride or oxynitride materials have been well known to be resistant to heat and mechanical stress. Data of mechanical hardness and elasticity of α- and β-SiAlON are shown in Table 6.1. These ceramic samples are as strong as Al_2O_3 against stress. The data suggest that the framework built by the connection of $[SiN_4]$ or $[(Si,Al)N_4]$ tetrahedra in a phosphor host is very rigid. Particularly in the $CaAlSiN_3$ lattice, two-thirds of all the N atoms are directly linked with three Si or Al atoms. In other words, three tetrahedra are joined together at a single corner, forming a rigid three-dimensional structure (see Figure 6.13). This can be the main reason of stiffness or small thermal quenching of the $CaAlSiN_3$:Eu^{2+} phosphor. The structure of $Ca_2Si_5N_8$ has three coordinated N atoms of 50% fraction, and the structure of $CaSiN_2$ has no such N atoms. The difference in rigidness among the three imidosilicate structures coincides with the order of thermal quenching shown in Figure 6.12.

6.3.2.2 Temperature-Dependent Concentration Quenching

If the absorption and emission processes are confined in a local cluster around an activator, thermal quenching should be independent of an activator concentration or imperfections in a host lattice. Actually, this is not always the case. For example, it was reported that $Y_3Al_5O_{12}$:Ce^{3+}, which is known to show considerable thermal quenching, maintains high efficiency even at 600 K, when the Ce concentration is as low as 3×10^{-1} mol% (Bachmann et al. 2009) (A commercial phosphor based on this compound is doped with 2–3 mol% Ce^{3+}.) This concentration-dependent thermal quenching occurs by energy transfer

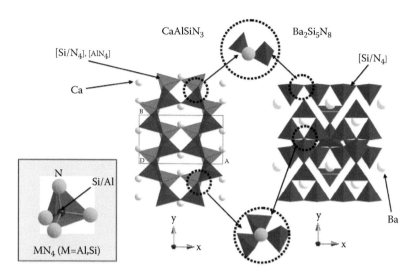

FIGURE 6.13

Comparison of the crystal structures of $CaAlSiN_3$ and $Ba_2Si_5N_8$. A tetrahedron of $(Si,Al)N_4$ at the lower left is shown as a dark or light grey tetrahedron in the structures. Nitrogen atoms coordinated with two or three Si/Al atoms are emphasized by the circles drawn by dotted lines.

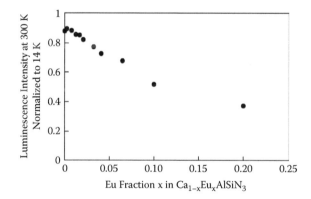

FIGURE 6.14

The ratio of the luminescence intensity at 300 K to that at 14 K for $(Ca_{1-x}Eu_x)$ $AlSiN_3$ as a function of nominal Eu fraction x.

between Ce^{3+} ions through spectral overlap of absorption and emission of photons. An increase in temperature broadens both spectra and increases the overlapped area. During the migration process of an excited state, the energy is lost at defects or impurities by nonradiative recombination with holes.

Thermal quenching of $CaAlSiN_3:Eu^{2+}$ also depends on the Eu concentration. Figure 6.14 shows luminescence intensity at 300 K normalized to the intensity at 14 K under 450 nm light excitation. At the optimum Eu concentration for the luminescence efficiency, about 0.01 fraction, or 1 atom%, thermal quenching is at a minimum. Meanwhile, in $Y_3Al_5O_{12}:Ce^{3+}$, considerable thermal quenching occurs at 1 mol% Ce. This result indicates that a weaker interaction between the activator ions in $CaAlSiN_3$ partially contributes to small thermal quenching at a Eu concentration of about 1 mol%.

6.3.2.3 Thermal Quenching by Photoionization or Autoionization Mechanism

Correlation between thermal quenching and the Stokes shift is not found for some groups of materials. In such a situation, it may be necessary to take into account the somewhat delocalized nature of an excited state. It was found that luminescence of Eu^{2+} or Yb^{2+} in wide-gap host crystals shows anomalous emission with strongly temperature-dependent efficiency. This phenomenon was explained by a process called autoionization or photoionization, in which a 5d electron is thermally excited to the conduction band, resulting in nonradiative processes (Pedrini et al. 2007). The process is schematically shown in Figure 6.15. For this process to take place, the conduction band has to overlap with a 5d level in energy or be located above the 5d level by energy separation narrow enough for thermal activation.

In a series of compounds expressed in a formula, $Ba_3Si_6O_{3(5-n)}N_{2n}:Eu^{2+}$ (n = 1, 2, and 3), the compound with n = 2 has very low efficiency even at low temperature, while the other two are efficient phosphors. The autoionization process was proposed for this irregular nature, because $Ba_3Si_6O_9N_4$ was shown to have a narrow band gap compared with the others (Mikami et al. 2009).

The autoionization process creates free electrons in the conduction band, which, in principle, can be detected by photoconductivity or photostimulated luminescence. Further works are required to provide evidence to demonstrate that this process may work in more compounds.

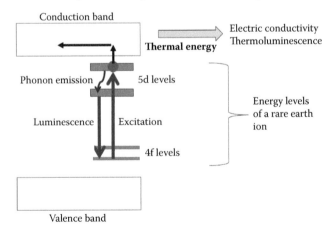

FIGURE 6.15
Schematic diagram of the autoionization mechanism.

6.4 Effects of Lattice Imperfections on Luminescence Properties

6.4.1 Multiple Luminescence Bands Observed for CaAlSiN$_3$:Eu^{2+}

Compared with many conventional phosphors, some nitride phosphors have multiple luminescence centers, which lead to somewhat complicated luminescence properties, such as multiple luminescence bands or nonexponential decay characteristics. At present, origins of the multiple centers are not known, but they can be ascribed to lattice imperfections contained in nitride lattices, e.g., cation or anion vacancies, dislocations, oxygen impurities, or disordered ion distribution. If such imperfections are located near an activator ion, they will change the local structure around the activator and modify absorption and luminescence spectra. As a result, they may give negative effects on luminescence, but in turn, they can be regarded as room left to improve phosphor performance. Below some experimental results are presented to show luminescence properties possibly induced by such imperfections.

Luminescence spectra of CaAlSiN$_3$:Eu^{2+} and CaS:Eu^{2+} phosphors at room temperature and 11 K are shown in Figures 6.16 and 6.18 in the unit of eV on the abscissa. The spectra of CaS:Eu^{2+} show a single band corresponding to a single crystallographic site for Eu^{2+}, i.e., Ca^{2+} site. As reported previously (Nakao 1980), the luminescence spectrum of CaS:Eu^{2+} at low temperature is composed of zero-phonon lines and a component extended to the lower photon energy, which arises by phonon emission processes. At higher temperature, the counterpart component by phonon absorption gains intensity, and at room temperature the spectral shape becomes nearly symmetrical. Meanwhile, CaAlSiN$_3$:Eu^{2+} (0.8 mol%) shows a side band at around 550 nm (2.3 eV), in addition to the main band peaking at 650 nm

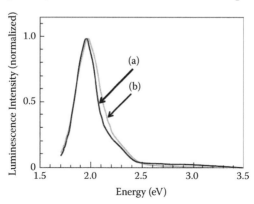

FIGURE 6.16
Luminescence spectra of CaAlSiN$_3$:Eu^{2+} (0.8 mol%) at (a) 11 and (b) 300 K. Excitation was made by 450 nm light. Two spectra are normalized at the peak.

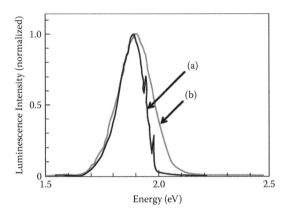

FIGURE 6.17
Luminescence spectra of CaS:Eu²⁺ at (a) 11 and (b) 300 K. Excitation was made by 450 nm light. Two spectra are normalized at the peak.

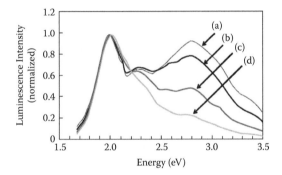

FIGURE 6.18
Luminescence spectra of CaAlSiN₃:Eu²⁺ (0.005 mol%) at (a) 11, (b) 100, (c) 200, and (d) 300 K. Excitation was made by 266 nm pulsed light with 6 ns duration, and the luminescence was detected at 100 ns delay time. The spectra are normalized at a peak of about 2 eV.

(1.9 eV) (see Figure 6.16). The spectral shape of the lower-photon-energy side does not change with an increase in temperature, but the higher-energy side and the side band grow with an increase in temperature in a way similar to that of CaS:Eu. In addition, weak bands are observed in the near-UV and blue region, or about 360–460 nm (2.7–3.4 eV). These bands are strongly observed when the Eu concentration and temperature are low, as shown in time-resolved spectra of a sample with a Eu concentration of 0.005 mol% and measured at a 100 ns delay time after the pulsed excitation was made (Figure 6.18). In this measurement, the excitation was made by the 266 nm pulsed light with a duration of 6 ns and a repetition rate of 10 Hz. The blue band appears strongly relative to the main orange band at 11 K, and its intensity decreases largely with an increase in temperature, though the main red band and the green side band maintain their intensities at high temperature.

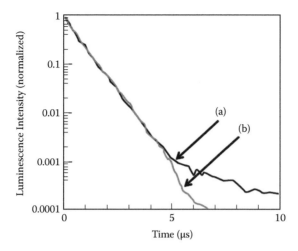

FIGURE 6.19
Luminescence decay curves of CaS:Eu^{2+} at 11 K. Excitation was made by 266 nm pulsed light with 6 ns duration, and the luminescence intensity was detected at (a) 680 and (b) 630 nm.

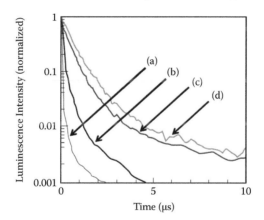

FIGURE 6.20
Luminescence decay curves of CaAlSiN$_3$:Eu^{2+} (0.8 mol%) at 11 K. Excitation was made by 266 nm pulsed light with 6 ns duration, and the luminescence intensity was detected at (a) 450, (b) 550, (c) 630, and (d) 680 nm.

Decay characteristics of CaS:Eu are shown in Figure 6.19 for a monitored wavelength of 630 nm, near the peak, and of 680 nm, at the tail of the luminescence band. The decay curves are fitted well with a single exponential form, with only a slight difference in low-intensity levels. Evidently luminescence is emitted by one kind of luminescence center. In contrast, decay curves of CaAlSiN$_3$:Eu^{2+} deviate largely from exponential forms and vary with monitoring wavelength, as shown in Figure 6.20. The green side band and the blue band, curves a and b, have shorter decay times than the main bands,

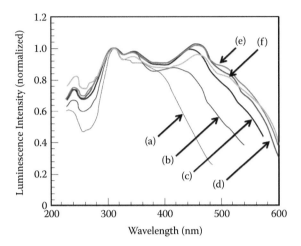

FIGURE 6.21
Excitation spectra of $CaAlSiN_3:Eu^{2+}$ (0.8 mol%) at room temperature. The monitored luminescence wavelength was changed to (a) 555, (b) 585, (c) 615, (d) 645, (e) 695, and (f) 745 nm. The luminescence peak is found at about 650 nm. All spectra are normalized at a 310 nm peak, which probably corresponds to the fundamental absorption edge.

c and d. And yet the luminescence at the main and the green side band decay in nearly the same time span of 10^{-1} to a few microseconds, indicating that the side band is originated from Eu^{2+} at a site different from the normal Ca^{2+} site. On the other hand, the blue band shows much faster decay. Accordingly, the blue band may be related to an unknown defect, although the possibility of Eu^{2+} luminescence from an anomalous site cannot be ruled out.

Dependence on the monitoring wavelength of luminescence is observed in the excitation spectra of $CaAlSiN_3:Eu^{2+}$, while not in those of $CaS:Eu^{2+}$. The results are shown in Figures 6.21 and 6.22. In the spectra of $CaS:Eu^{2+}$, the peak at about 255 nm and the shoulder around 285 nm correspond to the direct and indirect band gaps, respectively. The flat band at a wavelength longer than 420 nm is due to 4f-5d transitions of Eu^{2+}. Similarly, $CaAlSiN_3:Eu^{2+}$ shows the peak due to the absorption edge at 310 nm, and broad bands by Eu^{2+} absorption below the absorption edge. Again, the simple behavior of $CaS:Eu^{2+}$ shows that the luminescence is brought about by Eu^{2+} at a single site. For $CaAlSiN_3:Eu^{2+}$, the complicated dependence on the monitoring wavelength is observed in the region of Eu^{2+} absorption, particularly in the wavelength where the main band and the side band overlap with each other. This phenomenon is most probably caused by multiple luminescence bands.

6.4.2 Complicated Nature Observed for Other Nitride Phosphors

Dependence on the luminescence wavelength is also found in excitation spectra of $CaSiN_2:Eu^{2+}$, $Ca_2Si_5N_8:Eu^{2+}$, $Sr_2Si_5N_8:Eu^{2+}$, and AlN:Eu,Si. As an

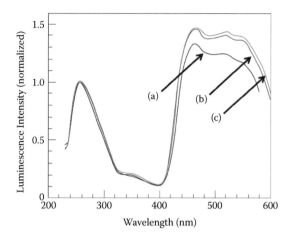

FIGURE 6.22
Excitation spectra of $CaS:Eu^{2+}$ at room temperature. The monitored luminescence wavelength was changed to (a) 600, (b) 650, and (c) 700 nm. All spectra are normalized at a 260 nm peak, which probably corresponds to the direct band gap.

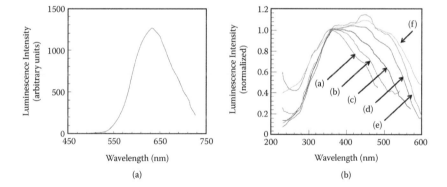

FIGURE 6.23
(a) A luminescence spectrum of $CaSiN_2:Eu^{2+}$ (1 mol%) excited by 420 nm at room temperature.
(b) Excitation spectra of $CaSiN_2:Eu^{2+}$ (1 mol%) at room temperature and at a monitored luminescence wavelength of (a) 530, (b) 560, (c) 590, (d) 640, (e) 690, and (f) 740 nm. The luminescence peak is found at about 640 nm.

example, luminescence and excitation spectra of $CaSiN_2:Eu^{2+}$ are shown in Figures 6.23a and 6.23b. When $CaSiN_2:Eu^{2+}$ is excited at a longer wavelength within the luminescence band shown in Figure 6.23a, it shows a monotonous shift of an excitation spectrum, as well as structures of the broad excitation band, to a longer-wavelength side. This behavior looks very much like that of the excitation spectra of $CaAlSiN_3:Eu^{2+}$ shown in Figure 6.21.

Excitation spectra were measured on some conventional phosphors as well, but evident dependence on luminescence wavelength was not confirmed. For example, for Eu^{2+} luminescence in $SrAl_2O_4$, $BaAl_2O_4$, and $BaMgAl_7O_{17}$,

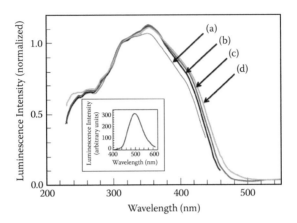

FIGURE 6.24
Excitation spectra of $BaAl_2O_4:Eu^{2+}$ (10 mol%) at room temperature and at monitored luminescence wavelengths of (a) 470, (b) 500, (c) 550, and (d) 600 nm. A luminescence spectrum is
shown in the inset.

excitation peaks were found to stay nearly at the same wavelength independent of the monitoring luminescence wavelength. Excitation spectra of
$BaAl_2O_4:Eu^{2+}$, monitored at varied wavelengths, are shown in Figure 6.24. This
difference between the two groups of phosphors seems to suggest that there
is a common reason for the variation of excitation spectra of nitride lattices.
However, it is necessary to extend the above measurements to a larger number
of samples to confirm the results and make clear imperfections in nitrides.

6.4.3 Application of Long-Persistent Luminescence

For some nitride phosphors, afterglow can be observed after cessation of
excitation. In fact, afterglow was reported for orange-emitting $Ba_2Si_5N_8:Eu^{2+}$
(Hoppe et al. 2000) and red-emitting $M_wAl_xSi_yN_{\{(2/3)w+x+(4/3)y\}}:Eu^{2+}$, where M
represents alkaline earth elements and $0.04 \leq w \leq 9$, $x = 1$, and $0.056 \leq y \leq 18$
(Hosokawa et al. 2006). Afterglow is an indication of trapping states formed
by impurity ions or lattice defects. If afterglow is strong and long lasting,
it can be applied for signage or displays.

As long-persistent phosphors, green-emitting $SrAl_2O_4:Eu^{2+},Dy^{3+}$ and blue-
emitting $CaAl_2O_4:Eu^{2+},Nd^{3+}$ have been developed and widely commercialized (Matsuzawa et al. 1996). However, a red-emitting phosphor has not
been as popular as the blue or green ones, probably because a red phosphor,
e.g., $Y_2O_2S:Eu^{3+},Mg,Ti$ (Murazaki et al. 1999) or $CaS:Eu^{2+},Tm^{3+}$ (Jia et al. 2000),
shows persistence luminance and a period not sufficient to wide application.
Red- or orange-emitting long-persistent phosphors with better performance
are therefore strongly demanded in the market.

Long-persistent phosphors are used under illumination or sunlight. It is
therefore necessary for them to be excited not just by UV light, but also by

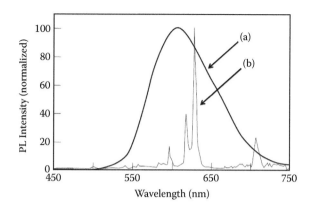

FIGURE 6.25
Luminescence spectra of (a) $Ca_2Si_5N_8$:Eu^{2+} (1.0 mol%), Tm^{3+} (0.2 mol%), and (b) Y_2O_2S:Eu^{3+},Mg,Ti at room temperature. Excitation was made by continuous irradiation of a violet LED emission peaked at 450 nm. The two spectra are normalized at the peaks. (Reprinted from Miyamoto, Y., Kato, H., Honna, Y., Yamamoto, H. and Ohmi, K., *J. Electrochem. Soc.*, 156, J235–J241, 2009. With permission.)

visible light. In this respect, nitride phosphors, which can be excited by blue LEDs, can be candidates of long-persistent phosphors.

With such a background, it was noticed that $Ca_2Si_5N_8$:Eu^{2+} shows relatively long persistence lasting for several seconds. In many long-persistent phosphors, a second impurity is introduced into a host crystal to form a trapping state, which is capable of releasing a trapped electron to the conduction band at room temperature. A typical example is Dy^{3+}, which replaces Sr^{2+} in $SrAl_2O_4$:Eu^{2+}. Among a series of rare earth ions, Tm^{3+} was found to work best to improve luminescence in intensity and persistence (Miyamoto et al. 2009; van den Eeckhout et al. 2009) without any change in luminescence color. The luminescence properties of this phosphor are described below, in comparison to a red phosphor, Y_2O_2S:Eu^{3+},Mg,Ti.

A typical luminescence spectrum of $Ca_2Si_5N_8$:Eu^{2+} (1.0 mol%) and Tm^{3+} (0.2 mol%) is compared with a spectrum of Y_2O_2S:Eu^{3+},Mg,Ti in Figure 6.25. It shows a broad band centered at 610 nm and exhibits an orange color. A red and saturated color of Y_2O_2S:Eu^{3+},Mg,Ti with the line spectrum peaked at 627 nm is more desirable to most applications. However, the luminescence color of $Ca_2Si_5N_8$:Eu^{2+},Tm^{3+} can be shifted to a longer wavelength by a partial substitution of Ca^{2+} with Sr^{2+} (Miyamoto et al., 2009). The excitation spectrum of $Ca_2Si_5N_8$:Eu^{2+},Tm^{3+} is extended to the visible region, as expected. This feature is an advantage compared to weak visible absorption of Y_2O_2S:Eu^{3+},Mg,Ti (see Figure 6.26). Afterglow luminance excited by 420 nm light is shown as a function of elapsed time in Figure 6.27. It was found that afterglow of $Ca_2Si_5N_8$:Eu^{2+},Tm^{3+} decays more slowly than that of Y_2O_2S:Eu^{3+},Mg,Ti. Except for the initial value, its luminance is higher than the intensity of Y_2O_2S:Eu^{3+},Mg,Ti in the case of $Ca_2Si_5N_8$:Eu^{2+},Tm^{3+} synthesized

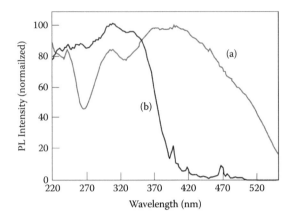

FIGURE 6.26
Excitation spectra of (a) $Ca_2Si_5N_8:Eu^{2+}$ (1.0 mol%), Tm^{3+} (0.2 mol%), and (b) $Y_2O_2S:Eu^{3+}$,Mg,Ti at room temperature. Luminescence was monitored at the peak wavelength of each sample.

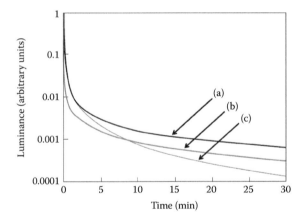

FIGURE 6.27
Afterglow luminance as a function of elapsed time after excitation by 420 nm light is made.
(a) $Ca_2Si_5N_8:Eu^{2+}$ (1.0 mol%) and Tm^{3+} (0.2 mol%) synthesized under 1 MPa N_2 and at 2,000°C.
(b) $Ca_2Si_5N_8:Eu^{2+}$ (1.0 mol%) and Tm^{3+} (0.2 mol%) synthesized under 0.1 MPa H_2/N_2 gas and at 1,400°C. (c) $Y_2O_2S:Eu^{3+}$,Mg,Ti.

under 1 MPa N_2 and at 1,800–2,000°C. Even $Ca_2Si_5N_8:Eu^{2+}$,Tm^{3+} synthesized under the atmospheric pressure and fired at 1,400°C, denoted as "low T&P firing," shows higher luminance than $Y_2O_2S:Eu^{3+}$,Mg,Ti in 8 min after the excitation light is turned off.

The effect of Tm^{3+} doping is evidenced by thermoluminescence glow curves, in which the glow peak temperature gives a measure of trap depth. Figure 6.28 shows glow curves of samples doped with Nd^{3+}, Dy^{3+}, and Tm^{3+}. In this result, doped ions generate new trapping states in $Ca_2Si_5N_8$. Among the three ions, Tm^{3+} produces thermoluminescence most intensely around

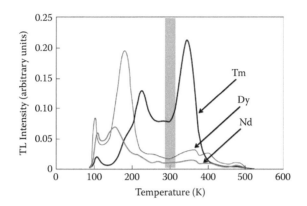

FIGURE 6.28
Thermoluminescence glow curves of $Ca_2Si_5N_8:Eu^{2+}$ (1.0 mol%) codoped with Nd^{3+}, Dy^{3+}, and Tm^{3+} ions, all of 0.2 mol nominal concentration. The glow curves were measured after irradiation of a D_2 lamp coupled with a glass filter transmitting 250–400 nm light for 25 min at about 80 K.

room temperature, with the peaks at 220 and 350 K. This fact proves that doping of Tm^{3+} is most effective in providing the strong afterglow.

The afterglow of this phosphor is, however, still not strong enough compared with the commercial phosphor, $SrAl_2O_4:Eu^{2+},Dy^{3+}$. Nevertheless, this could be meaningful as an example to explore new applications of nitride phosphors.

References

Akella, A., and Keszler, D. A. 1995. Structure and Eu^{2+} luminescence of dibarium magnesium orthoborate. *Mater. Res. Bull.* 30:105–111.

Bachmann, V., Ronda, C., and Meijerink, A. 2009. Temperature quenching of yellow Ce^{3+} luminescence in YAG:Ce. *Chem. Mater.* 21:2077–2084.

Bayer, E., and Schaack, G. 1970. Two-photon absorption of $CaF_2:Eu^{2+}$. *Phys. Stat. Sol.* 41:827–835.

Blasse, G., and de Vries, J. 1967. On the fluorescence of mixed metal oxides. *J. Electrochem. Soc.* 114:875–877.

Cother, N. E., and Hodgson, P. 1982. The development of sialon ceramics and their engineering applications. *Trans. J. Br. Ceram. Soc.* 8:141–144.

Diaz, A., and Keszler, D. A. 1996. Red, green, and blue Eu^{2+} luminescence in solid-state borates: A structure-property relationship. *Mater. Res. Bull.* 31:147–151.

Dorenbos, P. 2002. 5d-level energies of Ce^{3+} and the crystalline environment. IV. Aluminates and "simple" oxides. *J. Lumin.* 99:283–299.

Downer, M. C., Cordero-Montalvo, C. D., and Crosswhite, H. 1983. Study of new $4f^7$ levels of Eu^{2+} in CaF_2 and SrF_2 using two-photon absorption spectroscopy. *Phys. Rev. B* 28:4931–4943.

Hecht, C., Stadler, F., Schmidt, P. J., Schmedt auf der Günne, J., Baumann, V., and Schnick, W. 2009. SrAlSi$_4$N$_7$:Eu^{2+}—A nitridoalumosilicate phosphor for warm white light (pc)LEDs with edge-sharing tetrahedra. *Chem. Mater.* 21:1595–1601.

Hosokawa, M., Shinohara, T., Kameshima, M., Murazaki, Y., Takashima, M., and Tamaki, H. Nitride phosphors and light-emitting devices using the same (in Japanese). *JP* 2006–306982.

Hoppe, H. A., Lutz, H., Morys, P., Shnick, W., and Seilmeier, A. 2000. Luminescence in Eu^{2+}-doped Ba$_2$Si$_5$N$_8$: Fluorescence, thermoluminescence, and upconversion. *J. Phys. Chem. Solids* 61:2001–2006.

Jia, D., Zhu, J., and Wu, B. 2000. Trapping centers in CaS:Bi^{3+} and CaS:Eu^{2+},Tm^{3+}. *J. Electrochem. Soc.* 147:386–389.

Kaneko, Y., and Koda, T. 1988. New developments in IIa-VIb (alkaline-earth chalcogenide) binary semiconductors. *J. Crystal Growth* 86:72–78.

Kaneko, Y., Morimoto, K., and Koda, T. 1981. Growth of CaS, SrS single crystals by an arc image furnace and their optical properties. *Oyobutsuri (Appl. Phys.)* 50:289–294.

Kyocera Catalog Library. 2009. Mechanical & industrial ceramics catalog. Materials comparison chart 21. http://global.kyocera.com/prdct/fc/product/pdf/mechanical.pdf (accessed September 30, 2010).

Li, Y. Q., Van Steen, J. E. J., Van Krevel, J. W. H., Botty, G., Delsing, A. C. A., DiSalvo, F. J., de With, G., and Hintzen, H. T. 2006. Luminescence properties of red-emitting M$_2$Si$_5$N$_8$:Eu^{2+}(M=Ca, Sr, Ba) LED conversion phosphors. *J. Alloys Compds.* 417:273–279.

Matsuzawa, T., Aoki, Y., Takeuchi, N., and Murayama, Y. 1996. A new long phosphorescent phosphor with high brightness, SrAl$_2$O$_4$:Eu^{2+},Dy^{3+}. *J. Electrochem. Soc.* 143:2670–2673.

Mikami, M., and Kijima, N. 2010. 5d Levels of rare-earth ions in oxynitride/nitride phosphors: To what extent is the idea of covalency reliable? *Opt. Mater.*, 33:145–148.

Mikami, M., Shimooka, S., Uheda, K., Imura, H., and Kijima, N. 2009. New green phosphor Ba$_3$Si$_6$O$_{12}$N$_2$:Eu for white LED: Crystal structure and optical properties. *Key Eng. Mater.* 403:11–14.

Mitomo, M., Ishizawa, T., Ayusawa, N., Shironita, A., Takai, M., and Uchida, N. 1985. Development of alpha-sialon ceramics. *Shinagawa Technol. Rep.* 29:1–12.

Miyamoto, Y., Kato, H., Honna, Y., Yamamoto, H., and Ohmi, K. 2009. An orange-emitting, long-persistent phosphor, Ca$_2$Si$_5$N$_8$:Eu^{2+},Tm^{3+}. *J. Electrochem. Soc.* 156:J235–J241.

Murazaki, Y., Arai, K., Ichinomiya, K., and Oishi, T. 1999. A new long persistence red phosphor. In *Proceedings of IDW'99*, Sendai, Japan, 841.

Nakao, Y. 1980. Luminescent centers of MgS, CaS and CaSe phosphors activated with Eu^{2+} ion. *Jpn. J. Phys. Soc.* 48:534–541.

Pan, Z., Ning, L., Cheng, B.-M., and Tanner, P. A. 2006. Absorption, excitation and emission spectra of SrCl$_2$:Eu^{2+}. *Chem. Phys. Lett.* 428:78–82.

Pedrini, C., Joubert, M.-F., and McClure, D. S. 2007. Photoionization processes of rare-earth dopant ions in ionic crystals. *J. Lumin.* 125:230–237.

Reid, M. F., Van Pietersen, L., Wegh, R. T., and Meijerink, A. 2000. Spectroscopy and calculations for 4*fN*→4*fN*$^{-1}$5*d* transitions of lanthanide ions in LiYF$_4$. *Phys. Rev. B* 62:14744–14749.

Smet, P. F., Van Haecke, J. E., Loncke, F., Vrielinck, H., Callens, F., and Poelman, D. 2006. Anomalous photoluminescence in BaS:Eu^{2+}. *Phys. Rev. B* 74:035207.

Uheda, K., Hirosaki, N., and Yamamoto, H., 2006a. Host lattice materials in the system Ca_3N_2-AlN-Si_3N_4 for white light emitting diode. *Phys. Stat. Sol.* 203:2712–2717.

Uheda, K., Hirosaki, N, Yamamoto, Y. Naito, A., Nakajima, T., and Yamamoto, H. 2006b. Luminescence properties of a red phosphor. $CaAlSiN_3$:Eu^{2+}, for white light-emitting diodes. *Electrochem. Solid-State Lett.* 9:H22–H25.

Uheda, K., Yamamoto, H., Yamane, H., Inami, W., Tsuda, K., Yamamoto, Y., and Hirosaki, N. 2009. An analysis of crystal structure of Ca-deficient oxonitrido-aluminosilicate, $Ca_{0.88}Al_{0.91}Si_{1.09}N_{2.85}O_{0.15}$. *J. Ceram. Soc. Jpn.* 117:94–98.

van den Eeckhout, K., Smet, P. F., and Poelman, D. 2009. Persistent luminescence in rare-earth codoped $Ca_2Si_5N_8$:Eu^{2+}. *J. Lumin.* 129:1140–1143.

Wu, J. L., Gundiah, G., and Cheetham, A. K. 2007. Structure–property correlations in Ce-doped garnet phosphors for use in solid state lighting. *Chem. Phys. Lett.* 441:250–254.

Yamaga, M., Masui, Y., Sakuta, S., Kodama, N. and Kaminaga, K. 2005. Radiative and nonradiative decay processes responsible for long-lasting phosphorescence of Eu^{2+}-doped barium silicates. *Phys. Rev. B* 71:205102–205108.

Zhang, F.-L., Yang, S., Stoffers, C., Penczek, J., Yocom, P. N., Zaremba, D., Wagner, B. K., and Summers. C. J. 1998. Low voltage cathodoluminescence properties of blue emitting $SrGa_2S_4$:Ce^{3+} and ZnS:Ag,Cl phosphors. *Appl. Phys. Lett.* 72:2226–2228.

7

Applications of Nitride Phosphors in White LEDs

In Chapters 4–6, the crystal structure and photoluminescence properties of nitride phosphors were presented and discussed. Some of these phosphors show very broad excitation spectra, useful emission colors, high quantum efficiency, and small thermal quenching, so that they are very attractive for use as wavelength conversion materials in white LEDs. In comparison to traditional phosphors (aluminates, silicates, fluorides, oxysulfides, etc.), nitride phosphors exhibit superior photoluminescence properties and reliability, including abundant emission colors, stronger absorptions of blue light, higher chemical stability, and smaller degradation under thermal and irradiation attacks. These excellent properties of nitride phosphors allow for the production of white LEDs with tunable color temperatures, high color rendition, high luminous efficiency, small variations in chromaticity coordinates, and a long lifetime. In this chapter, white LEDs utilizing nitride phosphors will be introduced.

It has been over 30 years since the introduction of the first LED, and right now white LEDs are beginning to rival traditional lighting systems (incandescent and fluorescent lamps) in many architectural and small-area illumination applications. White LEDs have enjoyed tremendous growth over the last several years, with new applications ranging from small-area lighting to large-area lighting, from indicators to large-size backlight units, from flashlights to headlamps, etc. Much of this is due to the ever-increasing levels of efficiency being achieved with new wafer fabrication processes and phosphors, as well as advances in package and optics design. Due to the huge market and significant energy savings, the most important and ultimate applications of white LEDs are *general lighting* for replacing conventional incandescent and fluorescent lamps and *backlights* for substituting traditional cold-cathode fluorescent lamps (CCFLs) for liquid crystal displays (LCDs). Besides these, other applications, such as vehicle headlamps, street lighting, and in the medical field, are growing rapidly. For these different applications, it is necessary to select appropriate phosphors to meet the spectral requirements. In this chapter, Section 7.1 introduces white LEDs for general lighting, and Section 7.2 presents the results of white LEDs for LCD backlights.

7.1 White LEDs for Lighting/General Lighting

Traditional lighting sources, such as incandescent bulbs or fluorescent lamp tubes, have either problems of low energy conversion efficiency or environmental concerns. Consequently, incandescent bulbs are prohibited from sale in several countries (Brazil, Australia, Switzerland, Japan, etc.) and the European Union, and other nations (Russia, Canada, the United States, China, etc.) are planning to phase them out soon. This encourages the use of more energy efficient lighting alternatives, such as compact fluorescent lamps (CFLs) and white LEDs. Although cost and performance are two major bottlenecks for the use of white LEDs as general lighting, white LEDs are penetrating the lighting market at an extremely fast rate. Briefly looking at the history of a few different types of light sources helps provide some context for the LED's recent rapid progress. The incandescent bulb, which was developed in 1879, had an initial luminous efficacy of 1.5 lm/W, which improved to 16 lm/W over the next 130 years. Fluorescent bulbs, first developed in 1938, achieved a luminosity increase of 50 to 100 lm/W over the next 60 years. The progress of white LEDs is much more pronounced: since their commercialization in 1996, white LEDs' luminous efficacy has increased from 5 lm/W to today's commercial white LED of 150 lm/W, the highest luminous efficacy of all white light sources. The theoretical limit for white LEDs is about 260–300 lm/W.

Single-chip-based white LEDs (or phosphor-converted white LEDs) are superior to those using individual RGB LED chips in terms of cost, reliability, and lifetime, becoming the mainstream lighting sources. As phosphors play a key role in controlling the overall performance of white LEDs, such as luminous efficacy, color rendition, color temperature, and lifetime, the type of white LEDs (high efficiency or high color rendering) is thus determined by the combination of different kinds of phosphors.

7.1.1 Selection Rules of Phosphors

Selection of phosphors depends on the properties required for white LED lamps. There are generally two types of white LEDs for general illumination: high efficiency and high color rendition. High-efficiency white LEDs are usually fabricated by combining a blue LED chip with a yellow-emitting phosphor, and high-color-rendition white LEDs are produced by combining a blue LED chip with a multiphosphor blending, including green and red, or green, red, and yellow phosphors. For each phosphor used for white LEDs, the general selection rules are as follows:

- The phosphor should have strong absorption of the emitting light from LED chips (e.g., blue light).
- The phosphor should have high efficiencies of converting the absorbed light into visible light.

- The phosphor should emit useful colors, including green, yellow, and red, under the absorbed light excitation.
- The phosphor should exhibit a broad emission spectrum to achieve high color rendition.
- The phosphor should be stable against chemical attacks (e.g., H_2O, O_2, CO_2, and CO).
- The phosphor should have small degradation under thermal and irradiation attacks.
- The phosphor should have a small particle size ($<20\ \mu m$), a narrow particle size distribution, regular morphologies, and good crystallinity.
- The phosphor should be free of toxicity and environmentally friendly.

Among these requirements, the efficiency, emission color, and reliability of a phosphor are so important that they determine how good a white LED is and how long a white LED can be used. In addition, although the cost of a white LED is not mainly determined by the phosphors, the cheapness and high performance of phosphors are welcome.

7.1.2 High-Efficiency White LEDs

The first commercially available white LED in 1996 was fabricated by using the combination of a blue LED chip and a yellow-emitting YAG:Ce^{3+} phosphor (Nakamura and Fasol 1997). This type of device is also called a *one-phosphor-converted (1-pc) white LED*. Recently, Narukawa et al. (2007, 2010) at Nichia developed white LEDs with super-high luminous efficacy, achieving a value as high as 249 lm/W at 20 mA. This value nearly approaches the theoretical limit for white LEDs (260–300 lm/W). A high-power white LED with a luminous efficacy of 135 lm/W at 1 A was also developed, which has a higher flux than a 20 W-class fluorescent lamp and 1.5 times the luminous efficacy of a triphosphor fluorescent lamp (90 lm/W). On the other hand, the 1-pc white LED using YAG:Ce^{3+} usually has high color temperature (>5,000 K), due to the lack of a red component in the emission spectra of YAG:Ce^{3+} (see Figure 7.1). To achieve warm white light with low color temperature (~3,000 K), YAG:Ce^{3+} will combine with an additional red phosphor. Alternatively, red-shift of the emission band of YAG:Ce^{3+} by compositional tailoring (replacing Y^{3+} with Gd^{3+} or Tb^{3+}, codoping with Pr^{3+}, or substituting Si^{4+}-N^{3-} for Al^{3+}-O^{2-}) also leads to the fabrication of warm white LEDs. However, the former method would sacrifice the luminous efficacy, and the latter would reduce the lifetime of white LEDs.

As alternatives to YAG:Ce^{3+}, several yellow-emitting nitride and oxynitride phosphors have been developed, and were introduced in Chapter 4 (Xie et al. 2002, 2006a, 2006b; Li et al. 2008; Seto et al. 2009; Suehiro 2011). Using these

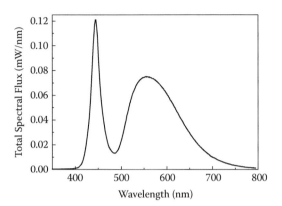

FIGURE 7.1
Emission spectrum of a white LED using a yellow YAG:Ce³⁺ (λ_{em} = 555 nm) and a blue LED chip (λem = 450 nm).

yellow phosphors, the fabrication of 1-pc white LEDs with varying color temperatures has been attempted, as will be discussed below.

7.1.2.1 Ca-α-sialon:Eu²⁺

As described in Chapter 4, Ca-α-sialon:Eu²⁺ is an orange-yellow phosphor that has a peak emission wavelength at 585 nm (Xie et al. 2002, 2004). It is thus expected that warm white LEDs can be achieved by using a single Ca-α-sialon:Eu²⁺ (Xie et al. 2004).

Sakuma et al. (2004) attempted to fabricate white LEDs using an orange-yellow Ca-α-sialon:Eu²⁺. The emission spectrum of the white LED is given in Figure 7.2. This white LED showed a color temperature of 2,750 K and chromaticity coordinates of x = 0.458 and y = 0.414. The luminous efficacy was reported to be 25.9 lm/W, which is 1.5 times higher than that of incandescent lamps (16–18 lm/W). In addition, with increasing temperature from 25°C to 200°C, the chromaticity coordinates (x,y) of the white LED using Ca-α-sialon:Eu²⁺ were varied from (0.503, 0.463) to (0.509, 0.464), whereas they were moved from (0.393, 0.461) to (0.383, 0.433) for white LED using YAG:Ce³⁺. This indicates that Ca-α-sialon:Eu²⁺ exhibits higher thermal stability than YAG:Ce³⁺, resulting in a very small variation in the color point of white LEDs.

By tailoring the chemical composition of Ca-α-sialon:Eu²⁺, we have blue-shifted the emission band, and obtained a short-wavelength Ca-α-sialon:Eu²⁺ phosphor. So it is also able to fabricate white LEDs with high color temperatures by employing the short-wavelength Ca-α-sialon:Eu²⁺. Figure 7.3 shows the emission spectrum of the white LED utilizing a Ca-α-sialon:Eu²⁺ with the peak emission wavelength of 573 nm. The correlated color temperature (CCT), chromaticity coordinates (x,y), and luminous efficacy of this white LED are 4000 K, (0.381, 0.377), and 58 lm/W, respectively. The effect of changes in

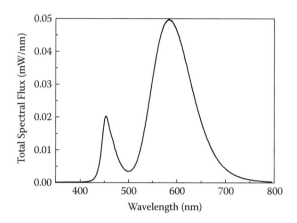

FIGURE 7.2
Emission spectrum of a white LED using an orange-yellow Ca-α-sialon:Eu²⁺ (λ_{em} = 585 nm) and a blue LED chip (λ_{em} = 450 nm). (Reprinted from Sakuma, K., Omichi, K., Kimura, N., Ohashi, M., Tanaka, D., Hirosaki, N., Yamamoto, Y., Xie, R.-J., and Suehiro, T., *Opt. Lett.*, 29, 2001–2003, 2004. With permission.)

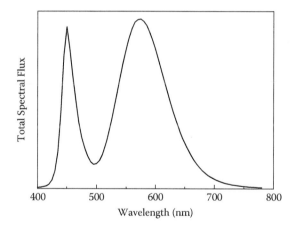

FIGURE 7.3
Emission spectrum of a white LED using a yellow Ca-α-sialon:Eu²⁺ (λ_{em} = 573 nm) and a blue LED chip (λ_{em} = 450 nm).

forward-bias current on the color point of white LEDs using YAG:Ce³⁺ and Ca-α-sialon:Eu²⁺ was investigated, and the results are given in Figure 7.4. It is seen that the color points of white LEDs shift upward with increasing current. The increment of chromaticity coordinates (Δx,Δy) is (0.0092, 0.046) for YAG:Ce³⁺ and (0.0081, 0.0144) for Ca-α-sialon:Eu²⁺. If we assume that the variation of the color point with current is the same for each LED chip, Ca-α-sialon:Eu²⁺ is more stable against changes in current than YAG:Ce³⁺, which is again due to the small thermal quenching observed in Ca-α-sialon:Eu²⁺.

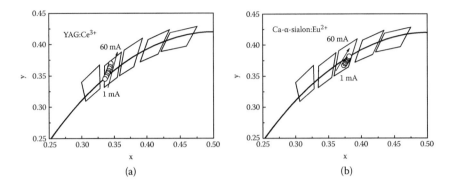

FIGURE 7.4
Variations of chromaticity coordinates with forward-bias current for white LEDs using
(a) YAG:Ce^{3+} and (b) Ca-α-sialon:Eu^{2+}.

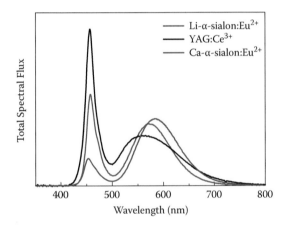

FIGURE 7.5
(See color insert.) Emission spectra of white LEDs using Li-α-sialon:Eu^{2+}, YAG:Ce^{3+}, and
Ca-α-sialon:Eu^{2+}.

7.1.2.2 *Li-α-sialon:Eu²⁺*

Compared to Ca-α-sialon:Eu^{2+}, Li-α-sialon:Eu^{2+} shows blue-shifted emission
spectra, and the peak emission wavelength varies from 573 to 577 nm under
460 nm excitation (Xie et al. 2006a). Li-α-sialon:Eu^{2+} has a maximal emission
wavelength 15–30 nm shorter than that of Ca-α-sialon:Eu^{2+}, which enables it
to fabricate white LEDs with high color temperatures.

Figure 7.5 shows the emission spectra of white LEDs using Li-α-sialon:Eu^{2+},
YAG:Ce^{3+}, and Ca-α-sialon:Eu^{2+}. The maximal emission wavelengths are
572, 567, and 585 nm for Li-α-sialon:Eu^{2+}, YAG:Ce^{3+}, and Ca-α-sialon:Eu^{2+},
respectively. The correlated color temperatures are 6,150, 7,600, and 2,750 K

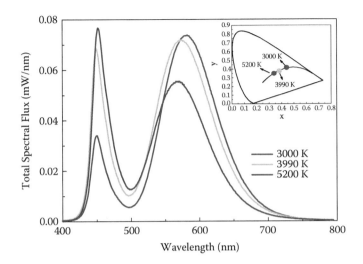

FIGURE 7.6
(See color insert.) Emission spectra and chromaticity coordinates of white LEDs using Li-α-sialon:Eu^{2+} with different emission wavelengths. (Reprinted from Xie, R.-J., Hirosaki, N., Mitomo, M., Sakuma, K., and Kimura, N., *Appl. Phys. Lett.*, 89, 241103, 2006a. With permission.)

for white LEDs employing Li-α-sialon:Eu^{2+}, YAG:Ce^{3+}, and Ca-α-sialon:Eu^{2+}, respectively. For the white LED using Li-α-sialon:Eu^{2+}, its chromaticity coordinates are x = 0.321, y = 0.308, the luminous efficacy is 43 lm/W, and the color rendering index Ra is 72 (Xie et al. 2006a).

In addition, by using Li-α-sialon:Eu^{2+} phosphors with different chemical compositions, we can achieve white LEDs with tunable color temperature. Figure 7.6 presents the emission spectra of white LEDs with different color temperatures. It is seen that the color temperature of a white LED can be tuned in a wide range by just utilizing a single Li-α-sialon:Eu^{2+} with different chemical compositions. Because the chemical composition (typically the n value) does not change the thermal quenching of Li-α-sialon:Eu^{2+} greatly, it is then expected that white LEDs using varying Li-α-sialon:Eu^{2+} will not change too much in thermal degradation.

7.1.2.3 CaAlSiN$_3$:Ce^{3+}

CaAlSiN$_3$:Ce^{3+} is a yellow phosphor having a maximum emission wavelength of 570 nm and a FWHM value of 134 nm (Li et al. 2008). A white LED was fabricated by combining a yellow CaAlSiN$_3$:Ce^{3+} and a blue LED chip, and its emission spectrum is given in Figure 7.7. This white LED has the following optical properties: correlated color temperature of 3,700 K, chromaticity coordinates of (0.397, 0.335), color rendering index of Ra = 70, and luminous efficacy of 51 lm/W at 20 mA.

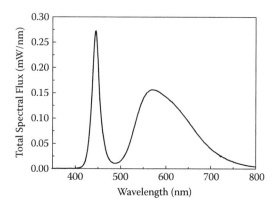

FIGURE 7.7

Emission spectrum of the white LED by combining a yellow-emitting CaAlSiN$_3$:Ce^{3+} (λ_{em} = 570 nm) and a blue LED (λ_{em} = 450 nm). (Reprinted from Li, Y. Q., Hirosaki, N., Xie, R.-J., Takeda, T., and Mitomo, M., *Chem. Mater.*, 20, 6704–6714, 2008. With permission.)

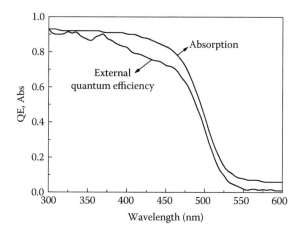

FIGURE 7.8

Absorption and external quantum efficiency of (Ca$_{0.95}$Eu$_{0.05}$)Si$_2$O$_2$N$_2$.

7.1.2.4 CaSi$_2$O$_2$N$_2$:Eu^{2+}

CaSi$_2$O$_2$N$_2$:Eu^{2+} is a yellow-green phosphor that has a broad emission band centered at 563 nm under 450 nm excitation (see Figure 4.31). Under blue light irradiation (λ_{em} = 450 nm), the absorption and external quantum efficiency of CaSi$_2$O$_2$N$_2$:Eu^{2+} (5 mol%) are 83 and 72% (see Figure 7.8), respectively, indicative of a highly efficient yellow phosphor. Furthermore, the luminescence of CaSi$_2$O$_2$N$_2$:Eu^{2+} (5 mol%) at 150°C is reduced by 8% of the initial intensity measured at room temperature (see Figure 7.9). These excellent photoluminescence properties imply that CaSi$_2$O$_2$N$_2$:Eu^{2+} (5 mol%) is a very interesting yellow phosphor for white LEDs.

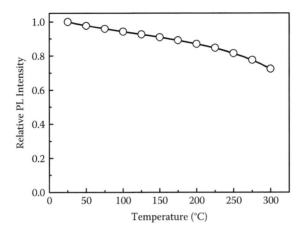

FIGURE 7.9
Thermal quenching of $(Ca_{0.95}Eu_{0.05})Si_2O_2N_2$ measured under the 450 nm excitation.

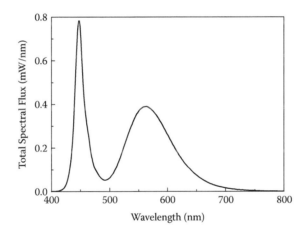

FIGURE 7.10
Emission spectrum of the white LED fabricated by using a yellow-emitting $CaSi_2O_2N_2:Eu^{2+}$ (λ_{em} = 563 nm) and a blue LED (λ_{em} = 450 nm).

Figure 7.10 shows the emission spectrum of the white LEDs using the yellow-emitting $CaSi_2O_2N_2:Eu^{2+}$ phosphor. This white LED exhibits cool white light with a correlated color temperature of 6,000 K and a luminous efficacy of 117 lm/W. The color rendering index Ra is 61, and the chromaticity coordinates are x = 0.323 and y = 0.328.

7.1.3 High-Color-Rendering White LEDs

In the previous section, several types of 1-pc white LEDs using different yellow phosphors were introduced. A common feature of these 1-pc white

LEDs is their high luminous efficacies, which is due to the fact that the proportion of light in the green part of the emission spectra of 1-pc white LEDs overlaps largely with the eye sensitivity curve (see Figure 1.7), which covers the spectral range of 370–750 nm and has a maximum at 555 nm. On the other hand, the color rendering index is low for 1-pc white LEDs because of the deficiency of red components in the white light. For example, the color rendering index Ra is about 60 for the white LED using $Ca-\alpha-sialon:Eu^{2+}$. Even using $CaAlSiN_3:Ce^{3+}$ or $YAG:Ce^{3+}$ with a much broader emission band, the white LED still has a color rendering index of less than 80. In fact, a color rendering index of greater than 85 is accepted for general lighting. Therefore, it is necessary to broaden the emission spectra of white LEDs by adopting two or more phosphors to achieve high color rendering.

In this section, several types of high-color-rendering white LEDs will be overviewed. These include 2-pc, 3-pc, and 4-pc white LEDs.

7.1.3.1 Two-Phosphor-Converted (2-pc) White LEDs

A simply way to achieve high color rendering is to employ a green and a red phosphor. This type of lamp is then named a two-phosphor-converted white LED, which contains three emission bands (blue, green, and red).

Xie et al. (2007) fabricated high-color-rendering white LEDs by employing a phosphor blend containing green $Ca-\alpha-sialon:Yb^{2+}$ (λ_{em} = 550 nm) and red $Sr_2Si_5N_8:Eu^{2+}$ (λ_{em} = 550 nm) phosphors. The phosphor blend was mixed with an appropriate amount of silicone resin, and then mounted on a blue LED chip (λ_{em} = 450 nm). The emission spectra of the 2-pc white LEDs with varying color temperatures are shown in Figure 7.11. The optical properties of these LEDs are summarized in Table 7.1. As seen, these white LEDs have a color rendering index of Ra = 83, and a luminous efficacy of 17–23 lm/W.

Fukuda et al. (2009) reported a new green-emitting oxynitride phosphor, $Sr_3Si_{13}Al_3O_2N_{21}:Eu^{2+}$. By combining this green phosphor and a red silicate phosphor $(Ca,Sr)_2SiO_4:Eu^{2+}$ with a blue LED, Fukuda et al. (2009) fabricated high-color-rendering white LEDs with varying color temperatures. The optical properties of white LEDs are summarized in Table 7.2. It is seen that the color rendering index varies in the range of 82–88, and the luminous efficacy is 56–62 lm/W for these white LEDs.

Yang et al. (2007) reported a high-color-rendering white LED. This white LED was fabricated by combining a blue LED (λ_{em} = 455 nm) with a green $SrSi_2O_2N_2:Eu^{2+}$ (λ_{em} = 538 nm) and a red $CaSiN_2:Ce^{3+}$ (λ_{em} = 642 nm). The three-band white LED has a luminous efficacy of 30 lm/W, chromaticity coordinates of (0.339, 0.337), a correlated color temperature of 5,206 K, and a color rendering index of 90.5. In addition, compared to the white LED prepared by using sulfide phosphors, the white LED using two nitride phosphors exhibits small variations in optical properties as the forward-bias current increases from 5 mA to 60 mA (see Table 7.3).

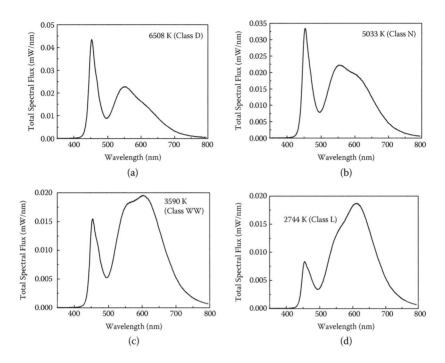

FIGURE 7.11

Emission spectra of white LEDs in (a)–(d), fabricated by combining the phosphor blend of green Ca-α-sialon:Yb²⁺ (λ_{em} = 550 nm) and red $Sr_2Si_5N_8$:Eu²⁺ (λ_{em} = 610 nm) with a blue LED (λ_{em} = 450 nm). (Reprinted from Xie, R.-J., Hirosaki, N., Kimura, N., Sakuma, K., and Mitomo, M., *Appl. Phys. Lett.*, 90, 191101, 2007. With permission.)

TABLE 7.1

Optical Properties of Two-Phosphor-Converted White LEDs (Ca-α-sialon:Yb²⁺ + $Sr_2Si_5N_8$:Eu²⁺)

Class	L	WW	W	N	D
CCT (K)	2,744	3,590	4,270	5,033	6,508
CIE x	0.461	0.402	0.372	0.344	0.313
CIE y	0.419	0.393	0.382	0.346	0.330
Ra	83	83	83	83	82
Luminous efficacy (lm/W)	17	20	21	22	23

Source: Data from Xie, R.-J., Hirosaki, N., Kimura, N., Sakuma, K., and Mitomo, M., *Appl. Phys. Lett.*, 90, 191101, 2007.

Muellar-Mach et al. (2005) fabricated a high-powder (input powder: 1 W) white LED by using green $SrSi_2O_2N_2$:Eu²⁺ and red $Sr_2Si_5N_8$:Eu²⁺ phosphors. This 2-pc white LED shows a correlated color temperature of 3,100 K and a color rendering index of Ra = 89. The luminous efficiency at 1 W input is 25 lm/W, which is almost twice the efficiency of any incandescent lamp.

TABLE 7.2

Optical Properties of Two-Phosphor-Converted White
LEDs ($Sr_3Si_{13}Al_3O_2N_{21}$:Eu^{2+} and $(Ca,Sr)_2SiO_4$:Eu^{2+})

Class	WW	W	N	D
CCT (K)	3,230	4,100	5,300	6,450
Ra	82	88	87	87
Luminous efficacy (lm/W)	56	57	62	61

Source: Data from Fukuda, Y., Ishida, K., Mitsuishi, I., and
Nunoue, S., *Appl. Phys. Express*, 2, 012401, 2009.

TABLE 7.3

Variations in Chromaticity Coordinates, Color Temperature, and Color Rendering
Index of White LEDs with Increasing the Forward-Bias Current from 5 mA to 60 mA

White LEDs	Chromaticity Coordinates		Color Temperature/ΔT	Color Rendering Index/ΔRa
	Δx	Δy		
$SrSi_2O_2N_2$:Eu^{2+} + $CaSiN_2$:Ce^{3+}	0.0022	0.0011	91 K	1.9
$SrGa_2S_4$:Eu^{2+} + $(Ca,Sr)S$:Eu^{2+}	0.0625	0.0299	4,144 K	4.6

Source: Data from Yang, C. C., Lin, C. M., Chen, Y. J., Wu, Y. T., Chuang, S. R., Liu, R. S., and
Hu, S. F., *Appl. Phys. Lett.*, 90, 123503, 2007.

TABLE 7.4

Color Rendering Indices of the 2-pc White LED and a Deluxe Fluorescent Lamp

	R9	R10	R11	R12	R13	R14	R15
2-pc white LED	56	85	81	68	93	96	—
Fluorescent lamp	11	50	71	54	97	71	—

Source: Data from Mueller-Mach, R., Mueller, G., Krames, M. R., Hoppe, H. A., Stadler, F.,
Schnick, W., Juestel, T., and Schmidt, P., *Phys. Stat. Sol. A*, 202, 1727–1732, 2005.

Table 7.4 lists the color rendering indices of both the 2-pc white LED and a
deluxe fluorescent lamp, indicating the higher color quality of 2-pc white
LED compared to traditional lighting tubes. In addition, the white LED
shows extraordinary stability with drive and temperature, due to the very
small thermal degradation of the nitride phosphors.

7.1.3.2 Three-Phosphor-Converted (3-pc) White LEDs

As stated above, the 2-pc white LEDs all show a color rendering index of
over 80, and some of them even have a color rendering index of close to 90
if the green and red phosphors are carefully selected. Since there is a great
gap between the emission spectra of the green and red phosphors, the color
rendering index can be further enhanced by adding a yellow phosphor in
between them. This leads to the three-phosphor-converted (3-pc) white LEDs
that contain four emission bands in total.

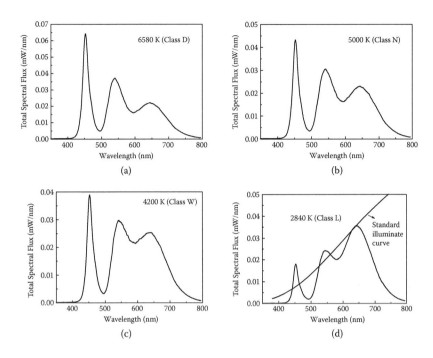

FIGURE 7.12
Emission spectra of white LEDs in (a)–(d), fabricated by combining the phosphor blend of green β-sialon:Eu^{2+} (λ_{em} = 550 nm), yellow Ca-α-sialon:Eu^{2+} (λ_{em} = 585 nm), and red CaAlSiN$_3$:Eu^{2+} (λ_{em} = 650 nm) with a blue LED (λ_{em} = 450 nm). (Reprinted from Sakuma, K., Hirosaki, N., Kimura, N., Ohashi, M., Xie, R.-J., Yamamoto, Y., Suehiro, T., Asano, K., and Tanaka, D., *IEICE Trans. Electron.*, E88-C, 2057–2064, 2005. With permission.)

Yamada et al. (2003) reported a red-enhanced high-color-rendering white LED by combining a green YAG:Ce^{3+}, a yellow YAG:Ce^{3+}, and a red nitride Sr-Ca-Eu-Si-N (SCESN) phosphor with a blue LED (λ_{em} = 460 nm). The peak wavelengths of the green and yellow YAG:Ce^{3+} and the red SCESN are 540, 570, and 655 nm, respectively. This 3-pc white LED showed better color rendering indices than the white LED using a single yellow YAG:Ce^{3+} phosphor, which is due to the enhancement of the proportions of light longer than 625 and 500 nm in the emission spectrum of 3-pc white LEDs by 1.3 and 2 times, respectively. The color rendering index Ra increased from 76.2 for the 1-pc white LED to 87.7 for the 3-pc white LED, and the R9 value was significantly improved, from –2.5 to 62.6. The luminous efficacy and color temperature were reported to be 25.5 lm/W at 20 mA and 4,670 K, respectively.

Sakuma et al. (2005) employed three (oxy)nitride phosphors: β-sialon:Eu^{2+} (λ_{em} = 535 nm), Ca-α-sialon:Eu^{2+} (λ_{em} = 585 nm), and CaAlSiN$_3$:Eu^{2+} (λ_{em} = 650 nm), to achieve high-color-rendering white LEDs. These LED lamps were prepared by combining these three nitride phosphors with a blue LED (λ_{em} = 450 nm). The emission spectra of the 3-pc white LEDs with varying color temperatures are presented in Figure 7.12. The color temperature varies from 2,840 K

TABLE 7.5

Optical Properties of Three-Phosphor-Converted White LEDs (β-sialon:Eu²⁺, Ca-α-sialon:Eu²⁺, and CaAlSiN₃:Eu²⁺)

Class	L	WW	W	N	D
CCT (K)	2,840	3,470	4,160	5,000	6,580
CIE x	0.449	0.407	0.373	0.345	0.311
CIE y	0.408	0.392	0.370	0.358	0.333
Ra	88	86	83	82	81
Luminous efficacy (lm/W)	25	24	27	25	28

Source: Data from Sakuma, K., Hirosaki, N., Kimura, N., Ohashi, M., Xie, R.-J., Yamamoto, Y., Suehiro, T., Asano, K., and Tanaka, D., *IEICE Trans. Electron.* E88-C, 2057–2064, 2005.

Note: L = incandescent light bulb, WW = warm white, W = white, N = neutral white, D = daylight.

TABLE 7.6

Color Rendering Indices of the Three-Phosphor-Converted White LED (CCT = 2,840 K)

R9	R10	R11	R12	R13	R14	R15
96	67	87	49	95	82	98

Source: Data from Sakuma, K., Hirosaki, N., Kimura, N., Ohashi, M., Xie, R.-J., Yamamoto, Y., Suehiro, T., Asano, K., and Tanaka, D., *IEICE Trans. Electron.*, E88-C, 2057–2064, 2005.

(equal to a incandescent lightbulb) to 6,580 K (equal to a fluorescent light tube), depending on the blending ratio of each phosphor. As see in Table 7.5, the color rendering index Ra of the 3-pc white LED is in the range of 81–88, and the luminous efficacy varies in the range of 24–28 lm/W (20 mA). Furthermore, the color rendering indices (R9–R15) of the 3-pc white LEDs are all enhanced for 3-pc white LEDs (see Table 7.6), even when compared to those of 2-pc white LEDs (see Table 7.4).

7.1.3.3 Four-Phosphor-Converted (4-pc) White LEDs

Super-high-color-rendering white LEDs, the color quality of which is equivalent to natural sunlight, are continuously pursued to satisfy general lighting and backlight needs. The addition of a yellow phosphor in between the green and red phosphors forms an extremely broad emission band that covers most of the visible light region, which dramatically increases the color rendering index up to ~90. However, the large distance between the sharp blue emission band and the green emission band leads to a deep "valley" lying in between them, breaking down the continuity of the overall emission spectrum of white LEDs. To solve this problem, a blue-light-excitable blue-green phosphor (λem = 480–520 nm) needs to be found and developed. By utilizing this

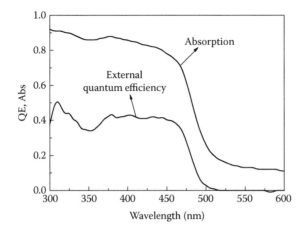

FIGURE 7.13
Absorption and external quantum efficiency of the blue-green-emitting $BaSi_2O_2N_2:Eu^{2+}$ phosphor, measured under varying excitation wavelengths.

blue-green phosphor together with green, yellow, and red phosphors, we can fabricate extra-high-color-rendering white LEDs. However, there are actually very few blue-green phosphors that can be excited by blue light efficiently because an extremely small Stokes shift is required for these phosphors. Alternatively, the use of ultraviolet or near-ultraviolet LEDs, rather than blue LEDs, is proposed to generate super-high-color-rendering white light.

As shown in Figure 4.31, $BaSi_2O_2N_2:Eu^{2+}$ is a blue-green phosphor having a maximum emission wavelength of ~500 nm. Moreover, as seen in Figure 7.13, it has a strong absorption of blue light (e.g., 78% at 450 nm) and exhibits relatively high external quantum efficiency (e.g., 41% at 450 nm). By employing $BaSi_2O_2N_2:Eu^{2+}$ and green β-sialon:Eu^{2+}, yellow Ca-α-sialon:Eu^{2+}, and red $CaAlSiN_3:Eu^{2+}$ phosphors, Kimura et al. (2007) prepared four-phosphor-converted (4-pc) white LEDs. The emission spectra of the 4-pc white LEDs are shown in Figure 7.14, which shows continuous spectral shapes that resemble the curves of standard illuminants. As summarized in Table 7.7, the color rendering index Ra for all LED samples is 95 or 96, which is very close to natural sunlight and extremely high for white LEDs using a blue LED as the primary lighting source. Most other color rendering indices are near 98 (see Table 7.8), much higher than those in 3-pc white LEDs. The luminous efficacy of these super-high-color-rendering white LEDs varies from 28 to 35 lm/W.

Takahashi et al. (2007) used a near-ultraviolet LED (λ_{em} = 405 nm) to fabricate super-high-color-rendering white LEDs. A blue-emitting oxynitride phosphor, JEM:Ce^{3+}, was first developed, which emits at 460–500 nm and has an external quantum efficiency of 50% under 405 nm excitation. 4-pc white LEDs were produced by combining the near UV LED with blue JEM:Ce^{3+}, green β-sialon:Eu^{2+}, yellow Ca-α-sialon:Eu^{2+}, and red $CaAlSiN_3:Eu^{2+}$

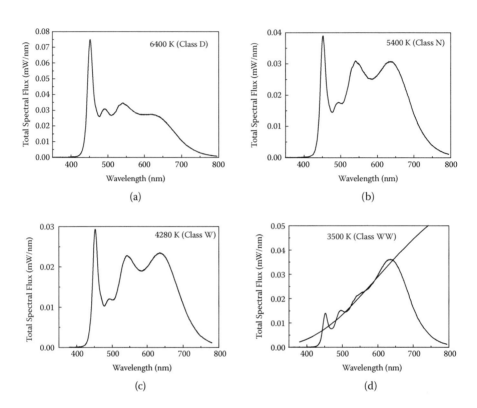

FIGURE 7.14

Emission spectra of white LEDs in (a)–(d), fabricated by combining the phosphor blend of blue-green $BaSi_2O_2N_2:Eu^{2+}$ (λ_{em} = 496 nm), green β-sialon:Eu^{2+} (λ_{em} = 550 nm), yellow Ca-α-sialon:Eu^{2+} (λ_{em} = 585 nm), and red $CaAlSiN_3:Eu^{2+}$ (λ_{em} = 650 nm) with a blue LED (λ_{em} = 450 nm). (Reprinted from Kimura, N., Sakuma, K., Hirafune, S., Asano, K., Hirosaki, N., and Xie, R.-J., *Appl. Phys. Lett.*, 90, 051109, 2007. With permission.)

TABLE 7.7

Optical Properties of Four-Phosphor-Converted White LEDs

Class	L	WW	W	N	D
CCT (K)	2,900	3,540	4,280	4,900	6,380
CIE x	0.446	0.406	0.371	0.348	0.315
CIE y	0.411	0.397	0.378	0.357	0.329
Ra	98	97	95	96	96
Luminous efficacy (lm/W)	28	30	31	33	35

Source: Data from Kimura, N., Sakuma, K., Hirafune, S., Asano, K., Hirosaki, N., and Xie, R.-J., *Appl. Phys. Lett.*, 90, 051109, 2007.

TABLE 7.8

Color Rendering Indices of the Four-Phosphor-Converted White LED (CCT = 2,900 K)

R9	R10	R11	R12	R13	R14	R15
89	98	98	87	99	98	97

Source: Data from Kimura, N., Sakuma, K., Hirafune, S., Asano, K., Hirosaki, N., and Xie, R.-J., *Appl. Phys. Lett.*, 90, 051109, 2007.

phosphors. The color rendering index is as high as 95–96 for these white LEDs with different color temperatures, and the luminous efficacy is 19–20 lm/W when measured at a forward-bias current of 20 mA.

7.2 White LEDs for LCD Backlights

Because liquid crystal displays (LCDs) do not produce light themselves, an illumination unit is required to produce a visible image. This illumination unit in LCDs is called *backlights*, which illuminate the LCD from the side or back of the display panel. In small LCDs backlights are used to improve readability in low light conditions, and in computer screens and LCD televisions backlights are used to produce light in a manner similar to that of a CRT display.

There are many types of light sources for backlights, including cold-cathode fluorescent lamps (CCFLs), external electrode fluorescent lamps (EEFLs), hot-cathode fluorescent lamps (HCFLs), and white LEDs. Among them, CCFLs are used for LCD backlighting and are currently the dominant technology for LCD backlighting. As there is much concern about backlight units with a larger color gamut, higher brightness, smaller volume, lower power consumption, high dimming ratio, and that are Hg-free, white LEDs are considered a promising alternative to CCFLs because LEDs are superior to CCFLs in terms of their (1) long life, (2) low-voltage operation, (3) fast response time, (4) wide color gamut, and (5) being mercury-free.

The LED backlight unit market has rapidly emerged in the thin film transistor (TFT) LCD industry, and momentum for this segment is expected to continue over the next 5 years. In the latest DisplaySearch *Quarterly LED Backlight Report* (Kwak and Uno 2010), it is reported that LED backlight units will surpass CCFL/EEFL backlights in large-area TFT LCD panels in 2011, and achieve 74% penetration in 2013. Large-area LED backlight demand for all applications will grow from 114 million units in 2009 to 770 million units in 2015 (see Figure 7.15). Although cost and performance remain bottlenecks

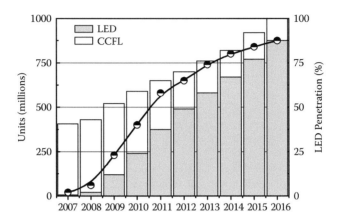

FIGURE 7.15
LED backlight unit penetration rate in 10-inch or more TFT LCDs.

for panel manufacturers for LED backlight units for monitor panels, the note-book PC segment has the highest LED backlight unit penetration rate, as the power savings benefit justifies the cost premium compared to CCFL back-light units. Meanwhile, the prices for side-view, high-intensity (1,900–2,200 mcd) white LEDs continues to fall. As a result, LED backlights will have an 84% share of notebook PC shipments in 2010, and will be close to 95% in 2011.

There are two types of LED backlights: RGB LEDs and phosphor-converted (pc) white LEDs. The RGB LEDs generate white light by combining individual red, green, and blue three-primary-color InGaN LEDs, which exhibit an excellent color gamut. However, the efficiency of green LEDs is still low, and the independent driving circuits for each LED lead to higher cost and much more complicated structures than phosphor-converted white LEDs. Furthermore, phosphor-converted white LEDs are more reliable and simpler than RGB LEDs. This is due to the fact that the temperature dependence of peak emission wave-lengths of phosphors is usually small, whereas that of semiconductor LED chips is so serious that a complicated feedback-driven system could be required.

In this section, phosphor-converted white LEDs used as LCD backlights will be introduced.

7.2.1 Selection Rules of Phosphors

Unlike general lighting applications, LCD backlights use RGB color filters to reproduce color images through the combination of the RGB primary col-ors. The light is not transmitted by the color filter because this would lead to energy loss and reduce the color gamut. The color gamut is determined by the combination of the spectral characteristics of the white light source and the spectral transmission factor of the RGB color filters. Therefore, in order to achieve a wide color gamut, the emission spectral features of phosphors should match well with the RGB color filters. Currently used phosphors for CCFLs are $BaMgAl_{10}O_{17}:Eu^{2+}$ (λ_{em} = 450 nm), $La(PO_4):Ce^{3+}$, Tb^{3+} (λ_{em} = 545 nm), and $Y_2O_3:Eu^{3+}$ (λ_{em} = 611 nm). The common feature of the green and red phosphors is that both of them have characteristic line spectra of Tb^{3+} or Eu^{3+}, and thus have high color purity. On the other hand, the blue light excitable nitride phosphors for white LEDs are usually doped with Eu^{2+} or Ce^{3+}, and therefore show broad emission spectra and relatively low color purity. It is necessary to optimize the combination of highly defined RGB colors with narrow emission spectra for white backlights by selecting suitable green and red phosphors.

Besides the general selection rules for phosphors described previously, additional requirements for phosphors in LED backlights include: (1) the emission spectra should match perfectly the spectral characteristics of RGB color filters, to minimize the color leak by lowering the cross-point of the color filter and the LED spectra, and (2) the emission spectra of proper phosphors should be as narrow as possible, to reduce the spectral overlap between the blue and the green emissions, as well as between the green and red emissions, which improves the color purity of the primary RGB colors.

7.2.2 Wide-Color-Gamut White LEDs

There are mainly two types of phosphor-converted white LED backlights: one is the 1-pc white LED using yellow YAG:Ce^{3+}, and the other is 2-pc (or even 3-pc) white LEDs using a green and a red phosphor. The 1-pc white LED exhibits a higher luminous efficacy but a poorer color gamut than the 2-pc one.

As shown in Figure 4.23, β-sialon:Eu^{2+} (Hirosaki 2005) is an attractive green phosphor for LED backlights because it has a very narrow emission band with a full width at half maximum (FWHM) of 58 nm. In addition, $CaAlSiN_3:Eu^{2+}$ is a highly efficient red phosphor (Uheda et al. 2006). Xie et al. (2009) then employed the green β-sialon:Eu^{2+} and red $CaAlSiN_3:Eu^{2+}$ phosphors to fabricate a 2-pc white LED for LCD backlights. For comparison, a 1-pc white LED using the yellow YAG:Ce^{3+} was also prepared.

Figure 7.16 shows the emission spectra of each phosphor as well as 1-pc and 2-pc white LEDs. The 1-pc white LED has a luminous efficacy of 59 lm/W and a color temperature of ~8,600 K, and the 2-pc white LED has a luminous efficacy of 38 lm/W at ~5,000 K. The color of illuminated objects is determined by the transmitted lights from LED backlights through RGB filters. The color gamut is calculated by multiplying the LED emission spectrum by the spectrum of each color filter. The transmission spectra of RGB color filters are shown in Figure 7.17. So, the calculated emission spectra for 1-pc and 2-pc white LEDs by applying the transmission spectra of color filters are given in Figure 7.18. It is seen that the cross-points between the red and green emission spectra, as well as between the blue and green emission spectra, are lowered in the 2-pc white LED compared to the 1-pc LED. Furthermore, the emission spectra of RGB primary colors are separated by a larger distance

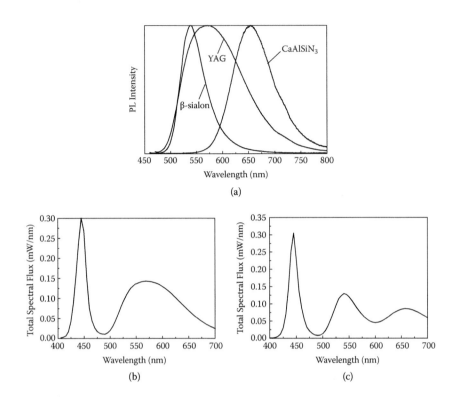

FIGURE 7.16
Emission spectra of (a) β-sialon:Eu^{2+}, YAG:Ce^{3+}, and CaAlSiN$_3$:Eu^{2+}; (b) the 1-pc white LED using only YAG:Ce^{3+}; and (c) the 2-pc white LED using β-sialon:Eu^{2+} and CaAlSiN$_3$:Eu^{2+}. (Reprinted from Xie, R.-J., Hirosaki, N., and Takeda, T., *Appl. Phys. Express*, 2, 022401, 2009. With permission.)

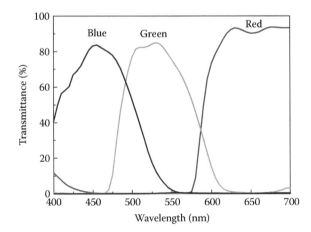

FIGURE 7.17
Typical transmission spectra of RGB color filters. (Reprinted from Xie, R.-J., Hirosaki, N., and Takeda, T., *Appl. Phys. Express*, 2, 022401, 2009. With permission.)

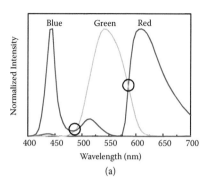
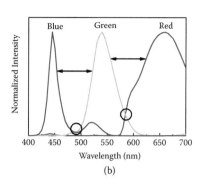

(a)   (b)

FIGURE 7.18
Calculated RGB spectra for the (a) 1-pc and (b) 2-pc white LEDs. (Reprinted from Xie, R.-J., Hirosaki, N., and Takeda, T., *Appl. Phys. Express*, 2, 022401, 2009. With permission.)

in the 2-pc LEDs. These spectral features lead to a higher color purity of each RGB primary color by using β-sialon:Eu^{2+} and $CaAlSiN_3$:Eu^{2+}.

A color gamut conforming to Commission International del'Eclairge (CIE) standard chromaticity coordinates can be calculated by using the following equations (Wyszecki and Stiles 1982):

$$X = \int S(\lambda)x(\lambda)d\lambda$$

$$Y = \int S(\lambda)y(\lambda)d\lambda \qquad (7.1)$$

$$Z = \int S(\lambda)z(\lambda)d\lambda$$

where X, Y, and Z are tristimulus values in the CIE 1931 standard colorimetric system. $S(\lambda)$ is the calculated spectrum distribution of the light source, as shown in Figure 7.18. $x(\lambda)$, $y(\lambda)$, and $z(\lambda)$ are CIE 1931 color-matching functions, as illustrated in Figure 1.9.

For both CIE 1931 and CIE 1976 standard systems, the chromaticity coordinates (x,y) and (u',v') are given by (Hunt 1991)

$$x = \frac{X}{X+Y+Z}$$

$$y = \frac{Y}{X+Y+Z} \qquad (7.2a)$$

$$u' = \frac{4X}{X+15Y+3Z}$$

$$v' = \frac{9Y}{X+15Y+Z} \qquad (7.2b)$$

The calculated chromaticity coordinates for 1-pc and 2-pc white LEDs are shown in Table 7.9. The chromaticity coordinates of National Television

TABLE 7.9

Chromaticity Coordinates of 1-pc, 2-pc White LEDs and the NTSC Standard

Chromaticity Coordinates		Red	Green	Blue
1-pc white LED	CIE 1931 (x,y)	(0.642, 0.344)	(0.335, 0.629)	(0.156, 0.092)
	CIE 1976 (u′,v′)	(0.439, 0.530)	(0.135, 0.573)	(0.165, 0.218)
2-pc white LED	CIE 1931 (x,y)	(0.656, 0.323)	(0.297, 0.664)	(0.155, 0.075)
	CIE 1976 (u′,v′)	(0.472, 0.522)	(0.115, 0.576)	(0.173, 0.188)
NTSC	CIE 1931 (x,y)	(0.670, 0.310)	(0.210, 0.710)	(0.140, 0.080)
	CIE 1976 (u′,v′)	(0.498, 0.519)	(0.076, 0.576)	(0.152, 0.196)

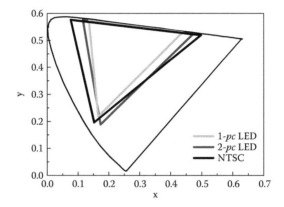

FIGURE 7.19

(See color insert.) CIE 1976 chromaticity coordinates of white LEDs. (Reprinted from Xie, R.-J., Hirosaki, N., and Takeda, T., *Appl. Phys. Express*, 2, 022401, 2009. With permission.)

Standards Committee (NTSC) are also included for comparison. Figure 7.19 plots the color triangles of the two types of white LEDs in a CIE 1976 chromaticity diagram. It is seen that the 2-pc white LED has a better color gamut than the 1-pc LED. The color gamut is 92% of NTSC's for 2-pc white LEDs, whereas it is only 76% of NTSC's for the 1-pc white LED. The wider color gamut, together with other excellent physical properties (e.g., reliability, longevity, free of mercury, compactness, light weight, energy saving), enables 2-pc white LEDs to be a very attractive light source for large-size LCDs (e.g., LCD televisions).

References

Fukuda, Y., Ishida, K., Mitsuishi, I., and Nunoue, S. 2009. Luminescence properties of Eu^{2+}-doped green-emitting Sr-sialon phosphor and its application to white light-emitting diodes. *Appl. Phys. Express* 2:012401.

Hirosaki, H., Xie, R.-J., Kimoto, K., Sekiguchi, T., Yamamoto, Y., Suehiro, T., and Mitomo, M. 2005. Characterization and properties of green-emitting β-SiALON:Eu^{2+} powder phosphors for white light-emitting diodes. *Appl. Phys. Lett.* 86:211905-1–211905-3.

Hunt, R. 1991. *Measuring colour*. 2nd ed. West Sussex, UK: Ellis Horwood.

Kimura, N., Sakuma, K., Hirafune, S., Asano, K., Hirosaki, N., and Xie, R.-J. 2007. Extrahigh color rendering white light-emitting diode lamps using oxynitride and nitride phosphors excited by blue light-emitting diode. *Appl. Phys. Lett.* 90:051109.

Kwak, K., and Uno, T. 2010. *Quarterly LED backlight report*. http://www.displaysearch. com/cps/rde/xchg/displaysearch/hs.xsl/quarterly_led_backlight_report.asp.

Li, Y. Q., Hirosaki, N., Xie, R.-J., Takeda, T., and Mitomo, M. 2008. Yellow-orange-emitting CaAlSiN$_3$:Ce^{3+} phosphor: Structure, photoluminescence, and application in white LEDs. *Chem. Mater.* 20:6704–6714.

Mueller-Mach, R., Mueller, G., Krames, M. R., Hoppe, H. A., Stadler, F., Schnick, W., Juestel, T., and Schmidt, P. 2005. Highly efficient all-nitride phosphor-converted white light emitting diode. *Phy. Stat. Sol. A* 202:1727–1732.

Nakamura, S., and Fasol, G. 1997. *The blue laser diode: GaN based light emitters and lasers*. Berlin: Springer.

Narukawa, Y., Ichikawa, M., Sanga, D., Sano, M., and Mukai, T. 2010. White light emitting diodes with super-high luminous efficacy. *J. Phys. D Appl. Phys.* 43:354002.

Narakuwa, Y., Narita, J., Sakamoto, T., Yamada, T., Narimatsu, H., Sano, M., and Mukai, T. 2007. Recent progress of high efficiency white LEDs. *Phys. Stat. Sol. A* 204:2087–2093.

Sakuma, K., Hirosaki, N., Kimura, N., Ohashi, M., Xie, R.-J., Yamamoto, Y., Suehiro, T., Asano, K., and Tanaka, D. 2005. White light-emitting diode lamps using oxynitride and nitride phosphor materials. *IEICE Trans. Electron.* E88-C:2057–2064.

Sakuma, K., Omichi, K., Kimura, N., Ohashi, M., Tanaka, D., Hirosaki, N., Yamamoto, Y., Xie, R.-J., and Suehiro, T. 2004. Warm-white light-emitting diode with yellowish orange SiAlON ceramic phosphor. *Opt. Lett.* 29:2001–2003.

Seto, T., Kijima, N., and Hirosaki, N. 2009. A new yellow phosphor La$_3$Si$_6$N$_{11}$:Ce^{3+} for white LEDs. *ECS Trans.* 25:247–252.

Takahashi, K., Hirosaki, N., Xie, R.-J., Harada, M., Yoshimura, K., and Tomomura, Y. 2007. Luminescence properties of blue La$_{1-x}$Ce$_x$Al(Si$_{6-z}$Al$_z$)(N$_{10-z}$O$_z$)(z ~ 1) oxynitrides phosphors and their application in white light-emitting diode. *Appl. Phys. Lett.* 91:091923.

Uheda, K., Hirosaki, N., Yamamoto, Y., Naoto, A., Nakajima, T., and Yamamoto, H. 2006. Luminescence properties of a red phosphor, CaAlSiN$_3$:Eu^{2+}, for white light-emitting diodes. *Electrochem. Solid State Lett.* 9:H22–H25.

Wyszecki, G., and Stiles, W. 1982. *Color science—Concepts and methods, quantitative data and formulae*. 2nd ed. New York: Wiley.

Xie, R.-J., Hirosaki, N., Kimura, N., Sakuma, K., and Mitomo, M. 2007. 2-Phosphor-converted white light-emitting diodes using oxynitride/nitride phosphors. *Appl. Phys. Lett.* 90:191101.

Xie, R.-J., Hirosaki, N., Mitomo, M., Sakuma, K., and Kimura, N. 2006a. Wavelength-tunable and thermally stable Li-α-sialon:Eu^{2+} oxynitride phosphors for white light-emitting diodes. *Appl. Phys. Lett.* 89:241103.

Xie, R.-J., Hirosaki, N., Mitomo, M., Takahashi, K., and Sakuma, K. 2006b. Highly efficient white-light-emitting diodes fabricated with short-wavelength yellow oxynitride phosphors. *Appl. Phys. Lett.* 88:101104.

Xie, R.-J., Hirosaki, N., Sakuma, K., Yamamoto, Y., and Mitomo, M. 2004. Eu^{2+}-doped Ca-alpha-SiALON: A yellow phosphor for white light-emitting diodes. *Appl. Phys. Lett.* 84:5404–5406.

Xie, R.-J., Hirosaki, N., and Takeda, T. 2009. Wide color gamut backlight for liquid crystal displays using three-band phosphor-converted white light-emitting diodes. *Appl. Phys. Express* 2:022401.

Xie, R.-J., Mitomo, M., Uheda, K., Xu, F. F., and Akimune, Y. 2002. Preparation and luminescence spectra of calcium- and rare-earth (R = Eu, Tb, and Pr)-codoped a-SiALON ceramics. *J. Am. Ceram. Soc.* 85:1229–1234.

Yamada, M., Naitou, T., Izuno, K., Tamaki, H., Murazaki, Y., Kameshima, M., and Mukai, T. 2003. Red-enhanced white light-emitting diode using a new red phosphor. *Jpn. J. Appl. Phys.* 42:L20–L23.

Yang, C. C., Lin, C. M., Chen, Y. J., Wu, Y. T., Chuang, S. R., Liu, R. S., and Hu, S. F. 2007. Highly stable three-band white light from an InGaN-based blue light-emitting diode chip precoated with (oxy)nitride green/red phosphors. *Appl. Phys. Lett.* 90:123503.

Index